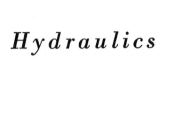

Hydraulics

H y d r a u l i c s

HORACE W. KING
Late Professor of Hydraulic Engineering
University of Michigan

CHESTER O. WISLER
Professor of Hydraulic Engineering
University of Michigan

JAMES G. WOODBURN
Professor of Hydraulic Engineering
University of Wisconsin

Fifth Edition

ROBERT E. KRIEGER PUBLISHING COMPANY
MALABAR, FLORIDA

Fifth Edition 1941
Reprint Edition 1980

Printed and Published by
ROBERT E. KRIEGER PUBLISHING COMPANY, INC.
KRIEGER DRIVE
MALABAR, FLORIDA 32950

Copyright © 1941 by
Horace W. King, Chester O. Wisler and James G. Woodburn
Reprinted by Arrangement with
John W. Wiley & Sons

Printed in the United States of America

Library of Congress Cataloging in Publication Data

King, Horace Williams, 1874-1951.
    Hydraulics.

    Reprint of the 5th ed. published by Wiley, New York.
    Includes bibliographical references.
    1. Hydraulics. I. Wisler, Chester Owen, 1881-
joint author. II. Woodburn, James Gelston, 1894-
joint author. III. Title.
TC160.K6 1980   620'.106   79-25379
ISBN 0-89874-106-8

Preface

THE purpose of this revision is (1) to improve the clarity and arrangement of the text and problems in the light of continued experience in the use of the book with classes in elementary hydraulics, (2) to add new problems illustrating applications of basic theory to certain engineering problems not previously covered, and (3) to expand the text material somewhat to include new developments in theory and practice which have been accepted in the past few years as an integral part of hydraulic engineering.

Changes to improve the wording and clarify the meaning have been made in many places throughout the book. Nearly all the problems of the fourth edition have been retained, although with some rearrangement to provide a more logical order. A number of new examples and problems have been added.

New text material is introduced on the variation of hydrostatic pressure with altitude in a compressible fluid, on the flow through gates and over dams, on the general consideration of the flow of liquids in pipes, on the analysis of flow in pipe networks, and on the resistance offered to motion of objects through a fluid. To give students an elementary knowledge in the field of hydraulic model testing, in which they are finding employment in increasing numbers, a chapter on hydraulic similitude and dimensional analysis has been added.

Although a quarter century has elapsed since this book was first prepared, the authors have no fault to find with the statements made in the preface to the first edition. The book is still intended to present to the engineering student or to the practicing engineer the fundamental principles relating to fluids at rest or in motion as they apply to engineering practice and to illustrate those principles with practical problems.

Thanks are due to many friends and associates whose suggestions and advice have greatly helped with the revision, in particular: A. T. Lenz, G. A. Rohlich, J. A. Borchardt, J. R. Villemonte, and A. C. Ingersoll of the University of Wisconsin; E. R. Dodge of Montana State College; E. F. Brater of the University of Michigan; R. D. Goodrich of the University of Wyoming; and F. W. Greve and W. E. Howland of Purdue University. Professor Howland read the manuscript and offered many valuable suggestions.

H. W. K.
C. O. W.
J. G. W.

Preface

TO FIRST EDITION

THIS book deals with the fundamental principles of hydraulics and their application in engineering practice. Though many formulas applicable to different types of problems are given, it has been the aim of the authors to bring out clearly and logically the underlying principles which form the basis of such formulas rather than to emphasize the importance of the formulas themselves.

Our present knowledge of fluid friction has been derived largely through experimental investigation, and this has resulted in the development of a large number of empirical formulas. Many of these formulas have necessarily been included, but, in so far as possible, the base formulas to which empirical coefficients have been applied have been derived analytically from fundamental consideration of basic principles.

The book is designed as a text for beginning courses in hydraulics and as a reference book for engineers who may be interested in the fundamental principles of the subject. Tables of coefficients are given which are sufficiently complete for classroom work, but the engineer in practice will need to supplement them with the results of his own experience and with data obtained from other published sources. Acknowledgment for material taken from many publications is made at the proper place in the text.

H. W. K.
C. O. W.

University of Michigan
April 1922

vii

Contents

ix

CHAPTER V
FUNDAMENTALS OF FLUID FLOW

CHAPTER VI
ORIFICES, TUBES, AND WEIRS

CHAPTER VII
PIPES

CHAPTER VIII

OPEN CHANNELS

CHAPTER IX

HYDRODYNAMICS

CHAPTER X

HYDRAULIC SIMILITUDE AND DIMENSIONAL ANALYSIS

GENERAL NOTATION

A area, total cross-sectional
A_m area, mean in reach
B width of canal bed
b breadth of weir
C coefficient of discharge; also Chezy coefficient
C_c coefficient of contraction
C_v coefficient of velocity
C_1 Hazen-Williams coefficient
D diameter
d depth
d_c depth, critical
d_m depth, mean in reach
E energy
F force
f pipe friction coefficient (Darcy-Weisbach)
G work, or power
g acceleration, gravitational
H head, total
H_0 head, above stream bed
h_f head lost by friction
h_v head, velocity
K coefficient of loss
k roughness, of pipe wall
L length along stream bed; also crest length of weir
m coefficient of roughness (Bazin)
N_F Froude's number
N_R Reynolds' number
N_W Weber's number
n coefficient of roughness (Kutter and Manning)

P wetted perimeter; also height of weir
P_m wetted perimeter, mean in a reach
p pressure, intensity of
p_a pressure, atmospheric
p_v pressure, vapor
Q discharge, total
q discharge, per unit width
R hydraulic radius, A/P
R_m hydraulic radius, mean in a reach
S slope of energy gradient
S_0 slope of channel bed
S_w slope of stream surface
s side slope (s horizontal to 1 vertical)
V velocity, mean, Q/A
V_c velocity, critical
V_m velocity, mean in a reach
V_w velocity of wave
W weight
w weight per unit volume
Z elevation above datum
α coefficient, unequal velocities in cross section
μ viscosity, dynamic (absolute)
ν viscosity, kinematic, μ/ρ
ρ density, mass per unit volume, w/g
σ surface tension

Chapter *I*

FUNDAMENTAL PROPERTIES OF FLUIDS

1. The Science of Hydraulics. Hydraulics is defined as that branch of science which treats of water or other fluid in motion. A prerequisite to the understanding of the motion of fluids, however, is a knowledge of the pressure exerted by fluids at rest. This study, called hydrostatics, is usually included in hydraulics. The field of hydraulics also includes hydrodynamics, which relates to the forces exerted by or upon fluids in motion.

2. Fluids. Fluids are substances capable of flowing, having particles which easily move and change their relative position without a separation of the mass. Fluids offer practically no resistance to change of form. They readily conform to the shape of the solid body with which they come in contact.

Fluids may be divided into liquids and gases. The principal differences between them are:

1. A liquid has a free surface, and a given mass of a liquid occupies only a given volume in a container, whereas a gas does not have a free surface, and a given mass occupies all portions of any container regardless of its size.

2. Liquids are practically incompressible and usually may be so considered without introducing appreciable error. On the other hand, gases are compressible and usually must be so treated.

The theory and the problems of this text deal mainly with fluids which may be considered incompressible. A few examples and problems require the use of the simple gas laws which give the relationship of pressure, volume, and temperature.

The distinctions between a solid and a fluid should be noted here:

1. A solid is deformed by a shearing stress, the *amount* of unit deformation up to a certain point being proportional to the unit stress; a fluid is also deformed by a shearing stress but at a *time rate* of deformation which is proportional to the stress.

2. If the elastic limit is not exceeded, the application of a given unit shearing stress to a solid produces a certain unit deformation

1

which is independent of the time of application of the force, and when the stress is removed the solid returns to its original form. On the other hand, if a given shearing stress is applied to a fluid, deformation continues to take place at a uniform rate with time, and when the stress is removed the fluid does not, through forces contained within itself, return to its original form.

The application of sufficient heat will change many solids into a fluid state. The hardest steel can be melted so that it will flow easily. A block of cold tar shows properties of a solid but if heated becomes fluid and can be poured into small cracks in concrete. The change from solid rock to molten lava is a well-known occurrence in nature. Relatively high temperatures are required for these changes. The change from solid ice to fluid water, however, occurs at 32° F.

The mechanics of the borderline condition in which a substance may be either a plastic solid or a very viscous fluid has not been as thoroughly studied in engineering as have the strictly solid and the strictly fluid states.

3. Units Used in Hydraulics. Engineering practice in the United States is generally based on the *foot-pound-second* system of units. In practically all hydraulic formulas these units are used, and if not otherwise stated they are understood. Frequently the diameters of pipes or orifices are expressed in inches, pressures are usually stated in pounds per square inch, and volumes may be expressed in gallons. Before applying such data to problems, conversion to the foot-pound-second system of units should be made. Care in the conversion of units is essential. Errors in hydraulic computations result more frequently from wrong use of units than from any other cause.

Since it is frequently necessary to interchange metric and foot-pound-second units, the relations of these systems of units are briefly reviewed here.

The fundamental equation relating force F, mass M, and acceleration a is

$$F = kMa$$

where k is a proportionality factor. The value of k is made equal to 1 by two different systems of defining units.

1. *Gravitational system*, in which k is made equal to 1 by defining the unit of mass. If a body of unit weight falls freely, unit force

is acting and the acceleration is g. Thus for unit force to produce unit acceleration the unit of mass must consist of g units of weight.

(a) Foot-pound-second system: 1 lb force = 1 slug mass $\times$ 1 ft per sec per sec, in which 1 slug mass = g pounds weight divided by g feet per second per second. An average, commonly used value of g is 32.2 ft per sec per sec.

(b) Metric system: 1 gram force = 1 unit of mass $\times$ 1 cm per sec per sec, in which a unit of mass = g grams weight divided by g centimeters per second per second. An average, commonly used value of g is 981 cm per sec per sec.

2. *Absolute system*, in which k is made equal to 1 by defining the unit of force.

(a) Foot-poundal-second system: Unit force is that force which, acting on a body of 1 lb mass, gives it an acceleration of 1 ft per sec per sec, and is called a poundal. Therefore, 1 poundal force = 1 lb mass $\times$ 1 ft per sec per sec. By 1 lb mass is meant an amount of matter equivalent to that in a block of metal, known as the standard pound, which is kept in Washington, D. C.

(b) Metric system: Unit force is that force which, acting on a body of 1 gram mass, gives it an acceleration of 1 cm per sec per sec, and is called a dyne. Therefore, 1 dyne force = 1 gram mass $\times$ 1 cm per sec per sec. The unit of mass is the gram, which is defined as 1/1000 of the mass of a block of platinum kept in Sèvres and known as the kilogram prototype.

The metric and foot-pound-second systems are related by the following units of length and weight:

$$1 \text{ meter} = 3.2808 \text{ ft}$$

$$1 \text{ kilogram} = 2.2046 \text{ lb}$$

EXAMPLE. How many dynes force are equivalent to 1 lb force?

Solution. 1 slug mass weighs g lb or 453.6 g grams. Also 1 ft per sec per sec = 30.48 cm per sec per sec. Therefore 1 lb force = 453.6 g $\times$ 30.48 = 444,800 dynes.

4. **General Properties of Fluids.** The properties of fluids which are of fundamental importance in the study of hydraulics are defined here.

Unit Weight w: The weight of a unit volume of a fluid. In foot-pound-second units, the unit weight is expressed in pounds per cubic foot.

Mass Density ρ *(rho):* The mass per unit of volume. Thus, in engineers' gravitational units,

$$\rho = \frac{w}{g} \quad \text{or} \quad w = \rho g$$

where g equals the acceleration due to gravity. In foot-pound-second units, mass density is thus slugs per cubic foot, or also

$$\frac{\text{lb/ft}^3}{\text{ft/sec}^2} = \frac{\text{lb sec}^2}{\text{ft}^4}$$

In the metric system density is measured in grams per cubic centimeter and is therefore numerically equal to specific gravity.

Specific Gravity s: The ratio of the unit weight of a fluid to the unit weight of water at 4° C (39.2° F).

Viscosity μ *(mu):* Viscosity is that property of a fluid which determines the amount of its resistance to a shearing stress. A perfect fluid would have no viscosity. There is no perfect fluid, but gases show less variation in viscosity than liquids. Water is one of the least viscous of all liquids, whereas glycerine, heavy oil, and molasses are liquids having comparatively high viscosities.

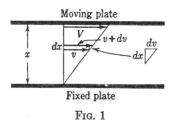

Moving plate

Fixed plate

Fig. 1

The viscosity of liquids decreases with increasing temperature, whereas the viscosity of gases increases with increasing temperature.

The mathematical basis of viscosity may be derived from Fig. 1. Consider two parallel plates of indefinite extent at distance x apart, the space between them being filled with a fluid. Consider further that one of these plates moves at velocity V parallel to the other plate. Three assumptions are made:

1. That the fluid particles in contact with a moving surface have the velocity of that surface.

2. That the rate of change of velocity is uniform in the direction perpendicular to the direction of motion.

3. That the shearing stress in the fluid is proportional to the rate of change of velocity.

By assumption 2, from similar triangles

$$\frac{V}{x} = \frac{dv}{dx} \tag{1}$$

But by assumption 3, the unit shearing stress

$$\tau \text{ (tau)} = \mu \frac{dv}{dx} \tag{2}$$

where μ is a proportionality factor called the coefficient of viscosity. Thence

$$\frac{V}{x} = \frac{\tau}{\mu} \tag{3}$$

and

$$\mu = \frac{\tau x}{V} \tag{4}$$

If the plates are unit distance apart and moving with unit relative velocity

$$\mu = \tau \tag{5}$$

In this case μ is known as the dynamic, or absolute, viscosity and is thus defined as the force required to move a flat surface of unit area at unit relative velocity parallel to another surface at unit distance away, the space between the surfaces being filled with the fluid.

The foot-pound-second units in which dynamic viscosity is expressed can be evaluated from equation 4. Unit shear τ is in pounds per square foot, distance x is in feet, and velocity v is in feet per second. Hence, the units of μ are

$$\frac{\text{lb/ft}^2 \times \text{ft}}{\text{ft/sec}} = \frac{\text{lb sec}}{\text{ft}^2} = \frac{\text{slug}}{\text{ft sec}}$$

In the metric system, the unit of viscosity is called the poise, 1 poise being 1 dyne sec per cm^2. A centipoise is 0.01 poise. It has been found experimentally that the dynamic viscosity of water at 68° F (20° C) is 1 centipoise. The ratio of the dynamic viscosity of any fluid to the dynamic viscosity of water at 68° F is termed the relative viscosity. Therefore, when expressed in centipoises, the dynamic viscosity and relative viscosity of any fluid are numerically equal.

Kinematic Viscosity ν (nu): The ratio of the dynamic viscosity of a fluid to its mass density. Thus

$$\nu = \frac{\mu}{\rho}$$

and the units are

$$\frac{\text{lb sec/ft}^2}{\text{lb sec}^2/\text{ft}^4} = \frac{\text{ft}^2}{\text{sec}}$$

In the metric system the unit of kinematic viscosity is called the stoke, 1 stoke being 1 sq cm per sec.

Viscosity of liquids is measured by means of an instrument called a viscometer, or viscosimeter. Viscometers are of various designs, of which the most widely used in the United States are the Saybolt and the Ostwald. Viscosity is determined, by equations or tables, from the time required for a certain volume of the liquid to flow through a capillary tube of specified design under standard conditions.[1]

Viscosity of air and gases has been determined by the capillary-tube method and also by determining the torsional force produced by rotating one cylinder inside a slightly larger cylinder, the space between the two being filled with the fluid the viscosity of which is desired.

Surface Tension σ (sigma): At any appreciable distance below the surface of a liquid the molecules are attracted toward each other by equal forces. The molecules near the surface, however, are subjected to an attraction downward that is not balanced by an upward attraction. This causes a film or skin to form on the surface and results in many interesting phenomena. A steel needle or razor blade may be made to float upon water so long as the surface film is not broken, but will sink immediately when the film is broken. Surface tension produces the spherical shape of dewdrops or drops of rain. The flow of liquids with a free surface is affected slightly by surface tension, particularly at low heads and small depths.

Capillary action is also explained by the phenomenon of surface tension combined with that of adhesion. In fact, the capillarity of a liquid has generally been used as a measure of its surface tension, although capillarity measures the adhesion of a liquid to a surface as well as the cohesion of the molecules of the liquid.

In Fig. 2, *A* illustrates an open tube of small diameter immersed in water or any other liquid that wets the tube. The liquid rises

[1] *Standards on Petroleum Products and Lubricants*, published annually by the American Society for Testing Materials.

in the tube higher than the level outside, the meniscus being concave upward. The tube B is immersed in mercury or some other liquid which does not wet the tube. In this case the meniscus is convex upward and the level of the liquid in the tube is depressed. The effect of capillarity decreases as the size of tube increases. The liquid in a tube $\frac{1}{2}$ inch in diameter is approximately at the same level as the outside liquid, but it is appreciably different in smaller tubes. Water-proofing liquids

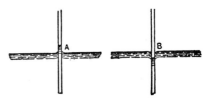

Fig. 2. Capillary action.

have been developed which, when applied to the inside of small glass tubes, greatly reduce the capillary rise of water in the tubes.

The dimensions of surface tension are pounds per foot, the surface film being considered of zero thickness. The surface tension of liquids decreases as their temperature rises.

5. Properties of Water. Various properties of water which are used in hydraulics are shown in the table on page 8. Certain properties are discussed briefly in this article.

Unit Weight. Water has its maximum unit weight at a temperature of 4° C (39.2° F). At this temperature pure water serves as a standard of specific gravity for all substances. Under atmospheric pressure at sea level, water freezes at 32° and boils at 212° F. The weight of pure water at its temperature of maximum density is 62.427 lb per cu ft.

As water occurs in nature, it invariably contains salts and mineral matter in solution. Silt or other impurities may also be carried in suspension. These substances, being heavier than water, increase its weight. The impurities contained in rivers, inland lakes, and ordinary ground waters do not usually add more than 0.1 lb to the weight per cubic foot. Ocean water weighs about 64 lb per cu ft. After long-continued droughts the waters of Great Salt Lake and of the Dead Sea have been found to weigh as much as 75 lb per cu ft.

Since the weight of inland water is not greatly affected by ordinary impurities or changes of temperature, an average weight of water may be used which will give results sufficiently accurate for ordinary purposes. In this book the weight of a cubic foot of water is ordinarily taken as 62.4 lb. In precise work, when the

PROPERTIES OF WATER

Temp. °F	Specific Gravity s	Unit Weight w lb/cu ft	Mass Density ρ slugs/cu ft	Dynamic Viscosity μ lb sec/sq ft	Kinematic Viscosity ν sq ft/sec	Surface Tension with Air σ lb/ft
32	0.9999	62.42	1.940	0.00003746	0.00001931	0.00518
35	0.9999	62.424	1.940	3536	1823	516
39.2	1.0000	62.427	1.941	3274	1687	514
40	0.9999	62.426	1.940	3229	1664	514
45	0.9999	62.42	1.940	2965	1528	511
50	0.9997	62.41	1.940	2735	1410	508
55	0.9994	62.39	1.939	2534	1307	506
60	0.9990	62.37	1.938	2359	1217	503
65	0.9986	62.34	1.937	2196	1134	500
70	0.9980	62.30	1.936	2050	1059	497
75	0.9973	62.26	1.935	1918	0991	494
80	0.9966	62.22	1.934	1799	0930	491
85	0.9958	62.17	1.932	1692	0876	488
90	0.9950	62.11	1.931	1595	0826	485
95	0.9941	62.06	1.929	1505	0780	482
100	0.9931	62.00	1.927	1424	0739	479
110	0.9909	61.86	1.923	1284	0667	473
120	0.9885	61.71	1.918	1168	0609	466
130	0.9860	61.55	1.913	1069	0558	460
140	0.9832	61.38	1.908	0981	0514	453
150	0.9802	61.20	1.902	0905	0476	447
160	0.9770	61.00	1.896	0838	0442	440
170	0.9738	60.80	1.890	0780	0413	433
180	0.9704	60.58	1.883	0726	0385	426
190	0.9667	60.36	1.876	0678	0362	419
200	0.9630	60.12	1.868	0637	0341	412
212	0.9583	59.83	1.860	0593	0319	403

Source: A.S.C.E. Manual 25, "Hydraulic Models"; and International Critical Tables. Mass density and kinematic viscosity were computed with $g = 32.174$ ft per sec per sec.

water temperature is indicated, the corresponding unit weight may be taken from the table.

Compressibility. Water is commonly assumed to be incompressible, but in reality it is about 100 times as compressible as mild steel. Upon release from pressure, water immediately regains its original volume.

Values of the modulus of elasticity of water as given by Daugherty are shown in the accompanying table. From a study of data obtained by several investigators, Daugherty concluded that for any water temperature the value of E increases uniformly with pressure; and that for any pressure E has its maximum value at a temperature of about 120° F.

VALUES OF MODULUS OF ELASTICITY OF WATER IN POUNDS PER SQUARE INCH

Pressure in pounds per square inch	Temperature				
	32° F	68° F	120° F	200° F	300° F
15	292 000	320 000	332 000	308 000	
1 500	300 000	330 000	342 000	319 000	248 000
4 500	317 000	348 000	362 000	338 000	271 000
15 000	380 000	410 000	426 000	405 000	350 000

Source: R. L. Daugherty, *Hydraulics*, McGraw-Hill Book Co., 1937, page 10; "Some Physical Properties of Water and Other Fluids," *Trans. Am. Soc. Mech. Engrs.*, Vol. 57, No. 5, July, 1935.

The compressibility of water usually affects the solution of practical problems in hydraulics only by changing its unit weight. Since pressures commonly encountered are relatively small, water may usually be considered incompressible without introducing any appreciable error.

Viscosity. As previously stated, the viscosity of pure water at 68° F is 0.01 poise. Dynamic and kinematic viscosities in foot-pound-second units at various temperatures are given in the table on page 8.

Surface Tension. The surface tension for water against air, as shown in this table, decreases with rise in temperature. Soap and certain other chemicals in water decrease the surface tension, permitting the formation of bubbles. It has been shown that water with its surface tension reduced will permeate a duck's feathers, thus reducing the bird's buoyancy and causing it to swim lower in the water.

6. Properties of Other Fluids. The table on page 10 shows properties of certain other liquids commonly encountered in engineering practice.

The unit weight of dry air at 32° F (491° F abs) and standard barometric pressure (2116 lb per sq ft, 14.7 lb per sq in.) is 0.0807

Temp. °F	Specific Gravity s	Mass Density ρ slugs / cu ft	Kinematic Viscosity ν sq ft / sec	Surface Tension σ lb / ft
A Light Dust-proofing Oil				
40	0.917	1.779	0.000809	0.00222
50	.913	1.772	565	220
60	.909	1.764	408	217
70	.905	1.757	306	214
80	.902	1.750	234	211
90	.898	1.742	185	209
100	.894	1.735	149	206
110	.890	1.728	122	203
A Medium Fuel Oil				
40	0.865	1.679	0.0000655	
50	.861	1.671	555	
60	.858	1.665	475	
70	.854	1.658	412	
80	.851	1.652	365	
90	.847	1.644	319	
100	.843	1.636	278	
110	.840	1.630	227	
A Heavy Fuel Oil				
40	0.918	1.782	0.00444	0.00219
50	.915	1.775	312	216
60	.912	1.769	221	214
70	.908	1.763	157	211
80	.905	1.756	114	208
90	.902	1.750	0836	205
100	.899	1.744	0627	202
110	.895	1.737	0480	200
A Regular Gasoline				
40	0.738	1.432	0.00000810	
50	.733	1.423	765	
60	.728	1.414	730	
70	.724	1.405	690	
80	.719	1.396	660	
90	.715	1.387	630	
100	.710	1.378	600	
110	.706	1.370	570	
A Cleaning Solvent				
40	0.728	1.413	0.0000161	
50	.725	1.407	148	
60	.721	1.399	137	
70	.717	1.392	126	
80	.713	1.384	117	
90	.709	1.376	110	
100	.705	1.368	103	
110	.702	1.362	096	

Source: L. H. Kessler and A. T. Lenz, Hydraulic Laboratory, University of Wisconsin, Madison.

lb per cu ft; and at any other absolute temperature T (° F) and absolute pressure p (pounds per square foot) is given by the equation

$$w = 0.0807 \times \frac{491}{T} \times \frac{p}{2116} = \frac{p}{53.3T}$$

This equation is based on Charles' and Boyle's laws and gives values which are correct within a fraction of 1 per cent except at high pressures or low absolute temperatures. The value 53.3 is the gas constant for air, the units being feet per degree Fahrenheit. In general the gas constant is denoted by R. Its value differs for different gases.

The viscosity of all gases increases with increasing temperature. The dynamic viscosity of dry air at 20° C as determined by Bearden[1] using the rotating-cylinder method is 0.0001819 poise = 0.000000380 lb sec per sq ft. The rate of increase at that temperature is 0.0000005 poise per degree Centigrade. The rate of change decreases slowly with increase in temperature.

7. Accuracy of Computations. Accuracy of computations is most desirable, but results should not be carried out to a greater number of significant figures than the data justify. Carrying them out too far implies an accuracy which does not exist and may give results that are entirely misleading.

EXAMPLE. Determine the theoretical horsepower available in a stream where the flow is 311 cfs (cubic feet per second) and the available head is 12.0 ft.

As shown later (Art. 50),

$$hp = \frac{QwH}{550}$$

where Q = the flow in cubic feet per second.
w = the unit weight of water in pounds per cubic foot.
H = the available head in feet.

Thus

$$\frac{311 \times 62.4 \times 12.0}{550} = 423.41 \text{ hp}$$

By referring to the table on page 8, it is seen that for the ordinary range of temperatures the weight of water may vary from 62.30 to 62.43

[1] *Physical Review*, Nov. 15, 1939, p. 1023.

lb per cu ft.   Furthermore, the statement that the flow is 311 cfs means merely that the exact value is more nearly 311 than 310 or 312.   In other words, the true value may lie anywhere between 310.5 and 311.5. Likewise, the fact that the head is given as 12.0 ft merely indicates that the correct value lies somewhere between 11.95 and 12.05.   Substituting the lower of these values,

$$\frac{310.5 \times 62.30 \times 11.95}{550} = 420.30 \text{ hp}$$

Substituting the higher of the possible values,

$$\frac{311.5 \times 62.43 \times 12.05}{550} = 426.06 \text{ hp}$$

The decimal 0.41 in the original answer 423.41, therefore, is unjustified, and the last whole number, 3, merely represents the most probable value, since the correct value may lie anywhere between 420.30 and 426.06.   The answer should, therefore, be given as 423.

It may be stated in general that in any computation involving multiplication or division, in which one or more of the numbers is the result of observation, the answer should contain the same number of significant figures as is contained in the observed quantity having the fewest significant figures.   In applying this rule it should be understood that the last significant figure in the answer is not necessarily correct but represents merely the most probable value.   To give in the answer a greater number of significant figures indicates a degree of precision that is unwarranted and misleading.

PROBLEMS

1. Water at 68° F has a dynamic viscosity of 1 centipoise.   Compute its dynamic viscosity and kinematic viscosity in foot-pound-second units.

2. The kinematic viscosity unit of 1 sq ft per sec is equivalent to how many stokes?

3. A volume of 15.5 cu ft of a liquid weighs 782 lb.   Compute the mass density of the liquid.

4. Compute the number of watts which are equivalent to 1 horsepower.   (1 watt = 10^7 cm-dynes per sec; 1 hp = 550 ft-lb per sec.)

5. A city of 6000 population has an average water use of 110 gallons per person per day.   Compute the average total rate of use in cubic feet per second and in gallons per minute.   (1 cu ft = 7.48 gal.)

6. Compute the numerical factor for converting foot-pound-second units of dynamic viscosity to poises.

7. Compute the unit weight and kinematic viscosity, in foot-pound-second units, of dry air at 15° C and an absolute pressure of 14.7 lb per sq in.

Chapter *II*

PRINCIPLES OF HYDROSTATIC PRESSURE

8. Unit Pressure, *p*. The unit pressure, meaning the intensity of pressure, at any point in a fluid is the amount of pressure per unit area.

If the unit pressure is the same at every point on any area, A, on which the total pressure is P,

$$p = \frac{P}{A}$$

If, however, the unit pressure is different at different points, the unit pressure at any point is equal to the total pressure on a small differential area surrounding the point divided by the differential area, or

$$p = \frac{dP}{dA}$$

Where there is no danger of ambiguity, the term pressure is often used as an abbreviated expression for unit pressure.

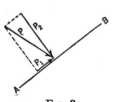

FIG. 3

The fundamental foot-pound-second unit for pressure is pounds per square foot, although the unit of pounds per square inch is often used.

9. Direction of Resultant Pressure. The resultant pressure on any plane in a fluid at rest is normal to that plane.

Assume that the resultant pressure P, on any plane AB (Fig. 3), makes an angle other than 90° with the plane. Resolving P into rectangular components P_1 and P_2, respectively parallel with and perpendicular to AB, gives a component P_1 which can be resisted only by a shearing stress. By definition, a fluid at rest cannot resist a shearing stress, and therefore the pressure must be normal to the plane. This means that there can be no static friction in hydraulics.

14

10. Pascal's Law. At any point in a fluid at rest, the pressure is the same in all directions. This principle is known as Pascal's law.

Consider an infinitesimally small wedge-shaped volume, BCD (Fig. 4), in which the side BC is vertical, CD is horizontal, and BD makes any angle θ with the horizontal. Let A_1, A_2, and A_3 and p_1, p_2, and p_3 represent, respectively, the area of these sides and the pressures to which they are subjected. Assume that the ends of the wedge are vertical and parallel.

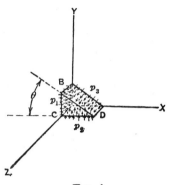

Fig. 4

Since the wedge is at rest, the principles of equilibrium may be applied to it. From Art. 9 it is known that the pressures are normal to the faces of the wedge. Choosing the coordinate axes as indicated in Fig. 4 and setting up the equations of equilibrium, $\Sigma F_x = 0$ and $\Sigma F_y = 0$, and neglecting the pressures on the ends of the wedge, since they are the only forces acting on the wedge which have components along the Z axis and therefore balance each other, the following expressions result:

$$p_1 A_1 = p_3 A_3 \sin \theta$$

$$p_2 A_2 = p_3 A_3 \cos \theta$$

But

$$A_3 \sin \theta = A_1 \quad \text{and} \quad A_3 \cos \theta = A_2$$

Therefore

$$p_1 = p_2 = p_3$$

Since BD represents a plane making any angle with the horizontal and the wedge is infinitesimally small so that the sides may be considered as bounding a point, it is evident that the unit pressure at any point must be the same in all directions.

11. Free Surface of a Liquid. Strictly speaking, a liquid having a free surface is one on whose surface there is absolutely no pressure. It will be shown later, however, that there is always some pressure on the surface of every liquid.

In practice the free surface of a liquid is considered to be a surface that is not in contact with the cover of the containing vessel.

Such a surface may or may not be subjected to the pressure of the atmosphere.

It may be shown that the free surface of a liquid at rest is horizontal. Assume a liquid having a surface which is not horizontal, such as $ABCDE$ (Fig. 5). A plane MN, inclined to the horizontal, may be passed through any liquid having such a surface in such manner that a portion of the liquid BCD lies above the plane. Since the liquid is at rest, BCD must be in equilibrium, but the vertical force of gravity would necessarily have a component along the inclined plane which could be resisted only by a shearing stress. As liquids at rest are incapable of resisting shearing stress it follows that the free surface must be horizontal.

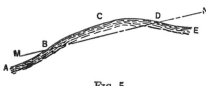

FIG. 5

12. Atmospheric Pressure. All gases possess mass and consequently have weight. The atmosphere, being a fluid composed of a mixture of gases, exerts a pressure on every surface with which it comes in contact. At sea level under normal conditions atmospheric pressure amounts to 2116 lb per sq ft, or about 14.7 lb per sq in., which is equivalent to 30 in. of mercury column (Art. 19).

VARIATION IN ATMOSPHERIC PRESSURE WITH ALTITUDE

Altitude above sea level in feet	Pressure in pounds per square inch	Altitude above sea level in feet	Pressure in pounds per square inch
0	14.69	5 280	12.08
1 000	14.17	6 000	11.76
2 000	13.66	7 000	11.32
3 000	13.16	8 000	10.89
4 000	12.68	9 000	10.48
5 000	12.21	10 000	10.09

The pressure and the unit weight of the atmosphere decrease with increase in altitude, theoretically in accordance with the mathematical laws which are discussed in Art. 27. The accompanying table gives values of the atmospheric pressure corresponding to different elevations above sea level.

13. Vacuum. A perfect vacuum, that is, a space in which there is no matter either in the solid, liquid, or gaseous form, has never been obtained. It is not difficult, however, to obtain a space containing a minute quantity of matter. A space in contact with a liquid, if it contains no other substance, always contains vapor from that liquid. In a perfect vacuum there could be no pressure.

In practice, the word " vacuum " is used frequently in connection with any space having a pressure less than atmospheric pressure, and the term " amount of vacuum " means the amount the pressure is less than the prevailing atmospheric pressure. The amount of vacuum is usually expressed in inches of mercury column or in pounds per square inch measured from atmospheric pressure as a base. For example, if the atmospheric pressure is standard and if the pressure within a vessel is reduced to 12 lb per sq in., which is equivalent to 24.5 in. of mercury column, there is said to be a vacuum of 2.7 lb per sq in. or 5.5 in. of mercury.

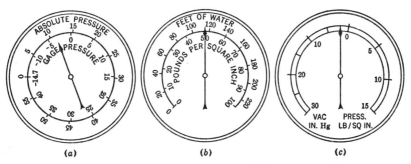

FIG. 6.   Pressure gage dials.

14. Absolute and Gage Pressure. The intensity of pressure measured above absolute zero is called *absolute* pressure. Obviously, a negative absolute pressure is impossible.

Usually pressure gages are designed to measure intensities of pressures above or below atmospheric pressure as a base. Pressures so measured are called relative or *gage* pressures. Negative gage pressures indicate the amount of vacuum, and at standard conditions at sea level pressures as low as (but no lower than) −14.7 lb per sq in. are possible. Absolute pressure is always equal to gage pressure plus atmospheric pressure.

Figure 6 shows various types of pressure gage dials. In (a) the inner circle shows the ordinary gage and vacuum scale in pounds per square inch. The outer scale is on a movable ring, which, when properly set, indicates the corresponding absolute pressures. In

(b) the inner circle reads in pounds per square inch, and the outer circle shows the corresponding heads in feet of water (Art. 16). In (c) the part of the circle to the left of zero shows vacuum in inches of mercury and the part to the right shows pressure in pounds per square inch.

Absolute pressures are sometimes measured in standard "atmospheres." Thus 1 atmosphere = 14.7 lb per sq in. abs = zero gage pressure; 3 atmospheres = 44.1 lb per sq in. abs = 29.4 lb per sq in. gage.

The term *pressure* as used in this book signifies gage pressure unless otherwise indicated.

15. Variation of Pressure with Depth in a Fluid. To determine the pressure at any point in a homogeneous fluid at rest or the variation in pressure in such a fluid, consider any two points such as 1 and 2 (Fig. 7) whose difference of elevation is h. Consider

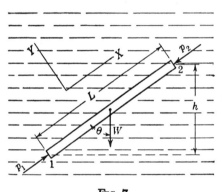

FIG. 7

that these points lie in the ends of an elementary prism of the fluid, having a cross-sectional area dA and length L. Since this prism is at rest, all the forces acting upon it must be in equilibrium. These forces consist of the fluid pressure on the sides and ends of the prism and the force of gravity.

Let X and Y, the coordinate axes, be respectively parallel with and perpendicular to the axis of the prism which makes an angle θ with the vertical. Also let p_1 and p_2 be the pressures at points 1 and 2, respectively, and let w be the unit weight of the fluid.

Considering forces acting to the left along the X axis as negative and remembering that the pressures on the sides of the prism are normal to the X axis and therefore have no X components, the following equation may be written:

$$\Sigma F_x = p_1 \, dA - p_2 \, dA - wL \, dA \cos \theta = 0 \qquad (1)$$

Since $L \cos \theta = h$, this reduces to

$$p_1 - p_2 = wh \qquad (2)$$

Therefore in any homogeneous fluid at rest the difference in pressure between any two points is the product of the unit weight of the fluid and the difference in elevation of the points.

If $h = 0$, $p_1 = p_2$, or, in other words, in any continuous homogeneous fluid at rest, the pressures at all points in a horizontal plane are the same. Conversely, in any homogeneous fluid at rest all points having equal pressures lie in a horizontal plane.

If the fluid is a liquid having a free surface in which point 2 (Fig. 7) is assumed to lie, p_2 becomes the pressure p_a at that surface. At any depth h the absolute pressure is

$$p_{abs} = wh + p_a \qquad (3)$$

The corresponding gage pressure is

$$p = wh \qquad (4)$$

16. Pressure Head. Equation 4 may be written in the form

$$\frac{p}{w} = h \qquad (5)$$

Here h, or its equivalent p/w, called in hydraulics the pressure head, represents the height of a column of homogeneous fluid of unit weight w that will produce an intensity of pressure p.

The pressure in a gas may be expressed in terms of hypothetical pressure head, which is the height of a column of the gas assumed to be of uniform unit weight required to produce the given pressure. A gas, of course, forms no free surface, its unit weight decreasing with altitude. (See Art. 27.)

Equation 2 may be written

$$\frac{p_1}{w} - \frac{p_2}{w} = h \qquad (6)$$

meaning that the difference in pressure heads at two points in a homogeneous fluid at rest is equal to the difference in elevation of the points.

Equations 1 to 6 may be applied only if the conditions of the problem are such that the fluid may be assumed to have constant unit weight. This is usually true of liquids. In respect to gases this assumption can be made without serious error if the differences in elevation involved are relatively small. Otherwise the theory relating to compressible fluids should be applied.

PROBLEMS

1. Determine the pressure on the face of a dam at a point 40 ft below the water surface, in (a) pounds per square foot gage; (b) pounds per square inch gage; (c) pounds per square foot absolute; (d) pounds per square inch absolute.

2. Determine the pressure in a vessel of mercury (sp gr 13.6) at a point 8 in. below the surface, expressing the answer in the same units as in problem 1.

3. Assuming sea water to be incompressible (w = 64.0 lb per cu ft), what is the pressure in tons per square foot 2 miles below the surface of the ocean?

4. What height of mercury column will cause a pressure of 100 lb per sq in.? What is the equivalent height of water column?

5. Derive and memorize the conversion factor for changing head in feet of water to pressure in pounds per square inch.

6. Derive and memorize the conversion factor for changing pressure in pounds per square inch to head in feet of water.

7. At what depth in a standpipe containing water is the pressure 30 lb per sq in.?

8. What height of a column of special gage liquid (sp gr 2.95) would exert the same pressure as a column of oil 15 ft high (sp gr 0.84)?

9. Assuming the unit weight of air constant at 0.0765 lb per cu ft, what is the approximate decrease in pressure in pounds per square inch corresponding to a rise in elevation of 1000 ft? Compare with the table on page 16. Express the decrease also in terms of equivalent head in inches of water.

10. If the pressure in a tank of oil (sp gr 0.80) is 60 lb per sq in., what is the equivalent head: (a) in feet of the oil; (b) in feet of water; (c) in inches of mercury?

11. What is the pressure in pounds per square inch 4 ft below the surface of a liquid of sp gr 1.50 if the gas pressure on the surface is 0.4 atmosphere?

12. The pressure in a gas tank is 2.75 atmospheres. Compute the pressure in pounds per square inch and the pressure head in feet of water.

13. A gage on the suction side of a pump shows a vacuum of 10 in. of mercury. Compute (a) pressure head in feet of water; (b) pressure in pounds per square inch; (c) absolute pressure in pounds per square inch if the barometer reads 29.0 in. of mercury.

14. A pressure gage on the discharge side of a pump reads 43.5 lb per sq in. Oil (sp gr 0.82) is being pumped. Compute the pressure head in feet of oil.

15. In a condenser containing air and water, the air pressure is 3.2 lb per sq in. abs. What is the gage pressure in pounds per square foot at a point 4.5 ft below the water surface?

16. A piece of wood 1 ft square and 10 ft long, weighing 40 lb per cu ft, is submerged vertically in a body of water, its upper end being flush with the water surface. What vertical force is required to hold it in position?

17. What is the absolute pressure in pounds per square inch 30 ft below the open surface in a tank of oil (sp gr 0.85) if the barometric pressure is 28.5 in. of mercury?

18. A glass tube 5 ft long and 1 in. in diameter with one end closed is inserted vertically, with the open end down, into a tank of water until the open end is submerged to a depth of 4 ft. If the barometric pressure is 14.3 lb per sq in., and neglecting vapor pressure, how high will water rise in the tube?

17. Transmission of Pressure. By writing equation 2 in the form

$$p_1 = p_2 + wh \tag{7}$$

it is seen that the pressure at any point, such as 1 (Fig. 7), in a liquid at rest is equal to the pressure at any other point, such as 2, plus the pressure produced by a column of the liquid the height of which, h, is equal to the difference in elevation between the two points. Any change in the pressure at point 2 would cause an equal change at point 1. In other words, a pressure applied at any point in a liquid at rest is transmitted equally and undiminished to every other point in the liquid.

This principle, which is also ascribed to Pascal, is made use of in the hydraulic jack by means of which heavy weights are lifted by the application of relatively small forces.

18. Vapor Pressure. Whenever the free surface of a liquid is exposed to the atmosphere, evaporation is continually taking place. If, however, the surface is in contact with an enclosed space, evaporation takes place only until the space becomes saturated with vapor. This vapor produces a pressure, the amount of which depends only upon the temperature and is entirely independent of the presence or absence of air or other gas within the enclosed space. The pressure exerted by a vapor within a closed space is called vapor pressure.

In Fig. 8, A represents a tube having its open end submerged in a liquid and having a stopcock at its upper end.   Consider the air within A to be absolutely dry at the time the stopcock is closed. At the instant of closure the surfaces inside and outside the tube stand at the same level.   Evaporation within the tube, however, soon saturates the space containing air and creates a vapor pressure,

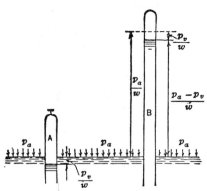

p_v, which causes a depression of the surface within the tube equal to p_v/w.

In the same figure, B represents a tube closed at the upper end.   Assume a perfect vacuum in the space above the liquid in the tube.   If this condition were possible the level in B would stand at an elevation p_a/w above the surface of the liquid outside. Vapor pressure within the

FIG. 8.   Vapor pressure.

vessel, however, causes a depression p_v/w equal to that produced within A, so that the maximum height of column possible under conditions of equilibrium in such a tube is $(p_a - p_v)/w$.   Vapor pressures increase with the temperature, as is shown in the accompanying table of pressure heads for water vapor.

WATER VAPOR PRESSURE HEADS IN FEET OF WATER

Temperature, F	$\dfrac{p_v}{w}$	Temperature, F	$\dfrac{p_v}{w}$	Temperature, F	$\dfrac{p_v}{w}$
−20°	0.02	60°	0.59	140°	6.63
−10	.03	70	0.83	150	8.54
0	.05	80	1.16	160	10.90
10	.08	90	1.59	170	13.78
20	.13	100	2.17	180	17.28
30	.19	110	2.91	190	21.49
40	.28	120	3.87	200	26.52
50	.41	130	5.09	212	33.84

19. The Mercury Barometer.   The barometer is a device for measuring intensities of pressure exerted by the atmosphere.   In 1643 Torricelli discovered that, if a tube (Fig. 9) more than 30 in. long and closed at one end is filled with mercury and then made to stand vertically with the open end submerged in a vessel of mercury, the column in the tube will stand approximately 30 in. above the surface of the mercury in the vessel. Such a device is known as a mercury barometer. Pascal proved that the height of the column of mercury depended upon the atmospheric pressure, when he carried a barometer to a higher elevation and found that the height of the column decreased as the altitude increased.

Although, theoretically, water or any other liquid may be used for barometers, two difficulties arise in using water. First, the height of water column necessary to balance the atmospheric pressure that usually occurs at sea level is about 34 ft, which height is too great for convenient use; and, second, as shown in Art. 18, water vapor collecting in the upper portion of the tube creates a pressure which partially balances the atmospheric pressure, so that the barometer does not indicate the total atmospheric pressure.

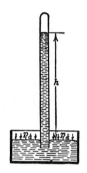

FIG. 9.
Barometer.

Since mercury is the heaviest known liquid, has a·very low vapor pressure, and does not freeze at ordinary air temperatures, it is more satisfactory for use in barometers than any other liquid.

PROBLEMS

1. At what height will water stand in a water barometer at an altitude of 5000 ft above sea level if the temperature of the water is 70° F? Under similar conditions what would be the reading of a mercury barometer, neglecting the vapor pressure of mercury?

2. A mercury barometer at the base of a mountain reads 28.95 in. At the same time, another barometer at the top of the mountain reads 23.22 in. Assuming w for air to be constant at 0.0765 lb per cu ft, what is the approximate height of the mountain?

3. A mercury barometer reads 26.45 in. (a) What would be the corresponding reading of a water barometer? (b) What is the atmospheric pressure in pounds per square inch? (c) What is the approximate elevation above sea level, assuming normal atmospheric conditions? Neglect vapor pressure.

4. On a mountain the barometric pressure is 24 in. of mercury at 32° F. (a) What is the pressure in pounds per square inch? (b)

Assuming normal atmospheric conditions, approximately how high is the mountain above sea level?

5. Compute " standard atmospheric head," which is defined as the hypothetical height of a column of air of uniform unit weight $w =$ 0.0765 lb per cu ft, with nothing above, that would produce standard atmospheric pressure.

20. Manometers. A manometer is a tube, usually bent in the form of a U, containing a liquid of known specific gravity, the surface of which moves proportionally to changes of pressure. Manometers are: (1) open type, with an atmospheric surface in one leg and capable of measuring gage pressures; and (2) differential type, without an atmospheric surface and capable of measuring only differences of pressure.

21. Piezometer. The simplest form of manometer is the piezometer, which is a tube tapped into the wall of a container or

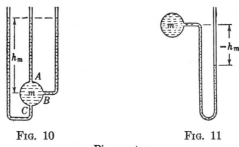

FIG. 10   FIG. 11
Piezometers.

conduit for the purpose of measuring the pressure. Figure 10 represents the cross section of a tank or pipe containing liquid under pressure. Piezometers are tapped into the top, side, and bottom at points A, B, and C, respectively. From the foregoing principles of pressure in a homogeneous liquid at rest, it is obvious that the level to which the liquid rises is the same in the three tubes and that the pressure at any point in the liquid is indicated by the height of the free surface in any one of the piezometers above that point. Thus, the pressure at m, the center of the pipe, is

$$p_m = wh_m$$

Conversely, h_m is the pressure head at m. Piezometers measure gage pressures, since the surface of the liquid in the tube is subjected to atmospheric pressure.

Negative pressures can be measured by means of the piezometer shown in Fig. 11. Since the pressure in a homogeneous liquid decreases as the elevation increases, the pressure at m is $-wh_m$ and the pressure head is $-h_m$.

Piezometers are also used to measure pressure heads in pipes where the liquid is in motion. Such tubes should enter the pipe in a direction at right angles to the direction of flow, and the connecting end should be flush with the inner surface of the pipe without burrs or roughness; otherwise the height of the column may be affected by the velocity of the liquid. (See Art. 55.) A vertical piezometer in the top of the pipe is simplest to construct. Moving liquids, however, frequently carry air which may collect along the top of the pipe, enter the piezometer, and affect the liquid level. Piezometer connections at the sides or bottom of a pipe are therefore more reliable.

The diameter of piezometer tubes at atmospheric surface should be large enough to prevent capillary action from affecting the height of the column of liquid. Usually $\frac{1}{2}$-in. diameter is sufficient for glass tubes. Smaller tubes have been found to give good results if a water-proofing liquid is applied to the inside glass surface. In order to damp fluctuations of liquid level a short length of capillary tube is frequently inserted between the pipe connection and the atmospheric surface. Such a restricted passage does not affect the pressure indication, provided that the pressure in the conduit remains approximately constant, except for momentary fluctuations above and below the average value. A continuously increasing or decreasing pressure is not measured accurately by a piezometer because the change in liquid level lags behind the change in pressure.

22. Open Manometer. A piezometer is limited in its range of pressure measurement since (1) large pressures in the lighter liquids require long tubes, and (2) gas pressures can not be measured because a gas forms no free atmospheric surface. These objections may be overcome by the use of tubes which are bent to contain one or more fluids of different specific gravities from that in which the pressure is desired. One arrangement of such a tube is shown in Fig. 12.

Liquids used in manometers must form a meniscus, that is, adjacent liquids must not mix. Liquids other than water that are frequently used are mercury, oil, salt solutions, carbon disulphide, carbon tetrachloride, bromoform, and alcohol.

It is not advisable to rely on general formulas for the solution of manometer problems.   Each problem should be considered individually and solved in accordance with funda-

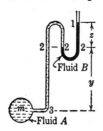

FIG. 12. Open manometer.

mental principles of variation of hydrostatic pressure with depth.   It is ordinarily easier to work in units of pressure head rather than pressure.   Suggested steps in the solution of open manometer problems are:

1. Draw a sketch of the manometer approximately to scale.

2. Decide on the fluid in feet of which the heads are to be expressed.

3. Starting with the atmospheric surface in the manometer as the point of known pressure head, number in order the levels of contact of fluids of different specific gravities.

4. Starting with atmospheric pressure head, proceed from level to level, adding or subtracting pressure heads as the elevation decreases or increases, respectively, with due regard for the specific gravities of the fluids.

EXAMPLE.   The pressure at point m in Fig. 12 is to be measured by the open manometer shown.   Fluid A is oil (sp gr 0.80).   Fluid B is mercury.   Height $y = 30$ in.   Height $z = 10$ in.

Solution.   1. In terms of gage pressure head in feet of oil:

At 1 p/w = 0.0 ft of oil

From 1 to 2:

Pressure head increase $= \dfrac{10}{12} \times \dfrac{13.6}{0.80} = 14.2$ "

All points 2 have the same pressure.   Why?

From 2 to 3:

Pressure head increase $= \dfrac{30}{12}$ $= 2.5$ "

At m: p/w = 16.7 "

and $p = 5.78$ lb per sq in.

2. In terms of gage pressure:

At 1 p = 0.0 lb per sq in.

From 1 to 2:

Pressure increase $= \dfrac{10}{12} \times \dfrac{13.6 \times 62.4}{144} = 4.92$ "

From 2 to 3:

Pressure increase $= \dfrac{30}{12} \times \dfrac{0.80 \times 62.4}{144} = 0.86$ "

At m: $p = 5.78$ "

PROBLEMS

1. In the piezometers of Fig. 10, liquid stands 4.50 ft above m. What is the pressure at m in pounds per square inch if the liquid is (a) water, (b) oil (sp gr 0.90), (c) mercury, (d) molasses (sp gr 1.50)? What is the pressure head in feet of each liquid?

2. How high will liquid rise in the piezometers of Fig. 10 if the pressure at m is 10 lb per sq in. and the liquid is (a) water, (b) oil (sp gr 0.85), (c) mercury, (d) brine (sp gr 1.15)?

3. The fluid in Fig. 11 is water and $h_m = 23.5$ ft. Assuming that atmospheric pressure is standard, compute gage pressure and absolute pressure at m.

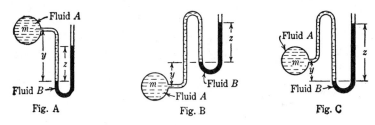

Fig. A Fig. B Fig. C

4. In Fig. A find the pressure head and the pressure at m when:

(a) Fluid A is water, fluid B is mercury, $z = 15$ in., $y = 30$ in.

(b) Fluid A is oil (sp gr 0.80), fluid B is a calcium chloride solution (sp gr 1.25), $z = 12$ in., $y = 8$ ft.

(c) Fluid A is gas ($w = 0.04$ lb per cu ft), fluid B is water, $z = 5$ in., $y = 15$ in. How much does the value of y affect the result?

5. In Fig. A, what is the height z if fluid A is water, fluid B is mercury, the gage pressure at m is 20 lb per sq in., and $y = 5$ ft?

6. In Fig. B, find the pressure head at m in feet of fluid A when:

(a) Fluid A is water, fluid B is mercury, $z = 15$ in., $y = 30$ in.

(b) Fluid A is sea water (sp gr 1.03), fluid B is bromoform (sp gr 2.87) $z = 12$ in., $y = 8$ ft.

7. In Fig. C, find the pressure at m when:

(a) Fluid A is water, fluid B is carbon tetrachloride (sp gr 1.60), $z = 22$ in., $y = 12$ in.

(b) Fluid A is oil (sp gr 0.915), B is water, $z = 8.5$ in., $y = 42$ in.

8. In Fig. D, assuming standard atmospheric pressure, compute the absolute pressure at m when:

(a) Fluid A is water, fluid B is mercury, $z = 15$ in., $y = 3.5$ ft.

(b) Fluid A is oil (sp gr 0.82), fluid B is a salt solution (sp gr 1.10), $z = 21.5$ in., $y = 7.5$ in.

9. In Fig. D the distance $y + \frac{1}{2}z = 4.0$ ft. When fluid A is water, fluid B is mercury, and the pressure at m is a vacuum of 5.4 lb per sq in., compute z.

10. In Fig. E, find the pressure head and pressure at m when:

(a) Fluid A is oil (sp gr 0.90), fluid B is carbon tetrachloride (sp gr 1.50), fluid C is air, $z = 24$ in., $y = 42$ in.

(b) Fluid A is water, fluid B is mercury, fluid C is air, $z = 13$ in., $y = 4.2$ ft.

11. In Fig. F, fluids A and C are air, fluid B is water, $z = 14$ in. Find gage pressure at m and absolute pressure assuming standard atmospheric conditions.

12. In Fig. G, fluids A and C are air, fluid B is mercury, $z = 5$ in. Compute gage and absolute pressures at m.

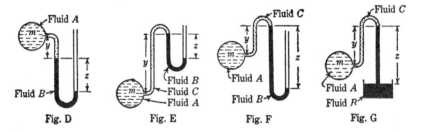

Fig. D   Fig. E   Fig. F   Fig. G

23. Single-tube Manometers. The open manometers described in Art. 22 usually require readings of fluid levels at two or more points, since a change in pressure causes a rise of liquid in one tube and a drop in the other. If, however, a reservoir having a large cross-sectional area compared to the area of the tube is introduced into one leg of the manometer, as illustrated in Fig. 13, the change in liquid level in that leg is held to a small or negligible amount, and the pressure is indicated approximately by the height of the liquid in the other leg. The tube may be vertical as in (a) or, for the measurement of small pressures, inclined as in (b).

With atmospheric pressure in the reservoir and tube at level O, both surfaces of the gage liquid stand at that level, called normal position. With increase in pressure at the interface, the gage liquid drops distance Δy in the reservoir and rises distance h in the tube. If A and a are the cross-sectional areas of reservoir and tube respectively,

$$A \cdot \Delta y = ah \qquad (8)$$

If fluid A is water and fluid B has specific gravity s, by the principles of variation of pressure head with depth, starting at 1, in terms of feet of water,

$$0 + hs + \Delta y \cdot s - \Delta y - y = \frac{p_m}{w} \qquad (9)$$

Simultaneous solution of equations 8 and 9 gives p_m/w in terms of h, s, and a/A.

By increasing the size of the reservoir sufficiently, the ratio a/A can be made so small that Δy is negligible and height h is a measure

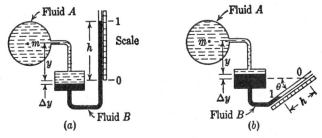

(a) (b)

FIG. 13. Single-tube manometers.

of the pressure head at the contact level in the reservoir. Or, if Δy is appreciable, the scale on which h is read can be so graduated as to correct for Δy so that only one reading of liquid level is required.

When the tube is inclined as in Fig. 13b,

$$A \cdot \Delta y = ah \qquad (8)$$

and

$$0 - (h \sin \theta)s - \Delta y \cdot s - y = \frac{p_m}{w} \qquad (10)$$

PROBLEMS

1. The diameters of reservoir and tube in Fig. 13a are 10 cm and 1 cm respectively. Fluid A is oil (sp gr 0.903); fluid B, a calcium chloride solution (sp gr 1.258). With the zero of the scale set opposite the liquid interface in the reservoir, the scale reading opposite point 1 is 1.615 ft and opposite m is 0.924 ft. Determine the pressure at m.

2. If the scale in problem 1 is to remain fixed with its zero at the normal level of fluid B, what should be the length of graduation on the scale in inches to represent a pressure change at m of 1 lb per sq in.?

3. In Fig. 13b, fluid A is gas, fluid B kerosene (sp gr 0.805). The slope of the tube is 6 horizontal to 1 vertical. The diameters of reservoir and tube are 2 in. and $\frac{1}{4}$ in., respectively. Compute the gage pressure head at m in inches of water when h equals 10.6 in.

4. In problem 3 what percentage error would result if the rise of oil in the reservoir were neglected?

24. Differential Manometer. Frequently in hydraulic problems the difference in pressure at two points in a pipe line or a system is desired rather than the actual pressure at the points. For this purpose a differential manometer can often be used. A manometer of this type is shown in Fig. 14.

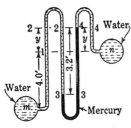

Fig. 14.   Differential manometer.

Again the computation is simpler if units of pressure head are used rather than of pressure. Suggested steps are:

1. Number the "strategic points" indicated by the levels of contact of the fluids. Some practice is needed in selecting the points which permit the simplest computation.

2. Starting with the unknown pressure head p/w at one of the end points, write a continued algebraic summation of heads, progressing from point to point, and equating the continued sum to the unknown head p/w at the other end point.

3. Solve the equation for the pressure-head difference and reduce to pressure difference if desired.

EXAMPLE.   Compute the pressure difference between m and n as shown by the differential manometer in Fig. 14.

Solution. One system of numbering the points is shown in the figure. Writing the continued sum of heads, in feet of water:

$$\frac{p_m}{w} - 4.0 - y + 3.2 - 3.2 \times 13.6 + y = \frac{p_n}{w}$$

from which

$$\frac{p_m}{w} - \frac{p_n}{w} = 4.0 - 3.2 + 43.5 = 44.3 \text{ ft of water}$$

$$p_m - p_n = 19.2 \text{ lb per sq in.}$$

PROBLEMS

1. In Fig. A, fluid A is water and fluid B has a specific gravity, s, greater than 1. Show by writing the step-by-step pressure-head equation that

$$\frac{p_m}{w} - \frac{p_n}{w} = (s - 1)\, z$$

2. In Fig. A, fluid A is water, fluid B is mercury, $z = 4.5$ ft. Compute pressure-head difference between m and n.

3. In Fig. A, when fluid A is water and fluid B is mercury, the differential gage reading is z. In what ratio is z magnified by changing from mercury to bromoform (sp gr 2.87)?

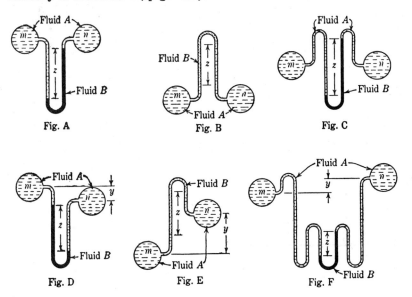

Fig. A   Fig. B   Fig. C

Fig. D   Fig. E   Fig. F

4. In Fig. B, fluid A is water, fluid B is oil (sp gr 0.80), $z = 14$ in. Compute pressure difference between m and n.

5. In Fig. C, fluid A is oil (sp gr 0.90), fluid B is a calcium chloride solution (sp gr 1.10), $z = 32$ in. Compute pressure difference between m and n.

6. In Fig. D, fluid A is water, fluid B is mercury, $z = 18$ in., $y = 3$ ft. Compute pressure-head difference between m and n in feet of water.

7. In Fig. E, fluid A is water, fluid B is oil (sp gr 0.85), $z = 27$ in., $y = 57$ in. Compute pressure difference between m and n.

8. In Fig. F, fluid A is water, fluid B is mercury, $z = 18$ in., $y = 5$ ft. Compute pressure difference between m and n.

25. Micromanometers.   The manometer reading in a differential gage can be increased for a given pressure difference by (1) using a gage liquid the specific gravity of which is closer to that of the liquid in which the pressure difference is desired, and (2) inclining the manometer tubes at an angle with the vertical.   Further magnification can be obtained by the micromanometers shown in Fig. 15.

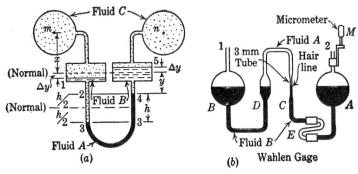

FIG. 15. Micromanometers.

If, in Fig. 15a, A and a are again the cross-sectional areas of reservoirs and tube, respectively,

$$A \cdot \Delta y = a \cdot \frac{h}{2} \tag{11}$$

Such micromanometers are generally used to measure the difference in gas pressures at m and n.   If fluid C is a gas, the effect of small differences of elevation in the gas can be neglected.   The equation for pressure head in feet of water is then

$$\frac{p_m}{w} + (y - \Delta y)s_B + hs_B - hs_A - (y + \Delta y)s_B = \frac{p_n}{w} \tag{12}$$

which can be solved simultaneously with equation 11.

The micromanometer illustrated in Fig. 15b was developed by Wahlen[1] and is said to be capable of indicating pressure differences as small as 0.0001 in. of water.   The bulb B and tubes C and D are fixed, whereas the jointed connector tube E permits bulb A to be raised or lowered.   The amount of travel of bulb A is measured by the micrometer M.

[1] *The Wahlen Gage,* bulletin, University of Illinois Engineering Experiment Station, Urbana.

The gage is first balanced with both pressure connections open by bringing the meniscus in the 3-mm tube to the hairline. The micrometer reading is then observed. With the pressure connections made at 1 and 2, a slight difference in pressure at these points produces a considerable change in elevation of the meniscus in the small tube. By raising or lowering bulb A the meniscus is brought back to the hairline and the micrometer again read. The difference in the micrometer readings in inches is the difference in pressure heads at 1 and 2 expressed in inches of fluid B.

If the two fluids A and B differ little in specific gravity, the gage is extremely sensitive. Fluid A is usually a kerosene-ligroin mixture and fluid B is alcohol which has been presaturated with kerosene and ligroin and colored with aniline dye. Best results have been obtained when the specific gravities of the fluids differed by about 0.0085.

26. Determination of Specific Gravity by U-tube. A glass U-tube open to the atmosphere at both ends is a convenient instru-

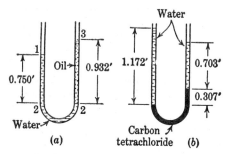

FIG. 16. Specific-gravity determination by U-tube.

ment for determining the specific gravity of a liquid provided that another non-miscible liquid of known specific gravity is available. The problem is best solved by writing the equation of heads, in feet of the liquid of known specific gravity, from one atmospheric surface to the other.

EXAMPLE. A quantity of distilled water (sp gr 1.000) is inserted in a glass U-tube, then a quantity of oil is inserted in one leg until the liquids stand as shown in Fig. 16a. Determine the specific gravity s of the oil.

Solution. Number the levels of contact as shown. Starting with zero gage pressure head at 1, write the equation of heads in feet of water, progressing through the tube to zero gage pressure head at 3:

$$0 + 0.750 - 0.932s = 0$$

from which

$$s = 0.805$$

PROBLEMS

1. In Fig. 15a, fluid A is a calcium chloride solution (sp gr 1.250), fluid B kerosene (sp gr 0.805), and fluid C air. The diameter of the reservoirs is 10 cm and that of the tube 5 mm. Compute h for a pressure-head difference between m and n of $\frac{1}{2}$ in. of water.

2. In Fig. 15b it is necessary to raise bulb A 0.3625 in. in order to maintain the meniscus at the hair line when the pressure at 1 is increased over the pressure at 2. The specific gravity of fluid A is 0.8100 and that of fluid B 0.8195. Compute the pressure-head difference between points 1 and 2 in inches of water.

3. Determine the specific gravity of the carbon tetrachloride in the U-tube shown in Fig. 16b.

27. Hydrostatic Relations for Compressible Fluids. The law of variation of pressure with depth in a fluid, as derived in Art. 15, applies to conditions under which the unit weight w of the fluid can be considered constant. If great changes in elevation in a gas are involved, the effect of compressibility may be appreciable since with compressible fluids w is a variable depending on the temperature and pressure as well as the physical characteristics of the gas.

If the prism of fluid in Fig. 7 is considered shortened to an infinitesimal length dL, the vertical projection of which is dh, the difference in pressure on the two ends of the prism becomes

$$dp = w\,dh \qquad (13)$$

In this equation dh is an increment in depth below some real or imaginary free surface. If it is desired to relate the change in pressure dp to a small change dz in altitude above some selected datum, the differential equation can be written

$$dp = -w\,dz \qquad (14)$$

The minus sign signifies that the pressure decreases as the altitude increases.

Equation 14 is the basic differential equation representing variation of pressure with altitude in a fluid at rest. If the unit weight w

of the fluid is variable it is necessary to express w in terms of p or z in accordance with the conditions of the problem.

EXAMPLE 1. Assuming isothermal conditions at 70° F, compute the barometric pressure in pounds per square inch at an altitude of 5000 ft if the pressure at sea level is 14.70 lb per sq in.

Solution. From Art. 6, the unit weight of air at 70° F, in terms of the absolute pressure p in pounds per square foot is

$$w = p/(53.3 \times 529)$$

Substituting in equation 14,

$$dp = -0.0000354p \, dz$$

Separating variables,

$$dp/p = -0.0000354 \, dz$$

Integrating,

$$\log_e p = -0.0000354z + C_1$$

where C_1 is a constant of integration. When $z = 0$, $p = 2116$ lb per sq ft. Hence $C_1 = \log_e 2116$, and

$$\log_e (2116/p) = 0.0000354z$$

$$\log (2116/p) = 0.4343 \times 0.0000354z = 0.00001540z$$

(The student should complete the evaluation of p.)

EXAMPLE 2. Assuming that the temperature decreases 3.57° F for each increase of 1000 ft in altitude, compute the pressure in pounds per square inch at an altitude of 25,000 ft, the pressure and the temperature at sea level being respectively 14.70 lb per sq in. and 70° F.

Solution. From Art. 6,

$$w = \frac{p}{53.3(529 - 0.00357z)} = \frac{p}{28,200 - 0.1903z}$$

$$\frac{dp}{p} = \frac{-dz}{28,200 - 0.1903z} = \frac{-5.25 \, dz}{148,200 - z}$$

$$\log_e p = 5.25 \log_e (148,200 - z) + C_1$$

When $z = 0$, $p = 2116$. Hence $C_1 = \log_e 2116 - 5.25 \log_e 148,200$. (The student should complete the evaluation of p.)

GENERAL PROBLEMS

1. A vertical pipe, 100 ft long and 1 in. in diameter, has its lower end open and flush with the inner surface of the cover of a box 2 ft square and 6 in. high. The bottom of the box is horizontal. Neglecting the weight of the pipe and box, both of which are filled with water, determine: (a) the total hydrostatic pressure on the bottom of the box; (b) the total pressure exerted on the floor on which the box rests.

2. A gas-holder at sea level contains gas under a pressure equivalent to 3.5 in. of water. Assuming the unit weights of air and gas to be constant and equal to 0.080 and 0.045 lb per cu ft, respectively, what will be the pressure head in inches of water in a distributing main at a point 800 ft above sea level?

3. The pressure head in a gas main at a point 400 ft above sea level is equivalent to 7.07 in. of water. Assuming the unit weights of air and gas are constant and equal to 0.075 and 0.035 lb per cu ft, respectively, what will be the pressure head in inches of water in a gas-holder at sea level?

4. A vertical tube, 1 in. in diameter and 4 ft long, having its upper end open to the atmosphere, contains equal volumes of water and mercury. If the tube is full, determine: (a) the gage pressure in pounds per square inch at the bottom of the tube, and (b) the weight of the liquids.

5. A U-tube with both ends open to the atmosphere contains mercury in the lower portion. In one leg, water stands 30 in. above the surface of the mercury; in the other leg, oil (sp gr 0.80) stands 18 in. above the surface of the mercury. What is the difference in elevation between the surfaces of the oil and water columns?

6. A and B are, respectively, the closed and open ends of a U-tube, both being at the same elevation. For a distance of 18 in. below A, the tube is filled with oil (sp gr 0.80); for a distance of 3 ft below B, the tube is filled with water, on the surface of which atmospheric pressure is acting. The remainder of the tube is filled with mercury. What is the absolute pressure at A in pounds per square inch?

7. A vertical U-tube, with both ends open, contains mercury to a depth of 10 in. in each tube above the bottom of the U. Water to a depth of 18 in. is added in one leg only. (a) What is the gage pressure at the bottom of the U? (b) What is the difference in level between the free surfaces in the two tubes?

8. The upper portion of an inverted U-tube is filled with oil having a specific gravity of 0.98. The remainder of the tube is filled with salt water having a specific gravity of 1.01. When there is a difference in level of 3 in. between the water surfaces in the two legs of the tube, what is the difference in pressure, in pounds per square inch, between two points of equal elevation at the base of the two legs?

9. A mercury gage connected with the air chamber of a condenser shows a reading of −10 in. Determine the absolute pressure in pounds per square inch at a point 4 ft below the water surface.

10. In Fig. A, what is the pressure in pounds per square inch at A?

11. In Fig. A, if the pressure at A is reduced 6 lb per sq in., what is the new difference in elevation of the mercury surfaces?

12. Compute the pressure head at m, in feet of fluid A, indicated by open manometer in Fig. B, when:

(a) Fluid A is oil (sp gr 0.856), the gage liquid is carbon tetrachloride (sp gr 1.60), $y = 30$ in., $z = 30$ in.

(b) Fluid A is molasses (sp gr 1.50), the gage liquid is water, $y = 3.6$ ft, $z = 1.2$ ft.

13. In Fig. C, fluids A and B have specific gravities of 0.915 and 2.95, respectively. If $y = 1.36$ ft and $z = 1.22$ ft, compute the pressure at m.

14. In problem 13, if the pressure at m is increased 1 lb per sq in., how many inches will fluid B rise in the $\frac{1}{2}$-in. tube?

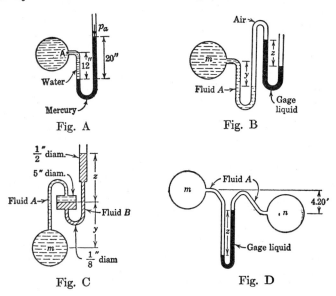

Fig. A   Fig. B

Fig. C   Fig. D

15. In Fig D, compute the pressure-head difference between m and n, in feet of fluid A, when:

(a) Fluid A is gasoline (sp gr 0.715), the gage liquid is mercury, and $z = 42$ in.

(b) Fluid A is oil (sp gr 0.908), the gage liquid is water, and $z = 26.6$ in.

(c) Fluid A is sea water ($w = 64.0$), the gage liquid has a specific gravity of 2.95, and $z = 57.2$ in.

16. In Fig. E, determine the pressure-head difference in feet of water between m and n, when z is: (a) 6.5 in.; (b) 21.2 in.; (c) 33.0 in.

17. In Fig. E, determine the value of z when the pressure at m is 10 lb per sq in. greater than at n.

18. In Fig. E, let $z = 10$ in. If, then, the pressure at m is increased by 5 lb per sq in., while the pressure at n remains constant, determine the new value of z.

19. In Fig. F, determine the pressure-head difference in feet of water between m and n, when z is: (a) 3.2 in.; (b) 10 in; (c) 15 in.

20. Determine the value of z in Fig. F if the pressure at n is 1.4 lb per sq in. greater than at m.

21. If, in Fig. F, the pressure at m is 20 lb per sq in., what is the corresponding pressure at n when $z = 8$ in.?

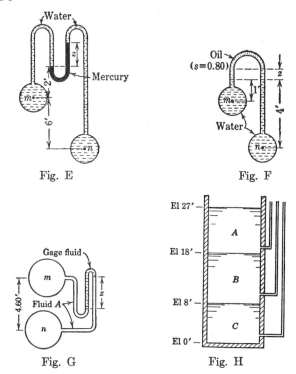

Fig. E   Fig. F

Fig. G   Fig. H

22. In Fig. G compute the pressure difference between m and n, in pounds per square inch, when:

(a) Fluid A is brine (sp gr 1.15), the gage fluid is oil (sp gr 0.92), and $z = 44$ in.

(b) Fluid A is water, the gage fluid is air, and $z = 12$ in.

23. Show that the sensitivity of the manometer in Fig. 15a, that is, the height h for a given pressure difference between m and n, increases as $(s_A - s_B)$ decreases and as a/A decreases.

24. Liquids A, B, and C in the container shown in Fig. H have specific gravities of 0.80, 1.00, and 1.60, respectively. With the liquid surfaces at the elevations shown what is the pressure in pounds per square inch on the bottom of the tank?   At what elevation will the liquid stand in each of the piezometer tubes?

25. In Fig. J assume that the piston and the weight W are at the same elevation, the face of the piston having an area of 2 sq in. and the face of the weight 20 sq in.   The intervening passages are filled with water. What weight W can be supported by a force P of 100 lb applied at the end of the lever as shown in the figure?

26. In Fig. J, the diameters of the two cylinders are 3 in. and 24 in., the face of the piston is 20 ft above the face of the weight W, and the intervening passages are filled with oil (sp gr 0.80).   What force P is required to maintain equilibrium if $W = 8000$ lb?

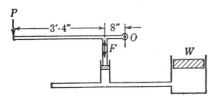

Fig. J.   Hydraulic jack.

27. Two 12-in. sections and a 2-in. section of pipe are connected by a 12-in. by 12-in. by 2-in. tee.   In each section there is a piston of the same diameter as the pipe.   Neglecting the weight of the water that fills the space between the pistons, what will be the total tensile stress in a steel rod connecting the 12-in. pistons if a force of 20 lb is applied to the 2-in. piston?

28. A 2-in. pipe is connected with the end of a cylinder having a diameter of 20 in.   There is a piston in the pipe and a piston in the cylinder, the space between being filled with water.   The larger piston is connected by a rod with a 2-in. piston in a third pipe, the two pipes and cylinder having their axes horizontal and collinear.   If a force of 20 lb is applied to the small piston in the first pipe, what will be the necessary intensity of pressure in the third pipe to maintain equilibrium?

29. Assuming normal barometric pressure, how deep is the ocean at a point where an air bubble, upon reaching the surface, has six times the volume that it had at the bottom?

30. A bottle, consisting of a cylinder 1 ft in diameter and 1 ft high, has a neck 2 in. in diameter and 1 ft long.   If this bottle, filled with air under atmospheric pressure, is inverted and submerged in water until the neck is just filled with water, find the depth to which the open end is submerged, neglecting vapor pressure.

31. A vertical tube 10 ft long, with one end closed, is inserted vertically, with the open end down, into a tank of water until the open end is submerged to a depth of 4 ft. Neglecting vapor pressure, how far will the water level in the tube be below the level in the tank?

32. A vertical tube 10 ft long, with one end closed, is inserted vertically, with the open end down, into a tank of water to such a depth that an open manometer connected with the upper end of the tube shows a reading of 6 in. of mercury. Neglecting vapor pressure, how far is the lower end of the tube below the water surface in the tank?

33. Assuming that at sea level the temperature is 59° F and the pressure is 14.7 lb per sq in., and that the temperature decreases 3.57° F for each 1000 ft increase in altitude, compute the temperature and barometric pressure at altitudes of 15,000 ft, 30,000 ft, 45,000 ft. Assuming that the temperature is constant above 45,000 ft, compute the pressure at 60,000 ft altitude.

Chapter *III*

HYDROSTATIC PRESSURE ON SURFACES

28. Total Pressure on Plane Surfaces. *The total hydrostatic pressure on any plane surface is equal to the product of the area of the surface and the unit pressure at its center of gravity.*

This rule may be proved as follows: Figure 17 shows, on two vertical planes normal to each other, projections of any plane

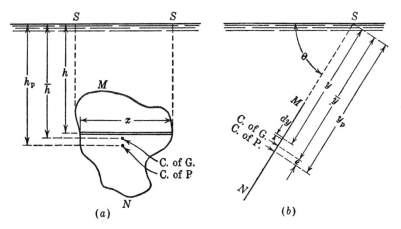

Fig. 17. Pressure on plane surface.

surface, *MN*, subjected to the full static pressure of a liquid with a free surface. Projection (*b*) is on a plane at right angles to *MN*. The surface *MN* makes any angle, *θ*, with the horizontal, and, extended upward, the plane of this surface intersects the surface of the liquid in the line *S–S*, shown as the point *S* in (*b*).

Consider the surface *MN* to be made up of an infinite number of horizontal strips each having an area *dA* and a width *dy* so small that the unit pressure on the strip may be considered constant. The liquid having a unit weight of *w*, the unit pressure on a strip at depth *h* below the surface and at distance *y* from the line *S–S* is

$$p = wh = wy \sin \theta$$

41

The total pressure on the strip is

$$dP = wy \sin \theta \, dA$$

and the total pressure on MN is

$$P = w \sin \theta \int y \, dA \qquad (1)$$

From the definition of center of gravity,

$$\int y \, dA = A\bar{y} \qquad (2)$$

where $\bar{y}$ is the distance from the line S–S to the center of gravity of A.   Hence

$$P = w \sin \theta \, A\bar{y} \qquad (3)$$

Since the vertical depth of the center of gravity below the surface is

$$\bar{h} = \bar{y} \sin \theta \qquad (4)$$

it follows that

$$P = w\bar{h}A \qquad (5)$$

where $w\bar{h}$ represents the unit pressure at the center of gravity of A.

29. Center of Pressure on Plane Surfaces.   Any plane surface subjected to hydrostatic pressure is acted upon by an infinite number of parallel forces the magnitudes of which vary with the depth, below the free surface, of the various infinitesimal areas on which the respective forces act.   These parallel forces may be replaced by a single resultant force P.   The point on the surface at which this resultant force acts is called the center of pressure.   If the total hydrostatic pressure on any surface were applied at the center of pressure the same effect would be produced on the surface, considered as a free body, as is produced by the distributed pressure.

The position of the horizontal line containing the center of pressure of a plane surface subjected to hydrostatic pressure may be determined by taking moments of all the forces acting on the surface about some horizontal axis in its plane.   In Fig. 17, the line S–S may be taken as the axis of moments for the surface MN. Designating by y_p the distance to the center of pressure from the axis of moments, it follows from the definition of center of pressure that

$$Py_p = \int y\, dP \tag{6}$$

$$y_p = \frac{\int y\, dP}{P} \tag{7}$$

But, as in article 28,

$$dF = wy \sin \theta\, dA$$

and

$$P = w \sin \theta\, A\bar{y} \tag{3}$$

Substituting in equation 7,

$$y_p = \frac{w \sin \theta \int y^2\, dA}{w \sin \theta\, A\bar{y}} = \frac{\int y^2\, dA}{A\bar{y}} \tag{8}$$

in which $\int y^2\, dA$ is the moment of inertia, I_S, of MN with respect to the axis S–S, and $A\bar{y}$ is the statical moment, S_S, of MN with respect to the same axis.   Therefore,

$$y_p = \frac{I_S}{S_S} \tag{9}$$

Problem. Apply the transfer formula of moment of inertia to equation 9 to show that

$$e = \frac{I_g}{S_S} \tag{10}$$

where e = perpendicular distance between horizontal axes through center of gravity and center of pressure, and lying in plane of area.

I_g = moment of inertia of area with respect to the horizontal axis through its center of gravity and lying in its plane.

S_S = moment of area with respect to the line in which the area extended cuts the surface of the liquid.

The above discussion refers only to the determination of the position of the horizontal line which contains the center of pressure; that is, y_p gives only the distance from the axis S–S, lying in the liquid surface, to the center of pressure.   For any plane figure such that the locus of the midpoints of the horizontal strips is a straight line, as, for instance, a triangle or trapezoid with base horizontal, the center of pressure falls on that straight line. It is with such figures that the engineer is usually concerned. For other figures, the horizontal location of the center of pressure may be found in a manner similar to that described above by taking moments about an axis, within the plane of the surface, at right angles to the horizontal axis of moments.

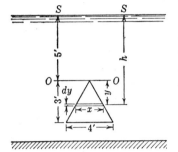

FIG. 18. Triangular gate.

EXAMPLE 1. Find the total pressure on the vertical triangular gate with water on one side as shown in Fig. 18, and locate the center of pressure.

Solution. (a) By integration: The total pressure dP on any thin horizontal strip at a distance y from the top of the gate equals the unit pressure, wh, times the area dA, or

$$dP = wh\, dA \quad \text{where} \quad h = 5 + y \quad \text{and} \quad dA = x\, dy$$

Since x varies with y it must be expressed in terms of y before integrating. From similar triangles,

$$\frac{x}{4} = \frac{y}{3} \quad \text{or} \quad x = \tfrac{4}{3}y$$

Thus

$$P = \int dP = \int_0^3 \tfrac{4}{3}w(5y + y^2)\, dy = 42w = 2620 \text{ lb}$$

Taking moments about the water surface axis, $S\text{--}S$,

$$Py_p = \int h\, dP$$

where h and dP have the same values as above. Thus

$$y_p = \frac{\displaystyle\int_0^3 \tfrac{4}{3}w(25y + 10y^2 + y^3)\, dy}{42w} = 7.07 \text{ ft}$$

measured from $S\text{--}S$.

It is often more convenient to locate the center of pressure by taking moments about some other axis, as, in this case, axis $O\text{--}O$ through the top of the gate.   Then

$$Py_p' = \int y\, dP$$

and

$$y_p' = \frac{\displaystyle\int_0^3 \tfrac{4}{3}w(5y^2 + y^3)\, dy}{42w} = 2.07 \text{ ft}$$

measured from O–O.

(b) By substitution in equations 5, 9, and 10,

$$P = w\bar{h}A = w \times 7 \times 6 = 42w = 2620 \text{ lb}$$

$$y_p = \frac{I_S}{S_S} = \frac{(4 \times 3^3/36) + 6 \times 7^2}{6 \times 7} = 7.07 \text{ ft}$$

$$e = \frac{I_g}{S_S} = \frac{4 \times 3^3/36}{6 \times 7} = 0.0714 \text{ ft}$$

Note that, with three-place accuracy of computation, the center of pressure can be located more accurately by equation 10 than by equation 9.

The horizontal location of the center of pressure in this case is on the median connecting the vertex with the base.

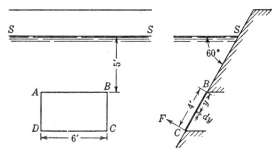

Fig. 19.   Inclined rectangular gate.

EXAMPLE 2.   Find the total pressure on the inclined rectangular gate with sea water on one side as shown in Fig. 19, and locate the center of pressure.

(a) By integration:

$$P = \int dP = \int wh\, dA = \int_0^4 w(5 + y \cos 30) \cdot 6\, dy = 162w = 10{,}370 \text{ lb}$$

Taking moments about the top of the gate,

$$y_p' = \frac{\int y\, dP}{P} = \frac{\int_0^4 w(5y + y^2 \cos 30) \cdot 6dy}{162w} = 2.17 \text{ ft from } B$$

(b) By equations 5, 9, and 10:

$$P = w\bar{h}A = w \times (5 + 2 \cos 30) \times 24 = 10{,}370 \text{ lb}$$

$$y_p = \frac{I_S}{S_S} = \frac{(6 \times 4^3/12) + 24(7.77)^2}{24 \times 7.77} = 7.94 \text{ ft from } S$$

$$e = \frac{I_g}{S_S} = \frac{6 \times 4^3/12}{24 \times 7.77} = 0.172 \text{ ft}$$

Both y_p and e are measured along the plane of the gate. The horizontal location of the center of pressure is 3 ft from either end of the gate.

EXAMPLE 3. In Example 2, what force F normal to the gate at its lower edge will be required to open it?

The total pressure, P, on the gate, and the location of the center of pressure being known, by taking moments about the upper edge which is the center of rotation,

$$4F = 2.17P = 22{,}450 \text{ ft-lb}$$

$$F = 5610 \text{ lb}$$

If this force were applied at the bottom of the gate, the gate would be in equilibrium and there would be no reaction on the supports along the lower edge or sides of the gate. Any force greater than 5610 lb would open the gate.

PROBLEMS

1. Water stands on one side of the vertical gates shown in Fig. A. Find by integration the total pressure on each gate and the location of the center of pressure.

2. Water stands on one side of the vertical gates shown in Fig. A, the water surface being 10 ft above the top of the gates. Find by integration the total pressure on each gate and the location of the center of pressure.

3. A gate 2 ft square lies in a plane making an angle of 30° with the vertical. Its upper edge is horizontal and 3 ft below the surface of the liquid (sp gr 3.0). Find by integration the total pressure on the gate and the location of the center of pressure.

4. A vertical circular gate 3 ft in diameter is subjected to pressure of molasses (sp gr 1.50) on one side. The free surface of the molasses is 8 ft above the top of the gate. Determine the total pressure and the location of the center of pressure.

5. A circular gate 5 ft in diameter is inclined at an angle of 45°. Sea water stands on one side of the gate to a height of 30 ft above the center of the gate. Determine the total pressure on the gate and the location of the center of pressure.

6. A vertical triangular surface has a horizontal base of 4 ft and an altitude of 9 ft, the vertex being below the base. If the center of pressure is 6 in. below the center of gravity, how far is the base below the liquid surface?

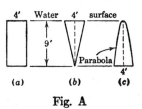

Fig. A

7. A vertical, trapezoidal gate in the face of a dike is subjected to seawater pressure ($w = 64.0$) on one side. The upper edge is in the water surface and is 5 ft long. Two edges are vertical and measure 6 ft and 9 ft each. Determine the total pressure on the gate and the location of the center of pressure.

8. A rectangular gate of height h, with upper and lower edges horizontal is inclined at any angle ($< 90°$) with the vertical. Liquid stands on one side of the gate, the upper edge of the gate being in the liquid surface. Show (and remember) that under these conditions the distance from the upper edge of the gate to the center of pressure is *two-thirds of h.*

9. A vertical rectangular gate 4 ft wide and 6 ft high, hinged at the top, has water on one side. What force applied at the bottom of the gate, at an angle of 45° with the vertical, is required to open the gate when the water surface is (*a*) at the top of the gate; (*b*) 3 ft above the top of the gate; (*c*) 3 ft below the top of the gate?

10. On one side, water stands level with the top of a vertical rectangular gate 4 ft wide and 6 ft high, hinged at the bottom; on the other side water stands 3 ft below the top. What horizontal force applied at the top of the gate is required to open it?

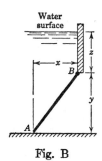

Fig. B

11. The rectangular gate in Fig. B is hinged at B and rests on a smooth floor at A, the horizontal component of the reaction at A therefore being zero. The gate is 5 ft wide perpendicular to the paper. Determine the vertical component of the reaction at A and the horizontal and vertical components of the reaction at B for the following sets of conditions: (*a*) $x = 6$ ft, $y = 8$ ft, $z = 0$ ft; (*b*) $x = 6$ ft, $y = 8$ ft, $z = 4$ ft; (*c*) $x = y = 8$ ft, $z = 10$ ft; (*d*) $x = 3$ ft, $y = 4$ ft, $z = 50$ ft.

12. The flashboard gate shown in Fig. C consists of a plane face *bd* resting in a groove at *d* and supported by the strut *ce* which is pinned at the ends.   Neglecting the weight of the gate, determine the greatest depth, h, which the water can have without causing the gate to collapse.

13. If, on the upstream side of the gate in Fig. C, water stands level with the upper edge b, and on the downstream side water stands level with the hinge c, neglecting the weight of the gate, what vertical force per foot length of crest must be applied at b in order to cause the gate to collapse?

14. The length of the gate in Fig. C, measured normal to the plane of the paper, is 10 ft.   Neglecting the weight of the gate, what vertical force must be applied at a, 4 ft from the hinge, to prevent collapse when h equals 20 ft?   What is the stress in ce?

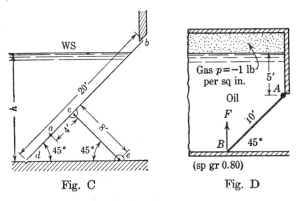

Fig. C   Fig. D

15. The gate in Fig. D is hinged at A and rests on a smooth floor at B. The gate is 10 ft square.   Oil stands on the left side of the gate to a height of 5 ft. above A.   Above the oil surface is gas under a gage pressure of -1 lb per sq in.   Determine the amount of the vertical force F applied at B that would be required to open the gate.

16. A triangular gate having a horizontal base 4 ft long and an altitude of 6 ft is inclined 45° from the vertical with the vertex pointing upward.   The base of the gate is 8 ft below the surface of the liquid, which has a specific gravity of 0.82.   What normal force must be applied at the vertex of the gate to open it?

30. Semigraphic Method of Location of Center of Pressure.

Semigraphic methods may be used advantageously in locating the center of pressure on any plane area whose horizontal dimension does not vary with the depth.   The rectangular surface $ABCD$, illustrated in Fig. 19, Example 2, is shown in perspective in Fig. 20a. BC, Fig. 20b, represents the projection of the rectangle on a vertical plane perpendicular to the plane of the surface.   The vertical depths of the top and bottom of the rectangle below the liquid surface are, respectively, h_1 and h_2. The unit pressure, wh_1, on the top of the rectangle is represented by the vectors $A'A$ and $B'B$ (Fig. 20a), and on the bottom of the rectangle the vectors $C'C$ and $D'D$ represent the unit pressure wh_2.

The trapezoid $BCC'B'$ may be divided into the rectangle $BCEB'$ and the triangle $B'EC'$, the locations of whose centers of gravity are known. By taking moments of each of these pressure areas

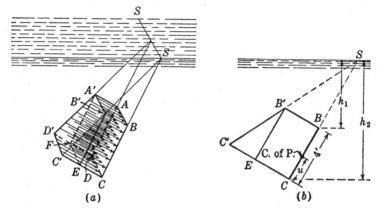

FIG. 20. Graphical representation of pressure distribution.

about $C'C$ and dividing the sum of these moments by the area of the trapezoid, the distance of the center of pressure from C is determined. Thus

$$B'B = 5w \quad \text{and} \quad C'C = (5 + 4 \cos 30)w = 8.46w$$

Therefore

$$C'E = 4 \cos 30w = 3.46w$$

Taking moments about $C'C$,

$$u = \frac{4 \times 5w \times 2 + \frac{4}{2} \times 3.46w \times \frac{4}{3}}{4 \times \frac{1}{2}[5w + 8.46w]} = 1.83 \text{ ft}$$

Example 3, page 46, can also be solved by taking moments about BB', as follows:

$$4F = 4 \times 5w \times 6 \times 2 + \frac{4}{2} \times 3.46w \times 6 \times \frac{8}{3}$$

from which

$$F = 5610 \text{ lb}$$

For areas having a variable width, $SB'C'$ is not a straight line and the center of gravity of the pressure area is not so easily located. For such areas it will probably be easier to use the analytical method described in Art. 29.

31. Position of Center of Pressure with Respect to Center of Gravity. If the unit pressure varies over any surface, the center of pressure is below the center of gravity. Applying the transfer formula for moment of inertia to equation 9 gives

$$y_p = \frac{I_g + A\bar{y}^2}{A\bar{y}} = \frac{I_g}{S_S} + \bar{y}$$

Since I_g/S_S must always be positive, y_p must be greater than $\bar{y}$.

This may also be seen from Fig. 20. The center of pressure on
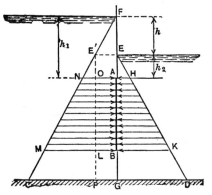
$ABCD$ is the normal projection on that plane of the center of gravity of the pressure volume $ABCDA'B'C'D'$. Evidently this projection must fall below the center of gravity of $ABCD$ since it would fall at the center of gravity if the unit pressure on the surface were uniform, in which case the pressure volume would be $ABCDA'B'EF$.

It also appears from the above discussion and from a study of Fig. 20 that for any area the greater its depth below the surface of the liquid the more nearly will the center of pressure approach the **center of gravity**. The two coincide at an infinite depth.

Fig. 21. Opposing hydrostatic pressures.

Under two conditions the unit pressure is constant over the area, and hence the center of pressure coincides with the center of gravity: (1) when the surface is horizontal and (2) when both sides of the area are completely submerged in liquids of the same unit weight.

As an illustration consider the gate AB (Fig. 21). The top of the gate is submerged h_1 feet on one side and h_2 feet on the other side. The distribution of pressure on the left is represented by the trapezoid $ABMN$ and on the right by the trapezoid $AHKB$.

The triangle GED is similar to CFG and equal to $CE'P$ by construction. The trapezoid of pressure $AHKB$ is therefore balanced by the trapezoid $ONML$. The resultant intensity of pressure on the gate is therefore constant, as represented by the rectangle $OABL$, and the center of pressure must coincide with the center of gravity of the gate. This is true regardless of the shape of the gate. The resultant intensity of pressure is wh, where h is the difference in level on the two sides.

In this latter case it should be observed that it is the center of the *resultant* total pressure that coincides with the center of gravity of the gate, since the center of gravity of either of the trapezoidal areas of pressure, considered alone, falls below the center of gravity of the gate.

32. Horizontal and Vertical Components of Total Hydrostatic Pressure on Any Surface. It is often more convenient to deal with the horizontal and the vertical components of the total pressure acting on a surface rather than with the resultant pressure. This is particularly true when dealing with pressures on curved surfaces.

Consider, for example, the liquid pressure acting on the curved surface AB shown in Fig. 22. The surface may have any length normal to the plane of the paper. Choosing the coordinate axes as shown, let BF rep-

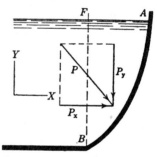

Fig. 22.   Pressure on curved surface.

resent the trace of a vertical plane normal to the XY plane. Consider the equilibrium of the volume of liquid the cross section of which, as shown in the figure, is ABF and whose ends are parallel with the XY plane. Since this volume of liquid is assumed to be in equilibrium, $\Sigma F_x = 0$ and $\Sigma F_y = 0$.

The only forces that have any components parallel with the X axis are the sum of the X components of the normal pressures acting on the surface AB and the normal total pressure on the vertical plane BF, which is the projection of the surface AB on a vertical plane normal to the X axis. These forces must be equal in magnitude. As the demonstration holds true independently of the location of the horizontal X axis, it may be stated that *the component, along any horizontal axis, of the total hydrostatic pressure on any surface is equal to the total pressure on the projection of that surface on a vertical plane which is normal to the chosen axis.* The *location* of the horizontal component is through the center of pressure of this projection.

In a similar manner consider the vertical forces acting on the volume of liquid whose cross section is *ABF* (Fig. 22). The only vertical forces are the force of gravity, represented by the weight of the liquid, and the sum of the vertical components of the pressures on the surface *AB*, which forces must therefore be equal in magnitude. In other words, *the vertical component of the total hydrostatic pressure on any surface is equal to the weight of that volume of the liquid extending vertically from the surface to the free surface of the liquid.* The location of the vertical component is through the centroid of this volume.

If the liquid is underneath the surface, the pressure acts upward on the surface, and the magnitude of the vertical component, as will be shown later (Art. 35), is equal to the weight of the *imaginary* volume of the liquid extending vertically from the surface to the level of the real or imaginary free surface of the liquid. The location of the vertical component is through the centroid of this imaginary volume.

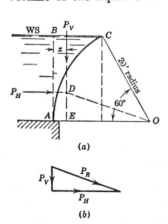

(a)

(b)

FIG. 23. Sector gate.

EXAMPLE. The sector gate shown in Fig. 23a consists of a cylindrical surface, of which *AC* is the trace, supported by a structural frame hinged at *O*. The length of the gate, perpendicular to the paper, is 30 ft. Determine the amount and location of the horizontal and the vertical components of the total hydrostatic pressure on the gate.

Solution. To determine the horizontal component of pressure on the gate, consider the gate projected onto a vertical plane of which *AB* is the trace. This projection is a rectangle with a width of 30 ft and a height of 20 cos 30 = 17.32 ft. Therefore

$$P_H = w\bar{h}A = 62.4 \times 8.66 \times 519.6 = 281,000 \text{ lb}$$

Since this projection is a rectangle with top edge horizontal and lying in the water surface, the center of pressure is at two-thirds the depth, or 11 55 ft below the water surface.

The vertical component of pressure on the gate is equal to the weight of the prism of water 30 ft long and having an end area ABC.

$$\text{Area } ABC = 17.32 \times 10 + \frac{1}{2} \times 17.32 \times 10 - \frac{\pi}{6} \times 20^2 = 50.4 \text{ sq ft}$$

Therefore

$$P_V = 50.4 \times 30 \times 62.4 = 94,400 \text{ lb}$$

The location of P_V is in line with the center of gravity of section ABC. This point can be located by summing up area moments with respect to line AB and dividing that sum by the area. (The centroid of a 60° circular sector is $2R/\pi$ from the center.) Thus

$$\Sigma M_{AB} = 173.2 \times 5 + 86.6 \times 13.33 - 209.4 \left(20 - \frac{40}{\pi} \cos 30 \right) = 141 \text{ ft}^3$$

$$x = 2.8 \text{ ft}$$

The location of P_V can be more easily determined in this problem by noting that the resultant of P_H and P_V must pass through O. (Why?) Triangle DEO is therefore similar to the force triangle shown in Fig. 23b. Since $DE = 5.77$ ft,

$$EO = 5.77 \times \frac{281,000}{94,400} = 17.2 \text{ ft}$$

and P_V is 2.8 ft from AB.

PROBLEMS

1. The curved surface represented by AB in Fig. A is the surface of the quadrant of a circular cylinder 10 ft long. Determine the horizontal and vertical components of total hydrostatic pressure on the surface, when (a) $r = 10$ ft, $z = 0$; (b) $r = 8$ ft, $z = 5$ ft; (c) $r = 5$ ft; $z = 50$ ft.

2. Locate the horizontal and vertical components of total pressure in problem 1. (Note: The center of gravity of a circular quadrant of radius r is $4r/3\pi$ distant from each straight side.)

3. The corner plate of the hull of a ship (AB in Fig. B) is curved on the arc of a circle with a 5-ft radius. With submersion in sea water as shown, compute for a 1-ft length perpendicular to the sketch the amount and location of the horizontal and vertical components of total pressure on AB. Determine graphically the amount and location of the resultant pressure.

4. In the crest gate on a dam shown in Fig. C, surface AB forms the arc of a circle of 10-ft radius subtending 45° at the hinge. The gate is 10 ft long. With water surface at B compute the amount and location of the horizontal and vertical components of total pressure on AB. Determine graphically the amount and direction of their resultant.

5. A pyramid weighing 4000 lb has a base 6 ft square and an altitude of 4 ft. The base covers an opening in the floor of a tank in which there is water 4 ft deep. Underneath the floor of the tank and on the water surface there is air at atmospheric pressure. What vertical force is required to lift the pyramid off the floor?

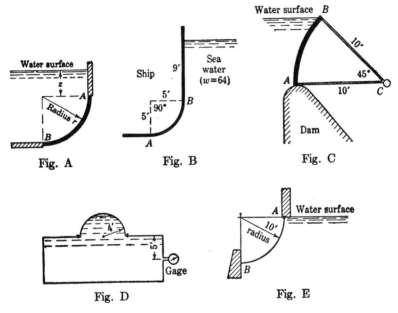

Fig. A   Fig. B   Fig. C

Fig. D   Fig. E

6. A hemispherical dome surmounts a closed tank as shown in Fig. D. If the tank and dome are filled with gasoline (sp gr 0.72) and the gage indicates 8 lb per sq in. gage pressure, what is the total tension in the bolts holding the dome in place?

7. The gate AB shown in Fig. E is hinged at A and rests against a perfectly smooth vertical wall at B. The gate is 20 ft long. With water as shown, determine the horizontal and vertical components of the reactions A and B.

33. Hoop Tension in Circular Pipes and Tanks. The circumferential stress, or hoop tension, in a pipe or tank of circular cross section is determined by applying the rule for horizontal component of pressure against a curved surface.

A cross section of a pipe of diameter D is shown in Fig. 24. If the pressure head h in the pipe is relatively large compared to the diameter of the pipe, it is customary to consider that the unit pressure p is uniform throughout the cross section.

Consider a semicircular segment, *AB*, of unit length, held in equilibrium by the two forces *T*. If the unit pressure is assumed to be uniform, *T* is the same at all points in the circumference. The sum of the horizontal components of the normal pressures acting on the semicircular segment is equal to the normal pressure on the projection of this segment (Art. 32). Calling this normal pressure *P*, since $\Sigma F = 0$,

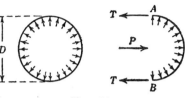

FIG. 24

$$2T = P = whA = pD \qquad (11)$$

It should be noted that equation 11 gives the tension per unit of length at the section where the pressure is *p*. If the pressure varies axially along a pipe or tank of uniform diameter, as for example in a vertical standpipe containing liquid, the hoop tension also varies.

EXAMPLE. Determine the tensile stress in the walls of a 24-in. steel pipe carrying water under a head of 1000 ft. Consider a section of pipe 1 in. long.

Solution. $p = 433$ lb per sq in.

$2T = pD = 10,400$

$T = 5200$ lb per in. length of pipe.

The required thickness of a steel pipe, using a working stress of 16,000 lb per sq in., would be

$$t = \frac{5200}{16,000} = 0.325 \text{ in.}$$

The foregoing theory takes account only of the static pressure in a pipe. Pipes in service are frequently required also to resist pressure due to water hammer (Art. 156) caused usually by too sudden closure of a valve. Moreover, it is necessary in practice to use pipe which will resist shocks due to rough handling and which will also have some reserve thickness to guard against breakage due to corrosion and other deterioration in service. The thickness of riveted steel pipe must also be increased in inverse ratio to the efficiency of the longitudinal riveted joints.

Empirical formulas are therefore in use to determine the proper thickness of pipe. For example, the New England Water Works Association formula for cast-iron pipe, assuming a working stress of 3300 lb per sq in., is

$$t = \frac{(p + p')d}{6600} + 0.25 \qquad (12)$$

where t = thickness of pipe wall in inches.

$\qquad p$ = static pressure in pounds per square inch.

$\qquad p'$ = allowance for water hammer in pounds per square inch.

$\qquad d$ = internal pipe diameter in inches.

The allowance for water hammer decreases as the size of the pipe increases according to the accompanying table.

d	p'	d	p'
3 to 10 in.	120 lb per sq in.	24 in.	85 lb per sq in.
12	110	30	80
16	100	36	75
20	90	42 to 60	70

PROBLEMS

1. Compute the wall stress in a 48-in. steel pipe $\frac{1}{4}$ in. thick under a head of 400 ft of oil (sp gr 0.82).

2. What is the minimum allowable thickness for a 24-in. steel pipe under an internal pressure of 125 lb per sq in. with a working stress in the steel of 10,000 lb per sq in.?

3. A wood-stave pipe is bound by steel rods which take the entire bursting stress. Find the proper spacing for 1-in. round steel rods for a 72-in. wood-stave pipe under a head of 200 ft of water if the working stress in the steel is 15,000 lb per sq in.

4. Find the proper thickness of 36-in. cast-iron pipe under a 300-ft head of water with a working stress of 3300 lb per sq in. (a) by theory (b) by N.E.W.W.A. empirical formula.

5. A cylindrical tank, having a vertical axis, is 6 ft in diameter and 10 ft high. Its sides are held in position by means of two steel hoops, one at the top and one at the bottom. What is the tensile stress in each hoop when the tank is filled with water?

6. A vertical, cylindrical tank, 5 ft in diameter and 12 ft high, is held together by means of two steel hoops, one at the top and one at the bottom. When molasses (sp gr 1.50) stands to a depth of 9 ft in the tank, what is the stress in each hoop?

34. Dams. Dams are built for the purpose of impounding water. Since the water level is raised on the upstream side, the dam is subject to hydrostatic forces which tend to (1) slide it horizontally on its foundation, and (2) overturn it about its downstream edge or toe. These tendencies are resisted by: (1) friction on the base of the dam, assisted in modern practice by keying the base of a solid masonry dam into the bedrock, and (2) gravitational forces which produce moment opposite in direction to the overturning moment.

Also because of the raised water level on the upstream face there is a tendency for the water to seep under the dam and escape at the lower level on the downstream side. Inasmuch as this seepage is under pressure, it exerts what is commonly called a hydrostatic uplift on the dam.

Depending upon the nature of the foundation material and upon the effectiveness of cut-off walls and foundation grouting in reducing the seepage, the uplift pressure head at the upstream edge, or heel, of the dam may vary from full hydrostatic head to a small fraction thereof. The intensity of this uplift pressure is usually assumed to vary uniformly from the heel to the toe, or downstream edge of the base, at which point it is approximately equal to wh', where h' is the depth of the base below the downstream water surface.

The hydrostatic uplift on a dam: (1) reduces the stability against sliding; (2) reduces the stability against overturning; and (3) reduces the total earth or rock pressure on the base, although it may increase the intensity of this pressure at the toe.

A typical application of the principles of the mechanics of a free body to the analysis of forces acting on the cross section of a gravity dam, both with and without hydrostatic uplift, is shown in the following examples. In designing important structures some of the assumptions herein made are modified in actual practice.

EXAMPLE 1. Analyze the forces acting upon the concrete dam subjected to water pressure as shown in Fig. 25 for:

(a) Amount and location of resultant reaction on base.
(b) Resistance to sliding.
(c) Resistance to overturning.
(d) Intensity of pressure on base at heel B and toe C.

The following numerical values are given. Area of section $ABCD =$ 1000 sq ft. Area of water section $OAB = 200$ sq ft. Weight of concrete $= 150$ lb per cu ft. Linear dimensions are shown in figure.

A section of dam 1 ft long will be considered to be in equilibrium under the action of the following forces:

> $W =$ the total weight of the section, acting through the center of gravity of the cross section $ABCD$.
>
> $P =$ resultant total hydrostatic pressure on the face AB.
>
> $R =$ reaction between the foundation and the base, BC, of the dam. This reaction must necessarily be equal and opposite to, and collinear with, the resultant of W and P.

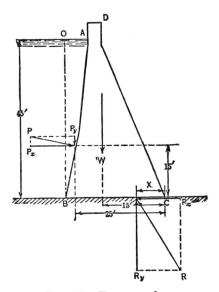

FIG. 25. Forces on dam.

Since the dam is in equilibrium when subjected to these three forces, the fundamental principles of equilibrium may be applied:

$$\Sigma F_x = 0, \quad \Sigma F_y = 0, \quad \Sigma M = 0.$$

Thus,

$$R_x = P_x = 62.4 \times 22.5 \times 45 = 63,200 \text{ lb}$$

and

$$R_y = P_y + W = 200 \times 62.4 + 1000 \times 150$$
$$= 12,500 + 150,000 = 162,500 \text{ lb}$$

The resultant reaction on the base is $\sqrt{R_x{}^2 + R_y{}^2} = 174,300$ lb.

For $\Sigma M = 0$, taking moments about an axis through C, normal to the plane of the section, X being the distance from C at which the resultant, R, intersects the base of the dam:

$$R_y X + 15 P_x - 25 P_y - 15W = 0$$

from which

$$X = 9.93 \text{ ft}$$

Thus the resultant reaction of 174,300 lb per linear foot of dam intersects the base 9.93 ft from the toe of the dam.

If the coefficient of friction is 0.4, the resistance to sliding is 162,500 × 0.4 = 65,000 lb, and the factor of safety against sliding is

$$\frac{65,000}{63,200} = 1.03$$

The factor of safety against overturning is

$$\frac{12,500 \times 25 + 150,000 \times 15}{63,200 \times 15} = 2.7$$

It might be stated in passing that this factor of safety against overturning would, in practice, be considered entirely satisfactory, whereas the above factor of safety against sliding could not be so considered.

The unit pressures on the foundation at B and C may be found by two methods:

(a) By geometric analysis, as indicated in Fig. 26a. The unit pressure is assumed to vary uniformly from x at B to $x + a$ at C. The diagram representing the distribution of pressure is in this case a trapezoid with its maximum ordinate at C, as indicated by the fact that if the diagram were a rectangle its center of gravity, that is, the location of R_y, would be 14 ft from C, whereas if it were a triangle the distance would be $9\frac{1}{3}$ ft.

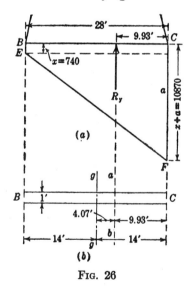

FIG. 26

Since the actual distance lies between these values the diagram must be a trapezoid.

The area of the pressure diagram represents the total vertical pressure on the base, or 162,500. Thus

$$28 \left(\frac{x + x + a}{2} \right) = 162,500$$

or
$$2x + a = 11,610$$

Taking moments about C,

$$162,500 \times 9.93 = 28 \times x \times 14 + \tfrac{1}{2} \times a \times 28 \times \tfrac{28}{3}$$

Solving the two preceding equations simultaneously,

$$x = 740 \text{ lb per sq ft}$$

and

$$x + a = 10,870 \text{ lb per sq ft}$$

(b) By the combined-stress formula from mechanics:

$$s = \frac{P}{A} \pm \frac{Mc}{I}$$

A plan view of the base of a section of the dam 1 ft long is shown in Fig. 26b. Let the vertical reaction R_y be considered applied along line ab, parallel to the toe of the dam and distant $e = 14 - 9.93 = 4.07$ ft from the parallel gravity axis gg. The terms in the combined-stress formula have the values

$$P = R_y = 162,500 \text{ lb}$$
$$A = 28 \text{ sq ft}$$
$$M = Pe = 661,000 \text{ ft-lb}$$
$$c = 14 \text{ ft}$$
$$I = \frac{1 \times 28^3}{12} = 1830 \text{ ft}^4$$

Substituting,

$$s = 5805 \pm 5065$$
$$= \begin{cases} 10,870 \text{ lb per sq ft at toe} \\ 740 \text{ lb per sq ft at heel} \end{cases}$$

EXAMPLE 2. Solve Example 1 assuming that the foundation material is pervious and that the uplift pressure head varies from full hydrostatic head at the heel to zero at the toe.

The intensity of the hydrostatic uplift at the heel is

$$62.4 \times 45 = 2810 \text{ lb per sq ft}$$

and the total uplift, U, is

$$\tfrac{1}{2} \times 2810 \times 28 = 39,300 \text{ lb}$$

Since $\Sigma F_y = 0$,

$$R_y = P_y + W - U = 12,500 + 150,000 - 39,300 = 123,200 \text{ lb}$$

Again with ΣM_c:

$$R_y X + (\tfrac{2}{3} \times 28)U + \tfrac{4.5}{3} P_z - 25P_y - 15W = 0$$

from which

$$X = 7.15 \text{ ft}$$

Since the resultant base pressure acts at a point 7.15 ft from C, or in other words, the center of gravity of the pressure diagram is at that distance horizontally from C, it follows that that diagram must be triangular as shown in Fig. 27 extending from C a distance equal to $3 \times 7.15 = 21.45$ ft. Since the total pressure equals 123,200 lb, the unit pressure at C is

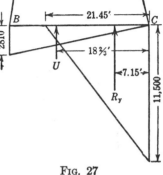

Fig. 27

$$\frac{123,200}{21.45} \times 2 = 11,500 \text{ lb per sq ft}$$

The resistance to sliding is now $123,200 \times 0.4 = 49,300$ lb, and, since this is less than the horizontal pressure, failure from sliding must occur.

The factor of safety against overturning is

$$\frac{12,500 \times 25 + 150,000 \times 15}{63,200 \times 15 + 39,300 \times \tfrac{2}{3} \times 28} = 1.52$$

as compared with 2.7 when there was no seepage under the dam.

PROBLEMS

1. A masonry dam of trapezoidal cross section, with one face vertical, has a thickness of 2 ft at the top and 10 ft at the bottom. It is 22 ft high and has a horizontal base. The vertical face is subjected to water pressure, the water standing 15 ft above the base. The weight of the masonry is 150 lb per cu ft. Where will the resultant pressure intersect the base, and what will be the intensity of pressure at the heel and at the toe assuming: (a) that there is no hydrostatic uplift; (b) that there is hydrostatic uplift which varies uniformly from that due to a full head of 15 ft at the heel to zero at the toe.

2. A masonry dam of trapezoidal cross section, with one face vertical, has a thickness of 2 ft at the top and 10 ft at the bottom. It is 22 ft high and has a horizontal base. The inclined face is subjected to water pressure, the water standing to a depth of 15 ft above the base. The weight of the masonry is 150 lb per cu ft. If there is no hydrostatic uplift, where will the resultant pressure intersect the base? Is this a good design?

35. Principle of Archimedes. The method of determining the vertical components of hydrostatic pressure on surfaces was discussed in Art. 32. It will now be shown that the so-called buoyant force acting on a body submerged in a fluid is merely the resultant of two vertical hydrostatic forces — one, the upward component of the total pressure exerted by the fluid on the under surface of the body; the other, the downward component of the total pressure exerted by the fluid on the upper surface. Since unit pressure increases with depth, the upward component is greater than the downward. The resultant is therefore an upward, or buoyant, force.

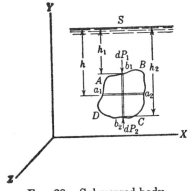

FIG. 28.   Submerged body.

In general, *any body immersed in a fluid is subjected to a buoyant force equal to the weight of the fluid displaced.* This is known as the Principle of Archimedes. It may be proved in the following manner.

Consider a body *ABCD* (Fig. 28) submerged in a fluid of constant unit weight *w* having a free surface *S*. If the fluid is a gas the free surface is imaginary; nevertheless the following analysis applies approximately if the dimensions of the body are small compared to its depth below the imaginary free surface. The body is referred to the coordinate axes *X*, *Y*, and *Z*.

Consider the small horizontal prism a_1a_2, parallel to the *X* axis, to have a cross-sectional area *dA*. The *X* component of the normal force acting on a_1 must be equal and opposite to the same force acting on a_2, each being equal to *wh dA*. There is, therefore, no tendency for this prism to move in a direction parallel to the *X* axis. Since the same reasoning may be applied to every other prism parallel to a_1a_2, it follows that there is no tendency for the body as a whole to move in this direction. The same reasoning applies to movement parallel to the *Z* axis or to any other axis in a horizontal plane. If, therefore, there is any tendency for the body to move it must be in a vertical direction.

Consider now the Y components of the hydrostatic presssure acting on the ends of any vertical prism b_1b_2 having a cross-sectional area, dA, so small that the intensity of pressure on either end of the prism may be considered uniform. The vertical component of the normal pressure on dA at b_1 is $wh_1\,dA$, acting downward; and the corresponding force at b_2 is $wh_2\,dA$, acting upward. The resultant of these two forces is upward and equal to $w(h_2 - h_1)\,dA$. But $(h_2 - h_1)\,dA$ is the volume of the elementary prism which, multiplied by w, gives the weight of the displaced fluid. Since the entire body, $ABCD$, is made up of an infinite number of such prisms, it follows that the resultant hydrostatic pressure on the body is an upward, or buoyant, force equal in magnitude to the weight of the displaced fluid.

If the weight of the body is greater than the buoyant force the body sinks unless prevented by external forces. If the weight of the body is less than the buoyant force, the body rises. If the fluid is a homogeneous liquid, the body rises to the free surface and floats there, displacing a volume of liquid having a weight equal to that of the body.

The resultant buoyant force acting on any vertical elementary prism of the submerged body is equal to the weight of the prism of displaced fluid. Since in a homogeneous fluid the weight of each prism is directly proportional to its volume, the center of gravity of the resultant buoyant force, called the *center of buoyancy*, coincides with the center of gravity of the displaced fluid.

PROBLEMS

1. A rectangular scow 15 ft by 32 ft, having vertical sides and ends, weighs 40 long tons (89,600 lb). What is its draft: (*a*) in fresh water; (*b*) in sea water?

2. If a rectangular scow 18 ft by 40 ft has a draft in fresh water of 5 ft, what is its weight in long tons?

3. A sphere 3 ft in diameter floats half submerged in a tank of oil (sp gr 0.80). (*a*) What is the total vertical pressure on the sphere? (*b*) What is the minimum weight of an anchor weighing 150 lb per cu ft that will be required to submerge the sphere completely?

4. A cubic foot of ice (sp gr 0.90) floats freely in a vessel containing water the temperature of which is 32° F. When the ice melts, will the water level in the vessel rise, fall, or remain stationary? Explain why.

5. In Fig. A a circular opening 4 ft in diameter is closed by a hemispherical shell weighing 2800 lb. (*a*) Neglecting friction, what force is required to lift the shell vertically? (*b*) With what force is the shell held against the seat?

6. An iceberg having a specific gravity of 0.92 floats in salt water having a specific gravity of 1.03. If the volume of ice above the surface is 700 cu yd, what is the total volume of the iceberg?

7. A cylinder 2 ft in diameter, 4 ft long, and weighing 75 lb floats in water with its axis vertical. An anchor weighing 150 lb per cu ft is attached to the lower end. Determine the total weight of the anchor if the bottom of the cylinder is submerged 3 ft below the water surface.

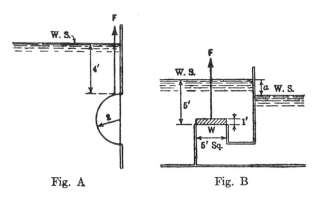

Fig. A   Fig. B

8. A concrete cube 2 ft on each edge, weighing 150 lb per cu ft in air, rests on the bottom of a tank in which sea water ($w = 64.0$) stands 16 ft deep. The bottom edges are sealed off so that no water is admitted under the block. Find the vertical pull required to lift the block.

9. A spherical balloon 20 ft in diameter is filled with gas weighing 0.035 lb per cu ft. In standard air weighing 0.0765 lb per cu ft, what is the maximum load, including its own weight, that the balloon can lift?

10. In Fig. B, if the weight W of the gate (in air) is 1000 lb and $a = 2$ ft, determine the force, F, required to lift the gate.

11. In Fig. B, if $a = 2$ ft, determine the value of W when a force of 3000 lb is required to lift the gate.

12. In Fig. B, if $W = 1000$ lb, determine the value of a when a force of 5000 lb is required to lift the gate.

36. Statical Stability of Floating Bodies. Any floating body is subjected to two systems of parallel forces: the downward force of gravity acting on each of the particles that goes to make up the body, and the buoyant force of the liquid acting upward on the various elements of the submerged surface.

In order that the body may be in equilibrium the resultants of these two systems of forces must be collinear, equal, and opposite. Hence the center of buoyancy and the center of gravity of the floating body must lie in the same vertical line.

Figure 29a shows the cross section of a ship floating in an upright position, the axis of symmetry being vertical. For this position the center of buoyancy lies on the axis of symmetry at B_0, which is the center of gravity of the area ACL. The center of gravity of the ship is assumed to be at G. If, from any cause, such as wind or wave action, the ship is made to heel through an angle θ, as

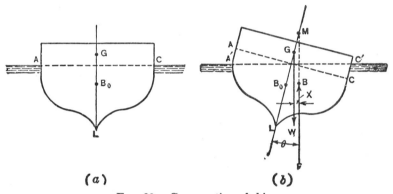

(a)   (b)

Fig. 29.   Cross section of ship.

shown in Fig. 29b, the center of gravity of the ship and cargo remaining unchanged, the center of buoyancy shifts to a new position, B, which is the center of gravity of the area $A'C'L$. The buoyant force F, acting upward through B, and the weight of the ship W, acting downward through G, constitute a couple WX which resists further overturning and tends to restore the ship to its original upright position.

If the vertical line through the center of buoyancy intersects the inclined axis of symmetry at a point M above the center of gravity, the two forces F and W produce a *righting moment*. If, however, M lies below G an *overturning moment* is produced. The point M is known as the *metacenter*, and its distance GM from the center of gravity of the ship is termed the *metacentric height*. The metacentric height is a measure of the statical stability of the ship. For small angles of inclination, as shown below, the position of M

does not change materially and the metacentric height is approximately constant.

37. Determination of Metacentric Height and Righting Moment. Figure 30 illustrates a ship having a displacement volume V. When the ship is tilted through the angle θ the wedge AOA' emerges from the water while the wedge $C'OC$ is immersed. If the sides AA' and $C'C$ are parallel, these wedges are similar and of equal volume, v, since the same volume of water is displaced by the ship whether in an inclined or upright position. The wedges therefore have the same length, and the water lines AC and $A'C'$ intersect on the axis of symmetry at O.

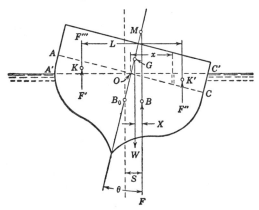

Fig. 30.   Inclined ship.

When the ship floats in an upright position a buoyant force F', equal to wv, acts upward through K, the center of gravity of the wedge AOA'. In the inclined position this force no longer acts, but an equal force F'' acts at K', the center of gravity of the wedge $C'OC$. It may be considered that a downward force F''', equal to F', has been introduced, the resultant of F''' and F' being zero. A couple has therefore been introduced equal to wvL, L being the horizontal distance between the centers of gravity of the wedges.

Because of the shifting of the force F' from K to K' the line of action of the buoyant force F acting on the entire ship is shifted from B_0 to B, a horizontal distance S such that

$$wVS = wvL$$

Hence

$$S = \frac{v}{V} L$$

But

$$S = \overline{MB_0} \sin \theta$$

Therefore

$$\overline{MB_0} = \frac{vL}{V \sin \theta} \tag{13}$$

For small angles,

$$\overline{MB_0} = \frac{vL}{V\theta} \text{ (approximately)} \tag{14}$$

Consider now a small prism of the wedge $C'OC$, at a distance x from O, having a horizontal cross-sectional area dA. For small angles the length of this prism $= x\theta$ (approximately). The buoyant force produced by this immersed prism is $wx\theta \, dA$, and the moment of this force about O is $wx^2\theta \, dA$. The sum of all these moments for both wedges must be equal to wvL or

$$w\theta \int x^2 \, dA = wvL = wVS$$

But for small angles $S = \overline{MB_0}\theta$ (approximately). Hence

$$\int x^2 \, dA = V \, (\overline{MB_0})$$

But $\int x^2 \, dA$ is the moment of inertia, I, of the water-line section about the longitudinal axis through O (approximately constant for small angles of heel). Therefore

$$\overline{MB_0} = \frac{I}{V} \tag{15}$$

The metacentric height

$$\overline{GM} = \overline{MB_0} \pm \overline{GB_0} \tag{16}$$

the sign being positive if G falls below B_0, and negative if above.

If I is the moment of inertia of the waterline section with the ship on an even keel, the resulting value of GM from equations 15 and 16 can be called the initial metacentric height and is a function of the geometrical form of the vessel and the location of the center of gravity for any particular condition of loading. This value is approximately constant for small angles of heel. For any angle, for a ship with straight sides, up to the angle at which the deck immerses or the bilge emerges, $\overline{GM}$ is given by equations 13 and 16.

The righting moment (see Figs. 29b and 30) is

$$\text{R.M.} = WX = W \cdot \overline{GM} \sin \theta \qquad (17)$$

EXAMPLE. The scow shown in Fig. 31 is 40 ft long, 20 ft wide, and 8 ft

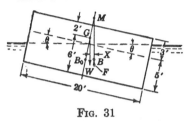

FIG. 31

deep. It has a draft of 5 ft when floating in an upright position. The center of gravity of the scow is on the axis of symmetry, 1 ft above the water surface. Compute (a) the initial metacentric height, and (b) the righting moment in fresh water when the angle of heel θ is 10°.

Solution. (a) By equations 15 and 16:

$$I = \frac{40 \times 20^3}{12} = 26,670 \text{ ft}^4 \quad V = 40 \times 20 \times 5 = 4000 \text{ ft}^3$$

$$\overline{MB}_0 = 6.667 \qquad \overline{GB}_0 = 3.50$$

Thus

$$\overline{GM} = 3.167 \text{ ft}$$

(b) $\overline{GM}$ is now determined by equations 13 and 16:

$$v = \tfrac{1}{2} \cdot 10 \tan 10° \cdot 10 \cdot 40 = 353 \text{ cu ft}$$
$$V = 4000 \text{ cu ft}$$
$$L = 13.33 \text{ ft}$$

$$\overline{MB}_0 = 6.77 \qquad \overline{GM} = 3.27$$
$$\text{R.M.} = 4000 \cdot 62.4 \cdot 3.27 \sin 10° = 142,000 \text{ ft-lb}$$

PROBLEMS

1. A ship of 4000 long tons displacement floats in sea water with its axis of symmetry vertical when a weight of 50 tons is midship. Moving the weight 10 ft towards one side of the deck causes a plumb bob, suspended at the end of a string 12 ft long, to move 9 in. Find the metacentric height.

2. A rectangular scow 30 ft wide, 50 ft long, and 12 ft high has a draft in sea water of 8 ft. Its center of gravity is 9 ft above the bottom of the scow. (a) Determine the initial metacentric height. (b) If the scow lists until one side is just on the point of submergence, determine the righting couple, or the overturning couple.

3. A cylindrical caisson having an outside diameter of 20 ft floats in fresh water with its axis vertical and with its lower end submerged 20 ft below the water surface. Its center of gravity is on the vertical axis and 8 ft above the bottom. Find: (*a*) the metacentric height; (*b*) the righting couple when the caisson is tipped through an angle of 10°.

4. A rectangular scow 30 ft wide and 50 ft long has a draft in fresh water of 8 ft. Its center of gravity is on the axis of symmetry, 15 ft above the bottom. Determine the height of the scow if, with one side just on the point of submergence, the scow is in unstable equilibrium.

5. A scow 50 ft long, 30 ft wide, and 15 ft high has a draft of 9 ft. Its center of gravity is at the center of the scow, both longitudinally and transversely. If the scow is tipped transversely until one side is just on the point of submergence, determine the righting couple.

6. A rectangular raft 10 ft wide and 20 ft long has a thickness of 24 in. and is constructed of solid timbers having a specific gravity of 0.60. If a man weighing 200 lb steps on the edge of this raft at the middle of one side, how much will the original water line on that side be depressed below the water surface?

7. Show that the value of $\overline{MB_0}$ for the scow in Fig. 31 is $6\frac{2}{3} \tan \theta / \sin \theta$. Make a table showing values of $\overline{MB_0}$, $\overline{GM}$, and the righting moment for angles of heel of 1°, 2°, 4°, 6°, 8°, and 10°.

GENERAL PROBLEMS

1. Water stands 40 ft above the top of a vertical gate which is 6 ft square and weighs 3000 lb. What vertical lift will be required to open the gate if the coefficient of friction between gate and guides is 0.3?

2. A plane parabolic gate with axis vertical and vertex down is submerged in oil (sp gr 0.80) to a depth of 9 ft. The width of the gate at the oil surface is 4 ft. Determine by integration the total pressure on the gate and the location of the center of pressure.

3. A vertical surface 6 ft square has its upper edge horizontal and on the water surface. At what depth must a horizontal line be drawn on this surface so as to divide it into two parts, on each of which the total pressure is the same?

4. A vertical triangular gate has a horizontal base 8 ft long and 6 ft below the water surface. Its vertex is 2 ft above the water surface. What normal force must be applied at the vertex to open the gate?

5. Find the horizontal and vertical components of the hydrostatic pressure per foot of length on the plane face of a dam inclined at an angle of 40° with the vertical if water stands 50 ft deep above the base?

6. The plane surface AB in Fig. A is rectangular in shape, 6 ft wide, and 10 ft long perpendicular to the paper. With oil on one side as shown, determine the total pressure on the gate, (a) by computing the horizontal and vertical components and their resultant; (b) directly by equation 5. Locate the center of pressure.

7. The rectangular gate in Fig. B is hinged at A and rests against a smooth vertical wall at B. The gate is 5 ft wide, perpendicular to the

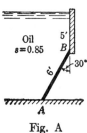

Fig. A

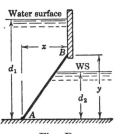

Fig. B

paper. Determine the horizontal and vertical components of the reactions at A and B when: (a) $x = 6$ ft, $y = 8$ ft, $d_1 = 12$ ft, $d_2 = 0$; (b) $x = 6$ ft, $y = 8$ ft, $d_1 = 12$ ft, $d_2 = 8$ ft; (c) $x = y = 6$ ft, $d_1 = 12$ ft, $d_2 = 10$ ft; (d) $x = y = 6$ ft, $d_1 = 6$ ft, $d_2 = 3$ ft.

8. The gate AB shown in Fig. C is rectangular and is 4 ft wide perpendicular to the paper. Compute the force F required to open the gate if: (a) it is hinged at A and rests against a smooth wall at B; (b) it is hinged at B and rests on a smooth floor at A.

9. The plane vertical side of a tank is formed as shown in Fig. D. When the tank is full of

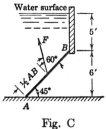

Fig. C

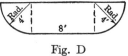

Fig. D

molasses (sp gr 1.50), compute the total pressure against the side and the location of the center of pressure.

10. An open flume of semicircular cross section is closed at the end by a semicircular gate hinged at the top diameter. The flume and the gate are 10 ft in diameter. When the flume is full of water, what moment applied to the hinge is required to keep the gate closed?

11. If, on the downstream side of the gate shown in Fig. C, page 48, water stands level with the hinge at c, neglecting the weight of the gate, determine the maximum height h to which water can rise on the upstream side without causing the gate to collapse.

12. A conical plug closes a circular opening in one end of a box as shown in Fig. E. The box is filled with water. If the absolute pressure at B is 40 lb per sq in., with what force is the plug held against the opening?

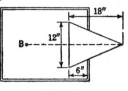

Fig. E

13. A funnel, in the shape of a cone having an area of base of 1 sq ft and an altitude of 1 ft, has a small hole at the vertex. If the base rests on the perfectly smooth platform of a scale and in this inverted position the funnel is filled with water, what downward force must be exerted upon it to maintain equilibrium and prevent the escape of water between the base of the funnel and the platform of the scale? Neglect the weight of the funnel. What is the least scale reading possible under these conditions?

14. The sides of a vertical cylindrical tank 12 ft in diameter and 12 ft high consist of vertical wood staves. The bursting stress is resisted by two horizontal steel hoops located at the quarter points of the sides. The lower half of the tank contains water, the upper half oil (sp gr 0.80). Compute the tension in each hoop.

15. A cubical box, 24 in. on each edge, has its base horizontal and is half filled with water. One of the sides is held in position by means of four screws, one at each corner. Find the tension in each screw due to the water pressure.

16. A cubical box, 24 in. on each edge, has its base horizontal and is half filled with a liquid the specific gravity of which is 1.5. The remainder of the box is filled with oil (sp gr 0.90). One of the sides is held in position by means of four screws, one at each corner. Find the tension in each screw due to the pressure of the liquids.

17. A cubical box, 6 ft on each edge, has its base horizontal and is half filled with water. The remainder of the box is filled with air under a gage pressure of 10 lb per sq in. One of the vertical sides is hinged at the top and is free to swing inward. To what depth can the top of this box be submerged in an open body of fresh water without allowing any water to enter?

18. A vertical, cylindrical cask, 1 ft in diameter and 2 ft high, contains mercury to a depth of 9 in., the remainder being filled with water. The cask is held together by means of two hoops, one at the top and one at the bottom. What is the stress in each hoop?

19. A cylindrical tank with its axis vertical is 9 ft high and 4 ft in diameter. It is held together by two steel hoops, one at the top and the other at the bottom. Three liquids, A, B, and C, having specific gravities of 1.0, 2.0, and 3.0, respectively, fill this tank, each having a depth of 3 ft. On the surface of A there is atmospheric pressure. Find the tensile stress in each hoop.

20. A 6-in. pipe line in which there is a horizontal 90° elbow contains water at rest under a gage pressure of 450 lb per sq in. Find the unit tensile stress in each of the eight $\frac{3}{4}$-in. bolts in the flanges by which the elbow is attached to the pipe.

21. The curved surface represented by AB in Fig. F is the surface of the quadrant of a circular cylinder 10 ft long, hinged at A and resting against a smooth wall at B. Determine the horizontal and vertical components of the reactions at A and B, when: (a) $r = 10$ ft, $d = 10$ ft; (b) $r = 8$ ft, $d = 16$ ft; (c) $r = 5$ ft, $d = 50$ ft.

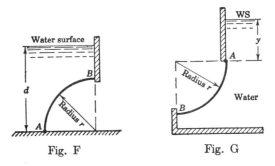

Fig. F   Fig. G

22. The curved surface represented by AB in Fig. G is the surface of the quadrant of a circular cylinder, hinged at A and resting against a smooth vertical wall at B. Considering a section 1 ft long perpendicular to the paper, compute the horizontal and vertical components of the reactions at A and B, when: (a) $r = 10$ ft, $y = 0$; (b) $r = 10$ ft, $y = 10$ ft; (c) $r = 5$ ft, $y = 25$ ft.

23. What is the tensile stress in pounds per square inch due to hydrostatic bursting pressure in the walls of a steel pipe 10 in. in diameter carrying gasoline ($w = 45$ lb per cu ft) under a pressure head of 400 ft of gasoline? The pipe wall thickness is $1/8$ in.

24. A masonry dam of trapezoidal cross section, with one face vertical and a horizontal base, is 24 ft high. It has a thickness of 2 ft at the top and 12 ft at the bottom. The weight of the masonry is 150 lb per cu ft. What is the depth of water on the vertical side if the resultant pressure intersects the base at the downstream edge of the middle third, or 2 ft from the middle of the base? Assume (a) that there is no hydrostatic uplift, and (b) that the uplift head varies uniformly from full hydrostatic head at the heel to zero at the toe.

25. A masonry dam of trapezoidal cross section is 50 ft high. It has a thickness of 5 ft at the top and 35 ft at the bottom. The upstream face has a batter of 10 ft in 50 ft. With the upstream water surface at the top of the dam, assuming no hydrostatic uplift, compute the distance from the toe to the point where the resultant pressure intersects the base, and the intensity of pressure at heel and toe.

26. A block of wood weighs 42.5 lb in air. A vertical force of 30 lb is required to keep the block submerged in water. Compute the unit weight and specific gravity of the wood.

27. A block of metal weighs 42.5 lb in air and 30 lb when suspended in water. Compute the unit weight and the specific gravity of the metal.

28. The inverted timber U-frame shown in Fig. H consists of a horizontal 12- by 12-in. timber 12 ft long, to the ends of which are attached two vertical 12- by 12-in. timbers 10 ft long, extending downward through openings in the bottom of a tank of water. Neglecting friction in the guides at the openings, determine the amount and direction (up or down) of the forces F, F, required to hold the frame in equilibrium. The timbers weigh 40 lb per cu ft.

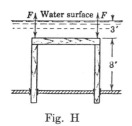

Fig. H

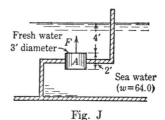

Fig. J

29. The solid cylindrical plug A in Fig. J is 3 ft in diameter and 2 ft high. It is made of concrete weighing in air 150 lb per cu ft. With fresh water on one side and sea water on the other as shown, and neglecting friction in the guides, find the vertical force F required to hold the plug in position.

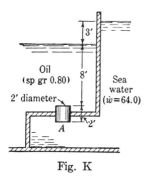

Fig. K

30. The solid cylindrical plug A in Fig. K is 2 ft in diameter and 2 ft high. With oil on one side and sea water on the other as shown, and neglecting friction in the guides, find the required weight of the plug so that it will be in equilibrium.

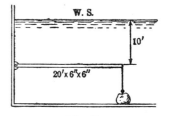

Fig. L

31. A 6-in. by 6-in. timber, 20 ft long, weighing 40 lb per cu ft, is hinged at one end and held in a horizontal position by an anchor at the other end as shown in Fig. L. If the anchor weighs 150 lb per cu ft, determine the minimum total weight it may have.

32. A cylinder weighing 100 lb and having a diameter of 3 ft floats in salt water (sp gr 1.03) with its axis vertical as shown in Fig. M.   The anchor consists of 10 cu ft of concrete weighing (in air) 150 lb per cu ft.   What rise in tide r will be required to lift the anchor off the bottom?

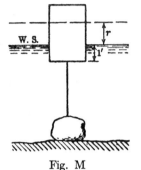

33. The timber shown in Fig. N is 12 in. square and has a specific gravity of 0.50.   A

Fig. M   Fig. N

man weighing 150 lb and standing at a point 2 ft from one end causes that end to be just submerged.   How long is the timber?

34. The timber shown in Fig. N is 12 in. square and 20 ft long and has a specific gravity of 0.50.   A man standing at a point 2 ft from one end causes that end to be just submerged.   Determine his weight.

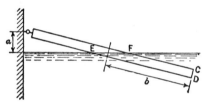

Fig. O

35. A solid cylindrical block of wood (sp gr 0.50) has a diameter of 12 in. and a length of 15 in.   Determine the position in which this block will float in water when in stable equilibrium.

36. A timber 6 in. square and 16 ft long has a specific gravity of 0.50. One end is hinged to a wall, the other end floating in the water as shown in Fig. O.   For values of a of 2 ft and 8 ft, find b, the length of timber submerged in the water.   Also find the reaction at the hinge. Assume that the center of gravity of the submerged trapezoid $CDEF$ is at a distance $b/2$ from the end CD.

37. A timber 6 in. square and 16 ft long has a specific gravity of 0.50. One end is hinged to a wall, the other end floating in the water as shown in Fig. O.   Determine the smallest value of a for which the timber will be in stable equilibrium in a vertical position.

38. A box, 1 ft square inside, 6 ft high, and made of wood (sp gr 1.00), having one end closed, is inserted vertically into water with the open end down, and floats with a block of 2 cu ft of concrete ($w = 150$ lb per cu ft) suspended from its lower end. Neglect the weight of the box. To what depth will the open end be submerged below the surface? What volume of concrete would cause the box to sink?

39. A rectangular pontoon 50 ft long, 24 ft wide, and 18 ft deep has a displacement of 180 short tons. Determine the location of the center of gravity, which is on the vertical axis of symmetry, if the maximum possible transverse angle of heel, without overturning, is 15°.

40. A rectangular pontoon 40 ft long, 16 ft wide, and 12 ft deep has a draft in fresh water of 7 ft. What load must be moved transversely from the center to a point 2 ft from the edge in order to cause the pontoon to list through an angle of 15°?

41. A cylinder, 40 ft long and having an inside cross-sectional area of 1 sq ft, is filled with water and submerged vertically so that the upper end stands 10 ft below the water surface in the tank. The upper end is closed and the lower end is open. Neglecting the weight of the cylinder, what maximum force will be required to lift it vertically out of the water.

42. A box 1 ft square and 6 ft high has its upper end closed and lower end open. If it is completely filled with water and submerged vertically with the open end down until the upper end is 10 ft below the water surface, neglecting the weight of the box, how many foot-pounds of work will be done in lifting it vertically out of the water?

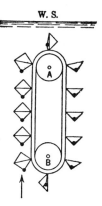

Fig. P

43. A and B, Fig. P, represent two pulleys free to rotate on their axes which are rigidly fastened to the inside vertical wall of a tank. Encircling these pulleys is a rubber hose to which are connected a number of cone-shaped metallic buckets. Connected with the base of each of these buckets is a flexible, rubber, cone-shaped diaphragm, in the vertex of which is a ball of mercury. At the point where each of these double-cone compartments is fastened to the hose, a small hole permits the flow of air between the hose and compartments. These compartments and hose form a closed system containing enough air to fill the hose and half of the compartments. When this device is completely submerged in water, a buoyant force will act upon the compartments to the left, which are filled with air, with no counterbalancing buoyant force acting on the closed compartments to the right. Since the magnitude of this buoyant force can be made as great as desired by increasing the size of the compartments, what will prevent the production of perpetual motion and the generation of an unlimited amount of power?

44. In Fig. Q is shown an endless belt composed of layers of cork with a central core of rope or wire cable. This belt passes around the four pulleys as shown, and through a stuffing-box in the side of the tank, which is filled with water. The portion of the belt that is submerged in water is subjected to a buoyant force equal to the weight of the water displaced, whereas the portion outside of the tank is not subjected to a similar counterbalancing force. What will prevent perpetual motion?

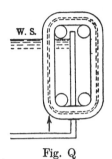

Fig. Q

Chapter IV

RELATIVE EQUILIBRIUM OF LIQUIDS

38. Relative Equilibrium Defined. In the preceding chapters liquids have been assumed to be in equilibrium and at rest with respect both to the earth and to the containing vessel. The present chapter treats of the condition where every particle of a liquid is at rest with respect to every other particle and to the containing vessel, but the whole mass, including the vessel, has a uniformly accelerated motion with respect to the earth. The liquid is then in equilibrium and at rest with respect to the vessel, but it is neither in equilibrium nor at rest with respect to the earth. In this condition a liquid is said to be in *relative equilibrium.* Since there is no motion of the liquid with respect to the vessel and no movement between the fluid particles themselves there can be no friction.

Hydrokinetics, which is treated in the following chapters, deals with liquids that are in motion with respect both to the earth and to their containers. In this case the retarding effects of friction must be considered.

Relative equilibrium may be considered as an intermediate state between hydrostatics and hydrokinetics. Two cases of relative equilibrium will be discussed.

39. Vessel Moving with Constant Linear Acceleration. If a vessel partly filled with any liquid moves horizontally along a straight line with a constant acceleration a, the surface of the liquid will assume an angle θ with the horizontal as shown in Fig. 32. To determine the value of θ for any value of a, consider the forces acting on a small mass of liquid M, at any point O on the surface. This mass is moving with a constant horizontal acceleration a, and the force producing the acceleration is the resultant of all the other forces acting upon the mass. These forces are the force of gravity W, acting verti-

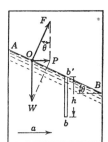

Fig. 32

77

cally downward, and the pressure of all the contiguous particles of the liquid. The resultant F of the pressure produced by these particles of liquid must be normal to the free surface AB. Since force equals mass times acceleration,

$$P = Ma = \frac{Wa}{g} \tag{1}$$

and from the figure

$$P = W \tan \theta \tag{2}$$

Solving these two equations simultaneously,

$$\tan \theta = \frac{a}{g} \tag{3}$$

which gives the slope that the surface AB will assume for any constant acceleration of the vessel.

Since O was assumed to be anywhere on the surface and the values of a and g are the same for all points, it follows that $\tan \theta$ is constant at all points on the surface, or, in other words, AB is a straight line.

The same value of θ will hold for a vessel moving to the right with a positive acceleration as for a vessel moving to the left with a negative acceleration or a retardation.

To determine the intensity of pressure at any point b at a depth h below the free surface, consider the vertical forces acting on a vertical prism bb' (Fig. 32). Since there is no acceleration vertically the only vertical forces acting are atmospheric pressure at b', gravity, and the upward pressure on the base of the prism at b. Hence, if the cross-sectional area is dA,

$$p_b \, dA = wh \, dA + p_a \, dA \tag{4}$$

or

$$p_b = wh + p_a \tag{5}$$

or, neglecting atmospheric pressure which acts throughout,

$$p_b = wh \tag{6}$$

Therefore, in a body of liquid moving with a horizontal acceleration the relative pressure at any point is that caused by the head of liquid directly over the point, as in hydrostatics. In this case, however, all points of equal pressure lie in an inclined plane parallel with the surface of the liquid.

In equation 3, if a were zero, tan θ would equal zero; or, in other words, if the vessel were moving with a constant velocity the surface of the liquid would be horizontal. Also if the acceleration were vertical, the surface would be horizontal.

To determine the relative pressure at any point b in a vessel with an acceleration upward, consider the forces acting on a vertical prism of liquid bb' of height h and cross-sectional area dA (Fig. 33). The force P, producing the acceleration, is the resultant of all the forces acting on the prism, consisting of gravity equal to $wh\,dA$, acting downward, and the pressure on the lower end of the filament at b, equal to $p_b\,dA$, acting upward. Therefore

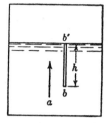

FIG. 33. Vessel with vertical acceleration.

$$P = p_b\,dA - wh\,dA = Ma = \frac{wh\,dA}{g}\,a$$

from which

$$p_b = wh + wh\,\frac{a}{g} \qquad (7)$$

This shows that the intensity of pressure at any point within a liquid contained in a vessel having an upward acceleration a is greater than the static pressure by an amount equal to wha/g. Evidently, if the acceleration were downward, the sign of the last term in the above expression would become negative, and if $a = g$, $p_b = 0$. In other words, if a vessel containing any liquid falls freely in a vacuum, so as not to be retarded by air friction, the pressure will be zero at all points throughout the vessel.

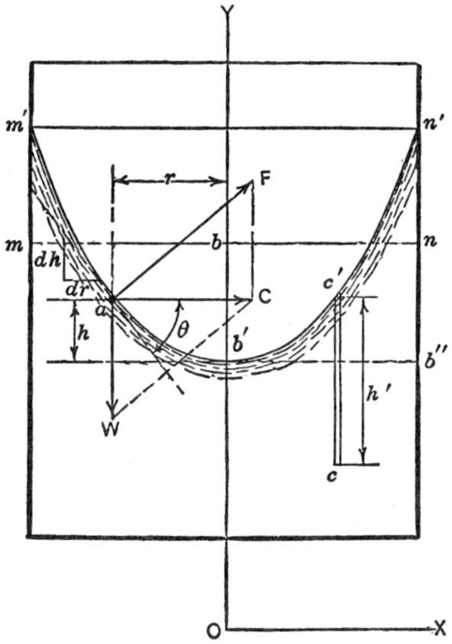

FIG. 34. Rotating vessel.

40. Vessel Rotating about a Vertical Axis. When the vessel shown in Fig. 34 is at rest, the surface of the liquid is horizontal and at mn. $m'b'n'$ represents the form of surface resulting from rotating the vessel with a constant angular velocity ω radians per second about its vertical axis OY. Consider the forces acting on a small mass of liquid M, at a, distant r from the axis OY.

Since this mass has a uniform circular motion it is subjected to a centripetal force, $C = M\omega^2 r$, which produces an acceleration directed toward the center of rotation and is the resultant of all the other forces acting on the mass. These other forces are the force of gravity, $W = Mg$, acting vertically downward, and the pressure exerted by the adjacent particles of the liquid. The resultant F of this liquid pressure must be normal to the free surface of the liquid at a.

Designating by θ the angle between the tangent at a and the horizontal,

$$\tan \theta = \frac{dh}{dr} = \frac{C}{W} = \frac{M\omega^2 r}{Mg} = \frac{\omega^2 r}{g}$$

or

$$dh = \frac{\omega^2 r}{g}\, dr \tag{8}$$

which, when integrated, becomes

$$h = \frac{\omega^2 r^2}{2g} \tag{9}$$

The constant of integration equals zero, because when r equals zero h also equals zero.

Since h and r are the only variables this is the equation of a parabola, and the liquid surface is a paraboloid of revolution about the Y axis. As the volume of a paraboloid is equal to one-half that of the circumscribed cylinder, and since the volume of liquid within the vessel has not been changed, $b'b = \frac{1}{2}b''n' = nn'$. The linear velocity at a is $v = \omega r$. Substituting v for ωr in 9,

$$h = \frac{v^2}{2g} \tag{10}$$

Expressed in words, this means that any point on the surface of the liquid will rise above the vertex of the paraboloid a height equal to the velocity head (see Art. 48) at that point.

To determine the relative pressure at any point c at a depth h' vertically below the surface at c' consider the vertical forces acting on the prism cc', having a cross-sectional area dA. As this prism has no vertical acceleration, $\Sigma y = 0$, and $p_c \, dA = wh' \, dA$. Hence

$$p_c = wh' \tag{11}$$

That is, the relative pressure at any point is that caused by the head of liquid directly over the point, as in hydrostatics. Therefore, the distribution of pressure on the bottom of the vessel is represented graphically by the vertical ordinates to the curve $m'b'n'$. It also follows that the total pressure on the sides of the vessel is the same as though the vessel were filled to the level $m'n'$ and were not rotating.

PROBLEMS

1. A vessel containing liquid moves horizontally along a straight line. What is the form of the liquid surface when the vessel moves with (a) a constant velocity of 10 ft per sec; (b) a constant acceleration of 10 ft per sec per sec?

2. A vessel partly filled with liquid and moving horizontally with a constant linear acceleration has its liquid surface inclined 45°. Determine its acceleration.

3. An open cylindrical vessel 2 ft in diameter and 3 ft high is two-thirds full of liquid. If the vessel is rotated about its vertical axis: (a) what is the greatest speed in revolutions per minute that it can have without causing any liquid to spill over the sides; (b) what speed must it have in order that the depth at the center be zero; (c) what speed must it have in order that there may be no liquid within 6 in. of the vertical axis?

4. An open cylindrical vessel, 2 ft in diameter, 3 ft high, and two-thirds filled with water, rotates about its vertical axis with a constant speed of 90 rpm. Determine: (a) the depth of water at the center of the vessel; (b) the total pressure on the cylindrical walls; (c) the total pressure on the bottom of the vessel.

5. An open cylindrical vessel 2 ft in diameter contains water 3 ft deep when at rest. If it is rotated about its vertical axis at a speed of 180 rpm, determine the least depth the vessel can have so that no water is spilled over the sides.

6. A vessel 12 in. in diameter and filled with water is rotated about its vertical axis with such a speed that the water surface at a distance of 3 in. from the axis makes an angle of 45° with the horizontal. Determine the speed in revolutions per minute.

7. A closed cylindrical vessel, 2 ft in diameter, 3 ft high, and completely filled with water, rotates about its vertical axis with a speed of

240 rpm. Assuming that the vessel is rigid and inelastic, determine the gage pressure, in pounds per square inch, just under the cover at the circumference and at the axis, under the following conditions: (a) with a small hole in the cover at the circumference; (b) with a small hole in the cover at the center.

8. A cylindrical vessel, 1 ft deep, is half filled with water. When it is rotated about its vertical axis with a speed of 150 rpm, the water just rises to the rim of the vessel. Determine the diameter of the vessel.

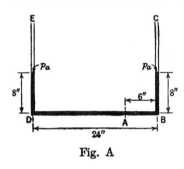

Fig. A

9. If the mercury U-tube shown in Fig. A is given an acceleration of 16.1 ft per sec per sec toward the right, determine the gage pressure at A in pounds per square inch.

10. If the mercury U-tube shown in Fig. A is rotated about a vertical axis through the leg BC, determine the height of mercury column in the leg DE, when the speed is (a) 40 rpm; (b) 60 rpm.

11. If the mercury U-tube shown in Fig. A is rotated about a vertical axis through A until there is no mercury in the leg BC, determine the lowest speed possible for this condition.

12. If the mercury U-tube shown in Fig. A is rotated about a vertical axis through the leg DE with a speed of 60 rpm, determine the gage pressure at A in pounds per square inch.

13. A conical vessel with vertical axis has an altitude of 3 ft and is filled with water. Its base, which is 2 ft in diameter, is horizontal and uppermost. If the vessel is rotated about its axis with a speed of 60 rpm, how much water will remain in it?

14. A conical vessel, with base uppermost, is rotated about its axis, which is vertical. Although the vessel was filled with water when at rest, after it is rotating at a speed of 60 rpm only 0.50 cu ft of water remains within it. Determine the ratio between the diameter of the base and the altitude.

15. A cylindrical bucket, 14 in. deep and 12 in. in diameter, contains water to a depth of 12 in. A man swings this bucket through a vertical plane, the bottom of the bucket describing a circle having a diameter of 7 feet. Assuming that the speed of rotation is contant, what is the lowest speed, in revolutions per minute, that the bucket can have without permitting any water to escape?

Chapter V

FUNDAMENTALS OF FLUID FLOW

41. Introduction. The principles relating to the behavior of water or other fluids at rest are based upon certain definite laws which hold rigidly in practice. In solving problems involving these principles it is possible to proceed by purely rational methods, the results obtained being free from doubt or ambiguity. Calculations are based upon a few natural principles which are universally true and simple enough to permit of easy application. In problems ordinarily encountered in hydrostatics, after the unit weight of the fluid has been determined, no other experimental data are required.

A fluid in motion, however, presents an entirely different condition. Though the motion undoubtedly takes place in accordance with fixed laws, the nature of these laws and the influence of the surrounding conditions upon them are very complex and have thus far defied complete expression in mathematical form. However, a great number of engineering problems involving fluid flow have been solved by combining mathematical theory with experimental data.

Although many of the laws governing the flow of fluids have been well known for centuries, it has been, as in other branches of science, only within comparatively recent times that applications of these laws have been widely extended. Most of the early knowledge of hydraulics applied only to water, as the name indicates. With modern use and transportation of oils, gasoline, chemicals, steam, and gases, it has been necessary to extend the laws of hydraulics both mathematically and experimentally to include these fluids. A great step in advance was made when it was discovered, largely through the research of Osborne Reynolds, that, from experiments on any particular fluid, it is often possible to predict, at least approximately, the characteristics of flow of any other fluid.

It cannot be emphasized too often, however, that there is still much to be learned regarding the behavior of all fluids in motion, even of water. Practical problems in hydraulics constantly arise which are apparently simple to explain or solve by theory or by laboratory technique but which prove extremely baffling because of lack of thorough determination of the fundamental properties of the fluids involved. As a result, many experimental data obtained with water and other fluids are conflicting, a condition due partly to the fact that records of temperature of the fluid were not kept.

42. Path Lines and Stream Tubes. A fluid in motion can be considered to consist of a great number of individual particles all of which move in the general direction of flow but usually not in parallel lines or even with continuous motion. The irregular rise of large volumes of smoke in air and the boiling and eddying of rivers are familiar phenomena which illustrate this point. Yet, at times a thin column of smoke in still air will rise in almost a straight vertical line, or a thin stream of dye injected into a liquid moving at slow velocity will continue to move in a straight line without being dispersed through the liquid.

These phenomena are cited to indicate the complex nature of fluid flow. It is possible, however, to arrive at a solution of many engineering problems by making certain simplifications. To aid in understanding these simplifications a brief statement of the nature of fluid flow with certain definitions is given here.

Any particle of a stream of fluid has at any given instant a certain velocity, v, which is a vector quantity and therefore possesses both magnitude and direction. At the next instant forces acting on the particle may cause it to have a velocity which is different in amount and direction. The path followed by a particle, called a *path line*, is ordinarily a curve in three dimensions. A two-dimensional projection of a typical path line is shown in Fig. 35.

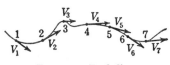

FIG. 35. Path line.

If path lines were drawn for all particles in a stream their composite effect would represent the motion of the entire stream. Most problems in applied hydraulics can be solved by considering the entire cross section of the stream. It is sometimes advisable, however, to consider only a small part of the cross section. For this purpose a small bundle of path lines called a *stream tube* can be used. Such a tube is bounded by an imaginary surface formed by surrounding layers of the fluid.

43. Laminar and Turbulent Flow. Flow is said to be laminar when the paths of the individual particles do not cross or intersect. With this type of flow in conduits having parallel sides the path lines are parallel. Moreover, the stream tubes are of constant cross section and have directions parallel to the sides of the conduit and to each other. The velocities in the various path lines are not the same, however, but increase with the distance from the walls of the conduit. (See Art. 93.)

In conduits having non-parallel sides, the path lines in laminar flow converge or diverge. The cross-sectional area of each stream tube varies proportionally with that of the conduit, but its relative position in the cross section does not change. Converging path lines result in increased velocity in each path line, whereas diverging path lines result in decreased velocity.

Any fluid will flow with laminar motion under certain limiting conditions. Conditions which tend to produce laminar flow are low velocity, small size of conduit, and high viscosity of fluid. (See Art. 92.) Laminar flow is frequently encountered in the flow of oil in pipes and in the flow of fluids through small tubes. The percolation of underground water or oil through sandbeds is an example of laminar flow.

Beyond these limiting conditions of laminar flow the flow becomes sinuous or turbulent. Flow is said to be turbulent when its path lines are irregular curves which continually cross each other and form a complicated network which in the aggregate represents the forward motion of the entire stream. The particles of a stream flowing with turbulent motion occupy successively various transverse positions without any regularity, and their paths are neither parallel nor fixed. A particle that at one instant is near the center of the conduit may an instant later be near the outer walls, and vice versa.

The laws governing laminar flow have been derived by mathematical theory, and the results agree closely with experimental data. However, the flow of water and other fluids in engineering problems is nearly always turbulent; and laws of turbulent flow have thus far baffled all attempts at complete mathematical derivation. Further studies are continually being made of the causes

and effects of the complex nature of turbulent flow. Pending a
more complete analysis than is now available, applications of the
hydraulics of turbulent flow must still be based on a combination
of mathematical analysis with experimental observation.

44. Discharge. The volume of fluid passing a cross section of
a stream in unit time is called the discharge. The symbol Q is used
to designate the discharge, the usual units being cubic feet per
second (cfs).

It is customary in certain lines of engineering to use other units,
for instance, cubic feet per minute (cfm) for measuring the flow
of air, gallons per minute (gpm) in connection with pumping ma-
chinery, and gallons per day (gpd) or millions of gallons daily
(mgd) in connection with municipal water supply.

If equal velocities at all points in the cross section of a stream
were possible there would be passing any section, every second, a
volume equivalent to that of a prism having a base equal to the
cross-sectional area of the stream and a length equal to the velocity.
Because, however, of the varying effects of friction, viscosity, and
surface tension, the individual particles in a stream have different
velocities. For this reason it is common in hydraulics to deal with
mean velocities. If V is the mean velocity in feet per second past
any cross section, and A is the cross-sectional area in square feet,

$$Q = AV \qquad (1)$$

and

$$V = \frac{Q}{A} \qquad (2)$$

45. Steady Flow. If the discharge Q passing a given cross
section of a stream is constant with time, the flow is steady at that
cross section. If Q at the cross section varies with time the flow is
unsteady.

Nearly all problems in this book deal with steady flow. Examples of unsteady flow are discharge through orifices under a falling
head (Art. 72) and sudden stopping of flow in pipe lines with the
resulting phenomenon called water hammer (Art. 156).

46. Uniform Flow. If, with steady flow in any length, or
"reach," of a stream, the average velocity at every cross section
is the same, the flow is said to be uniform in that reach. For
fluids considered incompressible this condition requires a stream of
uniform cross section. In streams where changes of cross section

and velocity occur, the flow is said to be non-uniform. With gases, strictly uniform flow seldom occurs, owing to the expansion resulting from the reduction in pressure that usually takes place along the path of flow.

Thus, steady flow involves permanency of conditions at any particular cross section, whereas uniform flow implies simultaneous uniformity of conditions at successive cross sections.

47. Continuous Flow. When, at any instant, the number of particles passing every cross section of the stream is the same, the flow is said to be continuous, or there is continuity of flow. Letting Q, A, and V represent, respectively, discharge, area, and mean velocity, with similar subscripts applying to the same cross section, continuity of flow with non-compressible fluids exists when

$$Q = A_1V_1 = A_2V_2 = A_3V_3, \text{ etc.} \tag{3}$$

Equation 3 applies when the number of particles of fluid per unit volume — that is, the density — can be considered constant. With gas flow the number of particles passing the given point depends not only on their mean velocity and the area of cross section but also on the density of the gas. The equation of continuity for compressible fluids thus becomes

$$\rho_1A_1V_1 = \rho_2A_2V_2 = \cdots, \text{ etc.} \tag{4}$$

or, since ρ is proportional to unit weight w,

$$w_1A_1V_1 = w_2A_2V_2 = \cdots, \text{ etc.} \tag{5}$$

However, the discharge past any section measured in volume per unit of time is still

$$Q = AV \tag{1}$$

If, between any two points of a stream, flow is added through a tributary or taken out through a distributary, the flow between the two points is not continuous and equations 3, 4, and 5 do not apply.

PROBLEMS

1. Compute the discharge of water through a 3-in. pipe[1] if the mean velocity is 8.5 ft per sec.

[1] The answers at the end of the book are computed on the basis of nominal pipe sizes as given in the problems. Actual diameters of commercial pipe vary slightly from the nominal sizes.

2. The discharge of air through a 24-in. pipe is 8600 cfm. Compute the mean velocity in feet per second.

3. The diameter of a 6-ft length of pipe decreases uniformly from 18 in. to 6 in. With a flow of 5 cfs of oil compute the mean velocity at cross sections 1 ft apart along the pipe. Plot velocity as ordinate against length as abscissa.

4. A pipe line consists of successive lengths of 15-in., 12-in., and 10-in. pipe. With a continuous flow through the line of 9 cfs of water compute the mean velocity in each size of pipe.

5. The diameter of a 6-ft length of pipe increases from 6 in. to 24 in. The diameter is to vary so that the mean velocity of liquid flowing through the pipe will decrease uniformly with distance from the 6-in. end. Compute the diameter at each foot along the pipe. Draw to scale the longitudinal profile of the pipe.

6. A city requires a flow of 25 mgd for its water supply. Compute the diameter of pipe required if the velocity of flow is to be: (a) 2 ft per sec; (b) 6 ft per sec.

7. What diameter of pipe is required to carry 10 gpm of gasoline at a velocity of 7 ft per sec?

8. A vertical circular stack 100 ft high converges uniformly from a diameter of 20 ft at the bottom to 16 ft at the top. Coal gas with a unit weight of 0.030 lb per cu ft enters the bottom of the stack with a velocity of 10 ft per sec. The unit weight of the gas increases uniformly to 0.042 lb per cu ft at the top. Compute the mean velocity every 25 ft up the stack.

48. Energy and Head. Since the principles of energy are applied in the derivation of fundamental hydraulic formulas, an explanation of such principles as will be used is here introduced.

Energy is defined as ability to do work. Both energy and work are measured in foot-pounds. The two forms of energy commonly recognized are kinetic energy and potential energy. Potential energy in fluids may in turn be subdivided into energy due to position or elevation above a given datum plane, and energy due to pressure in the fluid. The three forms of energy which must be considered in connection with flow of fluids are therefore usually stated as: 1. Kinetic energy. 2. Elevation energy. 3. Pressure energy. Other forms such as heat energy and electrical energy have little bearing on the laws governing flowing liquids, although thermodynamic effects are important in the flow of gases.

1. *Kinetic energy* is the ability of a mass to do work by virtue of its velocity. If, in any mass M, every individual particle has the

same velocity v, in feet per second, the kinetic energy of the mass is $\frac{1}{2}Mv^2$, and, since $M = W/g$,

$$\text{K.E.} = W\frac{v^2}{2g}$$

which reduces to $v^2/2g$ for a weight of unity. The expression $v^2/2g$ is of the form

$$\frac{(\text{ft/sec})^2}{\text{ft/sec}^2} = \text{ft}$$

and it therefore represents a linear quantity expressed in feet. It is the height through which a body must fall in a vacuum to acquire the velocity v. When applied to a moving mass it is called the *velocity head*. Although representing a linear quantity, the velocity head is directly proportional to the kinetic energy of any mass having a velocity v and is equal to the kinetic energy of 1 lb of any mass moving with that velocity.

2. *Elevation energy* is manifested in a fluid by virtue of its position or elevation with respect to some arbitrarily selected horizontal datum plane, considered in connection with the action of gravity. Elevation energy may be explained by considering a mass having a weight of W pounds the elevation of which above any horizontal datum plane is z feet. With respect to this plane the mass has Wz foot-pounds of energy. A mass weighing 1 lb will have z foot-pounds of energy. If a mass weighing 1 lb is placed z feet below the datum plane, its energy with respect to the plane will be $-z$ foot-pounds, being negative because this amount of energy will have to be imparted to the mass to raise it to the datum plane against the force of gravity. Here again the expression for energy, in this case z, represents a linear quantity called the *elevation head* of the mass, but it should be kept clearly in mind that z is also the energy expressed in foot-pounds contained in 1 lb of fluid by virtue of its position with respect to the datum plane.

It thus appears that the amount of energy of position possessed by a mass depends upon the elevation of the datum plane. In any particular problem, however, all references should be made to the same plane. In this way the relative amounts of energy contained in different masses or the relative amounts of energy in the same mass in different positions may be determined. Since all energy is relative, this is all that is required.

It is evident from the foregoing paragraphs that any mass of weight W, every particle of which is moving with the same velocity v, has an amount of kinetic energy equal to $Wv^2/2g$, regardless of all other conditions.   In a similar manner, any mass of weight W at a distance z above a datum plane has an amount of elevation energy equal to Wz, regardless of whether there is air, water, or an absolute vacuum between the mass and the datum plane.

3. *Pressure energy* differs fundamentally from kinetic and elevation energy, to the extent that no mass *per se* can have such energy.   Any mass having pressure energy acquires that energy only by virtue of contact with other masses having some form of energy.

Consider a reservoir of liquid (Fig. 36) from which a horizontal pipe leads to a valve which is closed so that there is no flow.   As-

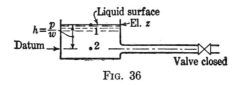

FIG. 36

sume that point 1 represents a small mass of liquid at the surface and that point 2 represents another small mass opposite the center line of the pipe.   Assume the datum through point 2.

It is evident that if the valve is opened liquid will flow from the open end of the pipe because of the elevation of the liquid surface in the reservoir.   The elevation head at point 1 is z feet, and the elevation head at point 2 is zero.

The pressure at point 2, however, assuming atmospheric pressure on the liquid surface, is $p = wh$ (Art. 15), so that the *pressure head* at point 2 is $p/w = h$ feet.   Thus from point 1 to point 2 the elevation head has decreased from z to zero, whereas the pressure head has increased from zero to h.

Since $z = h$ it is seen that the elevation head at point 1 is transformed into an equal amount of pressure head at point 2, and that this pressure head is the immediate cause of liquid flowing into the pipe on the opening of the valve.

Consider now that the reservoir is covered and that gas is forced into the space between the cover and the liquid surface so that the pressure on this surface is raised above atmospheric by an amount p_g.   Although the elevation head of point 1 is unchanged it is evident that if the valve is now opened the flow from the pipe will be greater than before, for the reason that the pressure head at point 2 has been increased to $p_g/w + h$. Moreover, if the pressure on the surface of the liquid is reduced below atmospheric pressure, the flow will be decreased.

Pressure energy should therefore be considered as energy transmitted to or through the mass considered. If pressure head is expressed in feet, it will also represent foot-pounds of energy per pound of fluid, as has been shown to be true for velocity head and elevation head.

The action of pressure energy is also illustrated by the piston and cylinder arrangement shown in Fig. 37, which is operated entirely by water under a gage pressure of p pounds per square foot. The area of the piston is A square feet. The cylinder is supplied with water through the valve R and may be emptied through the valve S.

At the beginning of the stroke the piston is at CD, the valve S is closed, and R is open. Water enters the cylinder and slowly drives the piston to the right against the force P. Neglecting friction, the amount of work done on the piston while it moves through the distance l feet is $Pl = pAl$ foot-pounds. The quantity of water required to do the work is Al cubic feet, and its weight is wAl pounds. The amount of work done per pound of water is therefore

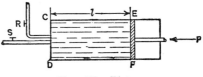

Fig. 37. Piston.

$$\frac{pAl}{wAl} = \frac{p}{w} \text{ foot-pounds}$$

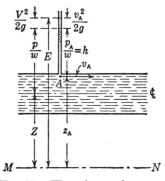

Fig. 38. Three forms of energy.

The three forms of energy which a fluid may have are illustrated in Fig. 38. At any point A in a stream of fluid where the velocity is v_A, the velocity head is $v_A^2/2g$, the pressure head is p_A/w, and the elevation head referred to the datum plane MN is z_A. Thus, with respect to the plane MN the total head at point A, or its equivalent, the total energy per pound of fluid at A, expressed in foot-pounds, is

$$E_A = \frac{v_A{}^2}{2g} + \frac{p_A}{w} + z_A \tag{6}$$

A practical solution of most hydraulics problems is obtained by considering the entire stream as a single stream tube, in a given cross section of which the average velocity is V. The pressure and the elevation heads are usually computed to the center line of the stream tube. Thus, the total head at *cross section* A of the stream tube shown in Fig. 38 is

$$E = \frac{V^2}{2g} + \frac{p}{w} + Z \tag{7}$$

49. Kinetic Energy in Stream with Non-uniform Distribution of Velocity. In equation 7 the term $V^2/2g$ represents the kinetic energy at a cross section of a stream at which all particles are assumed to have the same velocity. However, the velocities of the particles are ordinarily not equal.

The actual velocity head of a stream is the average of the velocity heads of the individual particles. Since the average of the squares of several numbers is always greater than the square of their average the actual velocity head of the stream is always greater than the velocity head computed from the mean velocity.

The usual method of correcting for non-uniform velocity in a cross section is to multiply the velocity head computed from the mean velocity by a coefficient α which is always greater than unity. The general expression for the kinetic energy per pound of fluid is therefore

$$\alpha \frac{V^2}{2g}$$

where V is the mean velocity in the cross section.

Usually no serious error is introduced by assuming $\alpha = 1$, but in some instances a knowledge of the value of α becomes important.

50. Power. It cannot be repeated too often that each term in equation 7 represents a linear quantity called a *head*, which is the amount of energy in foot-pounds contained in each pound of the fluid at that head. If the actual or potential flow is Q cubic feet per second and each cubic foot of fluid weighs w pounds, the weight of fluid which will pass any point per second is Qw pounds. Since the total energy at any point in the stream is E foot-pounds per pound of fluid, the power, or rate of doing work, at that point is

$$QwE \text{ foot-pounds per second}$$

the equation of units being

$$\frac{ft^3}{sec} \times \frac{lb}{ft^3} \times \frac{ft\text{-}lb}{lb} = \frac{ft\text{-}lb}{sec}$$

Equation 7 can thus be changed to units of power by multiplying through by Qw:

$$QwE = Qw\frac{V^2}{2g} + Qw\frac{p}{w} + Qwz \tag{8}$$

Horsepower is obtained by dividing each term by 550.

PROBLEMS

1. A fluid is flowing in a pipe 8 in. in diameter with a mean velocity of 10 ft per sec. The pressure at the center of the pipe is 5 lb per sq in., and the elevation of the pipe above the assumed datum is 15 ft. Compute the total head in feet if the fluid is (a) water, (b) oil (sp gr 0.80), (c) molasses (sp gr 1.50), (d) gas ($w = 0.040$).

2. A liquid (sp gr 2.0) is flowing in a 2-in. pipe. The total energy at a given point is found to be 24.5 ft-lb per lb. The elevation of the pipe above the datum is 10 ft, and the pressure in the pipe is 9.5 lb per sq in. Compute the velocity of flow and the horsepower in the stream at that point.

3. The jet of water from a nozzle discharging into air has a diameter of 6 in. and a mean velocity of 120 ft per sec. Compute the velocity head and the horsepower in the jet.

4. At a summit in a 12-in. pipe line in which 6.0 cfs of water is flowing the elevation above datum is 30 ft and the total head is 15 ft. Compute the absolute pressure in the pipe.

5. At point A where the suction pipe leading to a pump is 4 ft below the pump an open manometer indicates a vacuum of 7 in. of mercury. The pipe is 4 in. in diameter, and the discharge is 1.1 cfs of oil (sp gr 0.85). Compute the total head at point A with respect to a datum at the pump.

6. The cross section of a pipe was divided into ten equal areas by means of concentric circles. The mean velocities in the areas, beginning at the center of the pipe, were measured, in feet per second, as follows: 5.62, 5.58, 5.50, 5.38, 5.18, 4.90, 4.54, 4.04, 3.36, 2.52. Compute α.

51. Frictional Loss. A fluid in motion suffers a frictional loss, which is an expenditure of energy required to overcome resistance to flow. The expended energy is transformed into heat. After being so transformed it cannot, through the ordinary processes of nature, be reconverted into any of the useful forms of energy contained in a flowing fluid and is therefore often referred to as lost energy or *lost head*.

The exact manner in which this loss occurs is not completely known. The loss must not be thought of as caused by sliding friction between the stream and the walls of the conduit, since, when any fluid wets the walls of the conduit through which it flows, as generally occurs, the outermost particles of the fluid adhere to the wall and have no motion with reference to it. There can therefore be no friction between the fluid and the conduit.

The loss must be thought of rather as occurring within the stream itself. It may occur as the result of friction between the various fluid particles as they rub against one another, or it may be due to loss in kinetic energy resulting from the impact of molecules or masses moving with different velocities.

With laminar flow, as is shown later (Art. 98), the magnitude of the frictional loss is independent of the degree of roughness of the conduit. With turbulent flow, according to Prandtl and others, when the fluid wets the conduit walls there is a layer of the fluid adjacent to the walls in which the flow is laminar; but the roughness of the conduit wall does, however, have a direct effect upon the amount of frictional loss since turbulence increases with the degree of roughness.

52. Bernoulli's Energy Theorem. In 1738, Daniel Bernoulli, an eminent European mathematician and philosopher, demonstrated that in any stream flowing steadily without friction the total energy contained in a given mass is the same at every point in its path of flow. In other words, kinetic energy, pressure energy, and energy of position may each be converted into either of the other two forms, theoretically without loss. Thus if there is a reduction in the amount of energy contained in any one form there must be an equal gain in the sum of the other two.

In Fig. 39, *bcde* represents a stream tube in which all the particles in any cross section flow with the same velocity. For the present, frictional losses will be ignored. Every particle passing the section bc will, a little later, pass the section de, and no particles will pass the section de which have not previously passed bc.

Consider now the forces acting on this stream tube. On the section bc the area of which is a_1 there is a normal pressure in the direction of flow of intensity p_1 producing motion. On the section de the area of which is a_2 there is a normal intensity of pressure p_2 parallel with the direction of flow and resisting motion. On the lateral surfaces of the stream tube, indicated by the lines bd and ce, there is a system of forces, acting normal to the direction of

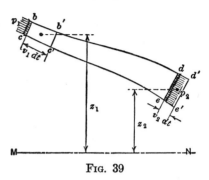

FIG. 39

motion, which have no effect on the flow and can therefore be neglected. The force of gravity, equal to the weight of the stream tube, acts downward. The work performed on the stream tube by the three forces will now be investigated.

Consider that in the time dt each of the particles at bc moves to $b'c'$ with a velocity v_1. In the same time interval each of the particles at de moves to $d'e'$ with a velocity v_2. Since there is continuity of flow,

$$a_1v_1 \, dt = a_2v_2 \, dt$$

The work G_1 done by the force acting on the section bc in the time dt is the product of the total force and the distance through which it acts, or

$$G_1 = p_1a_1v_1 \, dt \text{ foot-pounds} \tag{9}$$

Similarly the work done on the section de is

$$G_2 = -p_2a_2v_2 \, dt \text{ foot-pounds} \tag{10}$$

being negative because p_2 is opposite in sense to p_1 and resists motion.

The work done by gravity on the entire mass in moving from the position $bcde$ to $b'c'd'e'$ is the same as though $bcb'c'$ were moved to the position $ded'e'$ and the mass $b'c'de$ were left undisturbed. The force of gravity acting on the mass $bcb'c'$ is equal to the volume $a_1v_1 \, dt$ times the unit weight w. If z_1 and z_2 represent, respectively, the elevations of the centers of gravity of $bcb'c'$ and $ded'e'$ above the datum plane MN, the distance through which the force of gravity would act on the mass $bcb'c'$ in moving it to the position $ded'e'$ is $z_1 - z_2$, and the work done by gravity is

$$G_3 = wa_1v_1\,dt(z_1 - z_2) \text{ foot-pounds} \tag{11}$$

The resultant gain in kinetic energy is

$$\frac{Mv_2{}^2}{2} - \frac{Mv_1{}^2}{2} = \frac{wa_1v_1\,dt}{2g}\,(v_2{}^2 - v_1{}^2) \tag{12}$$

From fundamental principles of mechanics, the total amount of work done on any mass by any number of forces is equal to the resultant gain in kinetic energy. Therefore from equations 9, 10, 11, and 12,

$$p_1a_1v_1\,dt - p_2a_2v_2\,dt + wa_1v_1\,dt(z_1 - z_2) = \frac{wa_1v_1\,dt}{2g}\,(v_2{}^2 - v_1{}^2) \tag{13}$$

Dividing through by $wa_1v_1\,dt$ and transferring, and remembering that $a_1v_1 = a_2v_2$, there results

$$\frac{v_1{}^2}{2g} + \frac{p_1}{w} + z_1 = \frac{v_2{}^2}{2g} + \frac{p_2}{w} + z_2 \tag{14}$$

This is known as Bernoulli's energy equation. It is the mathematical expression of Bernoulli's energy theorem which is in reality the law of conservation of energy applied to fluids which may be considered incompressible. It may be stated as follows:

Neglecting friction, the total head, or the total amount of energy per unit of weight, is the same at every point in the path of flow.

Fluids in motion invariably suffer a loss of energy through friction. (See Art. 51.) If the direction of flow in the stream tube is from section 1 to section 2, the total energy at 2 must be less than at 1. In order to make equation 14 balance, a quantity, h_L, equal to the loss of energy, or what is equivalent, the loss of head due to friction between the two sections, must be added to the right-hand side of the equation.

The foregoing discussion is related primarily to the flow in any stream tube. Most problems involving continuous steady flow in pipes or open channels can be solved satisfactorily by considering the entire stream as a single stream tube. When the entire stream is so considered, capital letters are ordinarily used to denote velocity, elevation head, and lost head.

A statement of the energy equation for the entire cross section of a continuous stream including lost head then becomes

$$\frac{V_1^2}{2g} + \frac{p_1}{w} + Z_1 = \frac{V_2^2}{2g} + \frac{p_2}{w} + Z_2 + H_L \qquad (15)$$

The statement of this form of the energy theorem should be memorized: *With continuous, steady flow, the total head at any point in a stream is equal to the total head at any downstream point plus the loss of head between the two points.*

If energy is added to the stream between points 1 and 2, as for instance by a pump, the left side of the energy equation must include the added head H_U, and the complete energy equation then becomes

$$\frac{V_1^2}{2g} + \frac{p_1}{w} + Z_1 + H_U = \frac{V_2^2}{2g} + \frac{p_2}{w} + Z_2 + H_L \qquad (16)$$

If energy is given up by the stream to a turbine between points 1 and 2, the right side of the energy equation must include a term H_I to represent the head given up.

Equations 14, 15, and 16 apply to any stream of fluid which can be considered incompressible, regardless of the area of its cross section, if it is assumed that all particles in any cross section move with the same velocity. This assumption ordinarily gives results of sufficient accuracy. Therefore, in nearly all the problems and examples in this book the velocity head is computed from the mean velocity.

For streams in which the velocities at all points of a cross section cannot be considered the same, the energy equations may be written between any two points on the same path line, but in applying them to the entire cross section a corrective factor must be introduced. (See Art. 49.)

EXAMPLE 1. The fluid in Fig. 40 is water, with the surface 20 ft above the datum. The pipe is 6 in. in diameter, and the total loss of head between point 1 in the water surface and point 5 in the jet is 10 ft. Determine the velocity of flow in the pipe and the discharge Q.

Solution. Consider the entire stream from reservoir surface to jet as a stream tube having steady, continuous flow. Write the energy theorem from 1 to 5.

Total head at 1: Pressure head is atmospheric, therefore zero. Velocity head is zero since the reservoir is large and the water is practically stationary. Elevation head is 20 ft.

Total head at 5: Pressure head is zero since the jet is springing free in air. Velocity head is unknown but represented by $V^2/2g$, where V is

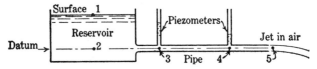

FIG. 40. Pipe discharging from reservoir.

the mean velocity of flow in the pipe and in the jet, which is the same diameter as the pipe. Elevation head is zero.

By the energy theorem:

$$0 + 0 + 20 = \frac{V^2}{2g} + 0 + 0 + 10$$

Therefore

$$\frac{V^2}{2g} = 20 - 10 = 10 \text{ ft.}$$

and $V = 25.3$ ft per sec. Since $A = 0.196$ sq ft, $Q = 5.0$ cfs.

QUESTION. If the liquid in the above example is oil (sp gr 0.80) instead of water, what, if any, changes must be made in the computation of Q?

Answer. No change, provided that the unit of head used is feet of oil. Thus, as the problem is stated, the elevation head at 1 is 20 ft of oil, the loss of head from 1 to 5 is 10 ft of oil, and the velocity head being a function of V and g only, is independent of the specific gravity of the fluid, but is measured in feet of oil.

EXAMPLE 2. In Fig. 41 is shown a siphon with its upper end immersed in a large reservoir of oil (sp gr 0.80). As long as the siphon is filled with air there is, of course, no tendency for flow to occur. If,

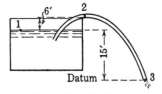

FIG. 41. Siphon.

however, air is exhausted by suction at the free end or by other means, atmospheric pressure will cause the liquid to rise in the upper end. If, after all the air has been exhausted and the siphon completely filled with liquid, the lower end of the siphon is opened, oil will be discharged from the reservoir.

The pipe is 6 in. in diameter. Compute the discharge and the pressure at point 2, if the loss of head from 1 to 2 is 5 ft and from 2 to 3 is 8 ft.

Solution. Writing the energy theorem from point 1 in the oil surface to point 3 in the jet,

$$0 + 0 + 15 = \frac{V_3^2}{2g} + 0 + 0 + 5 + 8$$

from which $V_3^2/2g = 2$ ft of oil, $V_3 = 11.3$ ft per sec, and, since $A_3 = 0.196$ sq ft, $Q = 2.22$ cfs.

The pressure at 2 can be found by writing the energy theorem either from 1 to 2 or from 2 to 3. It is advisable to do both in order to obtain a check on the results. Since the pipe is of uniform diameter, the velocity and the velocity head are the same at 2 as at 3.

From 1 to 2:

$$0 + 0 + 15 = 2 + \frac{p_2}{w} + 21 + 5$$

from which $p_2/w = -13$ ft of oil and $p_2 = -4.5$ lb per sq in. gage or $+10.2$ lb per sq in. absolute, assuming standard atmospheric conditions.

From 2 tc 3:

$$2 + \frac{p_2}{w} + 21 = 2 + 0 + 0 + 8$$

from which, as above, $p_2/w = -13$ ft of oil.

If the absolute pressure at the summit of a siphon should be found to be negative (an impossible condition), this result is obtained because the siphon does not flow full at the outlet as was assumed. Under such conditions the absolute pressure at the summit may be close to zero, and a portion of the siphon near the discharge end does not flow full.

The problems in this chapter are intended to illustrate applications of the energy theorem. A complete solution of such problems would include a determination of the head lost, but, in order not to complicate the problems unduly, reasonable losses of head have been predetermined and are given with the data. Methods of determining head losses in pipes are described in Chapter VII.

PROBLEMS

1. In Example 1 above, assume the following head losses: from 1 to 2, 0 ft; from 2 to 3, 2 ft; from 3 to 4, 7 ft; from 4 to 5, 1 ft. Make a table showing elevation head, velocity head, pressure head, and total head at each of the five points. How high above the center of the pipe will water stand in the piezometer tubes at 3 and 4?

2. A 12-in. pipe is connected by a reducer to a 4-in. pipe (Fig. A). Points 1 and 2 are at the same elevation. The pressure at 1 is 30

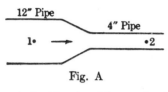

Fig. A

lb per sq in. $Q = 1$ cfs flowing from 1 to 2, and the energy lost between 1 and 2 is equivalent to 3 lb per sq in. Compute the pressure at 2 if the liquid is (a) water, (b) oil (sp gr 0.80), (c) molasses (sp gr 1.50).

3. In Fig. A, with 0.5 cfs of water flowing from 1 to 2, the pressure at 1 is 15 lb per sq in. and at 2 is 10 lb per sq in. Compute the loss of head between 1 and 2.

4. With 1 cfs of water flowing in Fig. A, what pressure must be maintained at 1 if the pressure at 2 is to be 10 lb per sq in. and the loss of head between 1 and 2 is 5 per cent of the difference in pressure heads at 1 and 2?

5. If the smaller pipe of Fig. A is cut off a short distance past the reducer so that the jet springs free into air as in Fig. B, compute the pressure at 1 if $Q = 5.0$ cfs of water, D_1 is 12 in., and D_2 4 in. Assume that the jet has the diameter D_2, that the pressure in the jet is atmospheric, and that the loss of head from point 1 to point 2 is 5 ft of water.

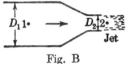

Fig. B

6. Compute the velocity head of the jet in Fig. B if D_1 is 3 in., D_2 is 1 in., the pressure head at 1 is 100 ft of the liquid flowing, and the lost head between points 1 and 2 is 5 per cent of the velocity head at point 2.

7. In Fig. C, with 1.2 cfs of sea water (sp gr 1.03) flowing from 1 to 2, the pressure at 1 is 15 lb per sq in. and at 2 is -2 lb per sq in. Point 2 is 20 ft higher than point 1. Compute the lost energy in pounds per square inch between 1 and 2.

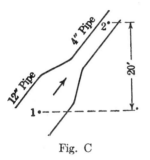

Fig. C

8. The diameter of a pipe carrying water changes gradually from 6 in. at A to 18 in. at B. A is 15 ft lower than B. What will be the difference in pressure, in pounds per square inch, between A and B, when 6.2 cfs is flowing, loss of energy being neglected?

9. The diameter of a pipe carrying water changes gradually from 6 in. at A to 18 in. at B. A is 15 ft lower than B. If the pressure at A is 10 lb per sq in. and that B is 7 lb per sq in. when 5.0 cfs is flowing, determine: (a) the direction of flow; (b) the frictional loss between the two points.

10. A horizontal pipe carries 30 cfs of water. At A the diameter is 18 in. and the pressure is 10 lb per sq in. At B the diameter is 36 in. and the pressure is 10.9 lb per sq in. Determine the head lost between the two points.

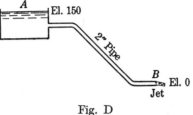

Fig. D

11. In Fig. D, a 2-in. pipe line leads downhill from a reservoir and discharges into air. If the loss of head between A and B is 145 ft, compute the discharge.

12. A 6-in. pipe line (Fig. E) conducts water from a reservoir and discharges at a lower elevation through a nozzle which has a discharge diameter of 2 in. The water

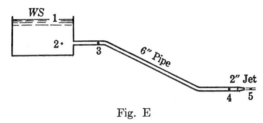

Fig. E

surface in the reservoir (1) is at elevation 100 ft, the pipe intake (2 and 3) at elevation 80 ft, and the nozzle (4 and 5) at elevation 0. The head losses are: from 1 to 2, 0 ft; from 2 to 3, 2 ft; from 3 to 4, 30 ft; from 4 to 5, 10 ft. Compute the discharge and make a table showing elevation head, velocity head, pressure head, and total head at each of the five points.

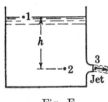

Fig. F

13. Water discharges through an orifice in the side of a large tank as shown in Fig. F. The orifice is circular in cross section and 2 in. in diameter. The jet is the same diameter as the orifice. The liquid is water, and the surface elevation is maintained at a height h of 12.6 ft above the center of the jet. Compute the discharge: (*a*) neglecting loss of head; (*b*) considering the loss of head to be 10 per cent of h. Make a table of heads at points 1, 2, and 3.

14. A pump (Fig. G) takes water from an 8-in. suction pipe and delivers it to a 6-in. discharge pipe in which the velocity is 8 ft per sec. At A in the suction pipe the pressure is -6 lb per sq in. At B in the discharge pipe, which is 8 ft above A, the pressure is $+60$ lb per sq in. What horsepower would have to be applied by the pump if there were no frictional losses?

15. A pump (Fig. G) draws water from an 8-in. suction pipe and discharges through a 6-in. pipe in which the velocity is 12 ft per sec. The pressure is -5 lb per sq in. at A in the suction pipe. The 6-in. pipe discharges horizontally into air at C. To what height h above B can the water be raised if B is 6 ft above A and 20 hp is delivered to the pump? Assume that the pump operates at 70 per cent efficiency and that the frictional loss in the pipe between A and C is 10 ft.

16. In Fig. H is shown a siphon discharging water from the reservoir A into the air at B. Distance a is 6 ft, b is 20 ft, and the diameter is 6 in. throughout. If there is a frictional loss

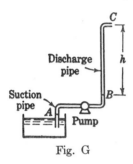

Fig. G

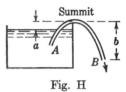

Fig. H

of 5 ft between A and the summit, and 5 ft between the summit and B, what is the absolute pressure at the summit in pounds per square inch? Also determine the rate of discharge in cubic feet per second and in gallons per minute.

17. Figure J shows a siphon discharging oil (sp gr 0.90). The siphon is composed of 3-in. pipe from A to B followed by 4-in. pipe from B to the open discharge at C. The head losses are: from 1 to 2, 1.1 ft; from 2 to 3, 0.7 ft; from 3 to 4, 2.5 ft. Compute the discharge, and make table of heads at points 1, 2, 3, and 4.

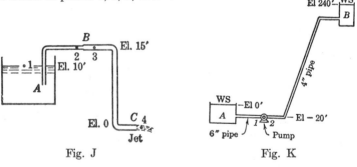

Fig. J   Fig. K

18. A pump draws water from reservoir A and lifts it to reservoir B as shown in Fig. K. The loss of head from A to 1 is 3 times the velocity head in the 6-in. pipe and the loss of head from 2 to B is 20 times the velocity head in the 4-in. pipe. Compute the horsepower output of the pump and the pressure heads at 1 and 2 when the discharge is: (a) 200 gpm; (b) 600 gpm.

19. The 24-in. pipe shown in Fig. L conducts water from reservoir A to a pressure turbine, which discharges through another 24-in. pipe into tailrace B. The loss of head from A to 1 is 5 times the velocity head in the pipe and the loss of head from 2 to B is 0.2 times the velocity head in the pipe. If the discharge is 25 cfs, what horsepower is being given up by the water to the turbine and what are the pressure heads at 1 and 2?

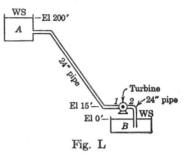

Fig. L

53. Venturi Meter. An illustration of the practical use of the energy equation is provided by the Venturi meter. This instrument, which is used for measuring the discharge through pipes, was invented by an American engineer, Clemens Herschel, and named by him in honor of the original discoverer of the principle involved.

A Venturi meter set in an inclined position is illustrated in Fig. 42. It consists of a short converging tube BC, connected to the approach pipe at the inlet end B, and ending in a cylindrical section CD, called the throat. Usually built as an integral part of a Venturi meter is the diverging section DE, connected to the pipe at the outlet end E. The angle of divergence is kept small to reduce the loss of head caused by turbulence as the velocity is reduced.

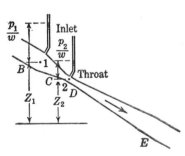

FIG. 42. Venturi meter.

Let V_1, p_1, and Z_1 represent the mean velocity, pressure, and elevation, respectively, at point 1 in the inlet. Also let V_2, p_2, and Z_2 represent the corresponding quantities at point 2 in the throat. Writing the energy equation between points 1 and 2, neglecting friction, and assuming uniform distribution of velocity in each cross section,

$$\frac{V_1^2}{2g} + \frac{p_1}{w} + Z_1 = \frac{V_2^2}{2g} + \frac{p_2}{w} + Z_2 \tag{17}$$

Transposing,

$$\frac{V_2^2}{2g} - \frac{V_1^2}{2g} = \left(\frac{p_1}{w} + Z_1\right) - \left(\frac{p_2}{w} + Z_2\right) \tag{18}$$

This equation shows that the increase in kinetic energy is equal to the decrease in potential energy, a statement which has been called the " Venturi principle." The decrease in potential head is the difference in levels of liquid in piezometer tubes connected to the inlet and throat, as in Fig. 42. It is commonly measured by means of a differential manometer connecting inlet and throat, as in Fig. 43.

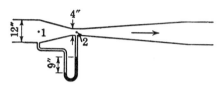

FIG. 43. Venturi meter with differential manometer.

With the decrease in potential head known, the only unknown terms in equation 18 are $V_1^2/2g$ and $V_2^2/2g$. These velocity heads are related, however, by the equation of continuity, $Q = AV$. Thus, for a given Q, V varies inversely as A. With circular cross sections, A varies directly as the square of the diameter D. Therefore V varies inversely as D^2, and the velocity head $V^2/2g$ varies inversely as the fourth power of the diameter D. Hence

$$\frac{V_2^2}{2g} = \left(\frac{D_1}{D_2}\right)^4 \times \frac{V_1^2}{2g} \tag{19}$$

By combining equations 18 and 19, the velocity head at either 1 or 2 and the corresponding velocity can be computed. With the area known, Q_t can be computed. This Q_t is the " theoretical " flow computed by neglecting the loss of head between inlet and throat. In the practical use of the Venturi meter this loss of head, though small, should not be neglected. It causes the actual flow, Q, to be less than the theoretical flow, Q_t. The correction is usually made by applying a factor less than 1 to the theoretical flow. This factor is called the meter coefficient. Thus, if C denotes the meter coefficient,

$$Q = C \times Q_t$$

The Venturi meter coefficient is best determined by measuring the actual flow Q through the meter by volume or by weight, computing the theoretical flow Q_t from the manometer readings and the meter dimensions, and finding the ratio of Q to Q_t. The value of C is affected by the design of the meter and by the roughness of its inner surface. It has been found that the coefficient of a standard Venturi meter usually has a fairly constant value between

0.96 and 0.98, although for relatively small flows it drops to somewhat lower values.

Venturi meters are usually installed in an approximately horizontal position. However, for a given discharge the difference between the elevations of the liquids in the two piezometers (Fig. 42), or the differential manometer reading (Fig. 43), will be the same regardless of whether the meter is horizontal or inclined. This is apparent from a consideration of equation 18. Since it is assumed that the rate of discharge remains the same, the increase in kinetic energy and likewise the decrease in potential energy must also remain unchanged, regardless of the position of the meter.

EXAMPLE. A Venturi meter having a throat 4 in. in diameter is installed in a horizontal 12-in. pipe line carrying a light oil (sp gr 0.82). A mercury U-tube connected as shown in Fig. 43 shows a difference in height of mercury columns of 9 in., the remainder of the tube being filled with oil. Find the rate of discharge, Q, in cubic feet per second, if $C = 0.975$.

Writing the energy equation between points 1 and 2, neglecting lost head,

$$\frac{V_1{}^2}{2g} + \frac{p_1}{w} + Z_1 = \frac{V_2{}^2}{2g} + \frac{p_2}{w} + Z_2$$

Since $Z_1 = Z_2$,

$$\frac{V_2{}^2}{2g} - \frac{V_1{}^2}{2g} = \frac{p_1}{w} - \frac{p_2}{w}$$

From the differential manometer reading,

$$\frac{p_1}{w} - \frac{p_2}{w} = \frac{9}{12} \times \frac{13.6}{0.82} - \frac{9}{12} = 11.7 \text{ ft of oil}$$

From the diameter ratio at points 1 and 2 and the equation of continuity,

$$\frac{V_2{}^2}{2g} = 3^4 \cdot \frac{V_1{}^2}{2g} = 81 \frac{V_1{}^2}{2g}$$

Substituting

$$80 \frac{V_1{}^2}{2g} = 11.7 \quad \text{and} \quad \frac{V_1{}^2}{2g} = 0.146 \text{ ft}$$

whence

$$V_1 = 3.06 \text{ ft per sec and } Q_t = 2.40 \text{ cfs}$$

Thus

$$Q = 0.975 \times 2.40 = 2.34 \text{ cfs}$$

PROBLEMS

1. A Venturi meter having a diameter of 6 in. at the throat is installed in a horizontal 18-in. water main. In a differential gage partly filled with mercury (the remainder of the tube being filled with water) and connected with the meter at the inlet and throat, the mercury column stands 15 in. higher in one leg than in the other. What is the discharge through the meter in cubic feet per second: (a) neglecting friction; (b) if the loss of head between inlet and throat is 1 ft of water? Compute meter coefficient in (b).

2. A Venturi meter having a diameter of 6 in. at the throat is installed in a horizontal 12-in. water main. In a differential gage partly filled with mercury (the remainder of the tube being filled with water) and connected with the meter at the inlet and throat, what would be the difference in level of the mercury columns if the discharge is 5.0 cfs? Neglect loss of head.

3. A 3-in. by $1\frac{1}{2}$-in. Venturi meter is installed in a 3-in. pipe to measure the flow of oil (sp gr 0.852). A differential gage connected with inlet and throat contains water in the lower part of the tube, the remainder of the tube being filled with oil. The difference z in height of water columns in the two legs of the tube is 1.832 ft. If the coefficient of the meter is 0.957, compute the discharge.

4. In a test to determine the discharge coefficient of a 2-in. by $\frac{1}{2}$-in. Venturi meter the total weight of water passing through the meter in 5.00 min was 768 lb. A mercury-water differential gage connected to inlet and throat of the meter showed an average mercury difference during that time of 1.18 ft. Determine the meter coefficient.

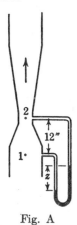

Fig. A

5. A Venturi meter is installed in a pipe line carrying air. The meter has a diameter of 24 in. at inlet and 18 in. at throat. A U-tube connected to inlet and throat contains water, the difference in levels in the two legs of the tube being 4 in. Considering the unit weight of air constant at 0.08 lb per cu ft, determine the approximate discharge in cubic feet per minute, neglecting friction.

6. A 12-in. by 6-in. Venturi meter is installed in a vertical pipe line carrying water, as shown in Fig. A. The flow is upward through the meter. A differential manometer containing carbon tetrachloride (sp gr 1.50) is attached to inlet and throat, the difference z in gage levels being 2.50 ft. Neglecting loss of head, write energy equation from inlet to throat of meter. Also write step-by-step equation of differential gage between inlet and throat, and combine the two equations to compute the discharge. How does the fact that the meter is vertical instead of horizontal affect the solution?

54. Nozzle. A nozzle is a converging tube attached to the end of a pipe or hose which serves to increase the velocity of the issuing jet. Figure 44 illustrates two types of nozzles in common use. Each of these has a cylindrical tip of such length that it will flow full. The converging part of the tube may be the frustum of a cone as in Fig. 44a, or the inside may be convex as in b. Each of these shapes gives an efficient stream.

A nozzle resembles the converging section of a Venturi meter and can therefore also be used for the measurement of flow. The base

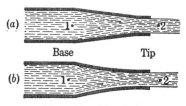

FIG. 44.   Nozzles.

of the nozzle at its connection with the pipe or hose corresponds to the inlet of the Venturi, and the tip of the nozzle corresponds to the throat. The flow conditions in a nozzle are therefore defined by writing the energy equation from the base to a point in the jet.

The pressure in the throat of a Venturi meter may have any value, whereas the pressure in an unconfined jet of liquid discharging into a medium of low density such as air or gas is the same as that in the medium surrounding the jet. This may be seen by investigating the conditions which would result from pressures greater or less than in the surrounding medium. If, for example, the internal pressure in a cross section of a jet were greater than the external pressure, there would be an unbalanced force along every radius of the section, and since an unconfined liquid is incapable of resisting stress this force would cause the jet to expand. In a similar manner if the internal pressure were less than the external pressure, the unbalanced force would cause the jet to contract. Since neither expansion nor contraction occurs, except as caused by acceleration due to gravity or air resistance, it follows that the pressure in the jet must be the same as that in the surrounding medium. If this medium is the atmosphere the gage pressure in the jet is zero.

The energy equation for a horizontal nozzle written between points 1 and 2 (Fig. 44) is

$$\frac{p_1}{w} + \frac{V_1^2}{2g} = \frac{V_2^2}{2g} + \text{lost head} \tag{20}$$

in which p_1 is the gage pressure at the base of the nozzle, V_1 is the mean velocity at the base, and V_2 is the mean velocity in the jet. The velocity heads are related as in the Venturi meter (equation 19). The lost head can be expressed as a percentage of the velocity head in the jet.

The nozzle is discussed further in Art. 75.

PROBLEMS

1. A $2\frac{1}{2}$-in. fire hose discharges water through a nozzle having a jet diameter of 1 in. The lost head in the nozzle is 4 per cent of the velocity head in the jet. If the gage pressure at base of nozzle is 60 lb per sq in.: (a) compute the discharge in gallons per minute; (b) what is the maximum horizontal range to which the stream can be thrown, neglecting air resistance?

2. A $2\frac{1}{2}$-in. fire hose discharges a $1\frac{1}{4}$-in. jet. If the head lost in the nozzle is 6 ft, what gage pressure must be maintained at the base of the nozzle to throw a stream to a vertical height of 100 ft, neglecting air resistance?

3. A power nozzle throws a jet of water which is 2 in. in diameter. The diameter of the base of the nozzle and of the approach pipe is 6 in.

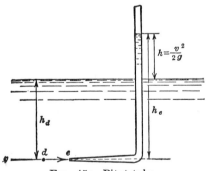

FIG. 45.   Pitot tube.

If the power in the jet is 42 hp and the pressure head at the base of the nozzle is 180 ft, compute the head lost in the nozzle.

55. Pitot Tube. A bent L-shaped tube with both ends open, similar to Fig. 45, is called a Pitot tube, after the French investigator who first used such a device for measuring the velocity of liquids.

When the tube is first placed in a moving stream in the position shown, the liquid enters the opening at e until the surface in the tube rises a distance h above the surface of the stream. A condition of equilibrium is then established, and the quantity of liquid in the tube remains unchanged as long as the flow remains steady.

A magnified sketch of flow conditions near the end of the tube is shown in Fig. 46. In the tube is a volume of motionless liquid, the upstream limit of which is not definitely known but is a surface which may be represented by some line such as abc or $ab'c$ or by the intermediate line $ab''c$. On the adjacent upstream side of this surface, particles are moving with an extremely low velocity.

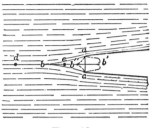

Fig. 46

Let e represent the apex of this surface, called a point of stagnation. Let d represent a point in the undisturbed stream, that is, far enough upstream so that the velocity is not affected by the presence of the tube. For simplicity, let d be on the axis of the tube and at the same elevation as e.

As a particle flows from d to e its velocity is gradually retarded from v to practically zero at e. The velocity head at e may therefore be called zero. Writing the energy theorem for path line de, the elevation heads being equal and friction neglected,

$$\frac{v^2}{2g} + \frac{p_d}{w} = 0 + \frac{p_e}{w} \tag{21}$$

From Fig. 45

$$\frac{p_e}{w} = h_e, \quad \frac{p_d}{w} = h_d, \quad \text{and} \quad h_e - h_d = h'$$

Thus from equation 21

$$\frac{v^2}{2g} = \frac{p_e}{w} - \frac{p_d}{w} = h$$

or

$$v = \sqrt{2gh} \tag{22}$$

Hence the velocity head at d is transformed into pressure head at e, and, because of this increased pressure inside the tube, a column h_e is maintained the height of which is $v^2/2g$ above the level outside.

Figure 47 illustrates several tubes immersed vertically in a stream. The upper ends of the tubes are open and exposed to the atmosphere. At the same depth, h_d, there is an opening in each tube which allows free communication between the tube and the stream.

Tube (a) is similar to the tube of Fig. 45, and $h = v^2/2g$. Tubes (b) and (c) are similar to (a), being bent through an angle of 90°, the tip of each tube being open. If the tube is placed with the open end directed downstream as in (b), or with its lower leg transverse to the stream as in (c), the pressure head at the opening is less than h_d and the surface of the liquid in the tube is a certain distance, h_1 or h_2, below the surface of the stream. Experiments by Darcy showed that h_1 is approximately 0.43 $v^2/2g$ and h_2 is approximately 0.68 $v^2/2g$.

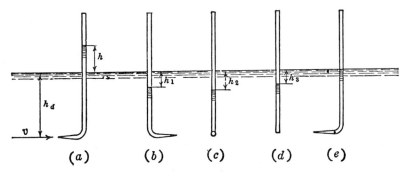

FIG. 47. Tubes extending into flowing liquid.

Similarly for (d), which is a straight tube open at each end, there is a depression, h_3, of the column in the tube. The conditions of flow affecting the height of water column in tube (d) are similar to those encountered when piezometer tubes (Art. 21) project through the conduit walls into the stream. Piezometer tubes are designed to measure pressure head only, and, in order that their readings may be affected a minimum amount by the velocity of the liquid, their ends should be set flush with the inner surface of the conduit and they should never project beyond this surface.

Tube (e) is the same as tube (a) except that the tip of the tube is closed and there is a small hole on each side of the lower leg. If this tube is held with the lower leg parallel to the direction of flow the surface in the tube remains at about the same elevation as the surface of the stream, thus measuring the static pressure at the depth of the opening.

Pitot tubes of the type shown in Fig. 47 are not practicable for measuring velocities because of the difficulty of determining the height of the column in the tube above the surface of the stream. In order to overcome this difficulty the tubes of Fig. 47a and b can be combined as shown in Fig. 48. The open end of one tube is directed upstream, whereas that of the other is downstream. The two tubes can be joined at their upper ends to a single tube connected with a suction pump and provided with a stopcock at A. By opening the stopcock and drawing some of the air from the tubes, both columns are raised an equal amount, since the pressure in their surfaces is reduced equally. The stopcock can then be closed and the difference in height of columns can be read.

FIG. 48. Pitot tube.

This difference is a function of the velocity head, or

$$h = K_p \frac{v^2}{2g} \tag{23}$$

The velocity measured by the tube is thus

$$v = C_p\sqrt{2gh} \tag{24}$$

where

$$C_p = \frac{1}{\sqrt{K_p}} \tag{25}$$

For tubes of the type shown in Fig. 48, h should be the sum of h and h_1 in Fig. 47a and b. Therefore, approximately, $K_p = 1.43$ and $C_p = 0.84$.

Forms of Pitot tubes adapted for measuring velocity in pipes are shown in Figs. 49 and 50. Provision must be made for inserting the tube into the pipe through a fluid-tight connection. A differential manometer is ordinarily used to measure the pressure-head difference.

The arrangement shown in Fig. 49a is similar in principle to that of Fig. 48 except that the horizontal part of the tubes is cut off short to permit insertion into the pipe through a corporation stop of 1 in. inside diameter. Figure 49b shows an arrangement in which the open end of the velocity or "kinetic" tube is directed upstream, while the open end of the "static" tube is normal to the direction of flow and is flush with the side of the "head" of the tube. The coefficients of such a tube should have values close to 1.00. Another arrangement is shown in Fig. 50, in

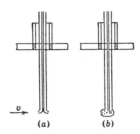

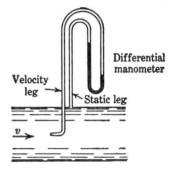

FIG. 49.   Forms of Pitot tube.   FIG. 50.   Pitot tube in pipe.

which the velocity tube is inserted into the pipe, and the static pressure is indicated by a piezometer in the pipe wall.

For accurate measurement a pitot tube must be calibrated or rated. This can be done by moving the tube through still water at a known velocity or by holding the tube at various positions in a pipe in which a known quantity of fluid is flowing. In the latter case correction should be made for the projected area of the tube.[1]

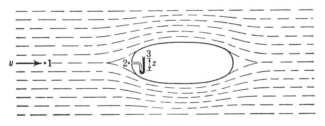

FIG. 51

The foregoing discussion has related to measuring the velocity of a stream of fluid past a fixed point. The principle of the Pitot tube can also be applied to indicate the speed of a body relative to a fluid through which it is moving.

Consider, for instance, a body moving with velocity v through a

[1] Cole, "Pitot-tube Practice," *Trans. Am. Soc. Mech. Engrs.*, 1935, p. 281.

fluid at rest. It is convenient and usually sufficiently accurate to consider the motion of the fluid relative to the body, as in Fig. 51.

At point 1 in the undisturbed stream of fluid where the velocity is v the pressure is p_1. Somewhere on the forward portion of the body, as at 2, there is a point of stagnation where the velocity is brought to zero. Writing the energy theorem from 1 to 2 assuming horizontal motion

$$\frac{v^2}{2g} + \frac{p_1}{w} = 0 + \frac{p_2}{w}$$

from which

$$v = \sqrt{2g \times \left(\frac{p_2}{w} - \frac{p_1}{w}\right)} \tag{26}$$

The usual case concerns measurement of the speed of a body through air. If the velocity is not too great, approximately correct results can be obtained by considering the unit weight of air, w_a, constant. For high velocities, however, appreciable error is made if effects of compressibility are not considered.

If the pressure at 2 is measured by means of an open manometer inside the body, with atmospheric pressure at 3,

$$\frac{p_2}{w_a} = \frac{p_3}{w_a} + z\frac{w_1}{w_a} \tag{27}$$

where w_1 is the unit weight of the gage liquid. If $p_1 = p_3$, from equations 26 and 27,

$$v = \sqrt{2g \times \left(z\frac{w_1}{w_a}\right)} = \sqrt{2gh} \tag{28}$$

where h is the head of air equivalent to height z of the gage liquid.

As with the Pitot tube (equation 24), an empirical coefficient C_p must usually be introduced in equation 26 or 28 to give the actual air speed of a body. More frequently the pressure at the point of stagnation on the forward portion of the body, as, for instance, on the leading edge of the wings of an airplane, is measured by a pressure gage calibrated to read air speed directly in miles per hour.

GENERAL PROBLEMS

1. The diameter of a pipe line is 6 in. at A and 18 in. at B. A is 11 ft lower than B. If the pressure at A is 10 lb per sq in. and at B 7 lb per sq in. when the flow is 2.5 cfs, determine the direction of flow and the frictional loss between A and B when the liquid is: (a) water; (b) a regular gasoline at 60° F.

2. A and B are two points in a pipe line. The diameter is 6 in. at A and 18 in. at B. A is 15 ft lower than B. Determine the discharge of water when the pressure is the same at the two points and the loss of head between A and B is 1.5 ft.

3. A pipe discharges 5.0 cfs of water into a reservoir at a point 6 ft below the water surface. At A the diameter is 10 in. and the center of the pipe is 4 ft above the water surface. At the discharge end the pipe is 12 in. in diameter. If the loss of head from A to the reservoir is 2.2 ft, determine the pressure head at A.

4. A pipe, 12 in. in diameter at A, discharges 4.0 cfs of a heavy fuel oil at 100° F into the air at B, where the diameter is 6 in. If B is 12 ft above A and the frictional loss between the two points is equivalent to 3.0 lb per sq in., determine the pressure at A in pounds per square inch.

5. A jet of liquid is directed vertically upward. At A its diameter is 3 in. and its velocity is 30 ft per sec. Neglecting air friction, determine its diameter at a point 10 ft above A.

6. A fire pump delivers water through a 6-in. main to a hydrant to which is connected a 3-in. hose, terminating in a 1-in. nozzle. The nozzle is 10 ft above the hydrant and 60 ft above the pump. Assuming a total frictional loss of 28 ft from the pump to the base of the nozzle, and a loss in the nozzle of 6 per cent of the velocity head in the jet, and neglecting air resistance, what gage pressure at the pump is necessary to throw a stream 80 ft vertically above the nozzle?

7. A fire pump delivers water through a 6-in. main to a hydrant to which is connected a 3-in. hose, terminating in a 1-in. nozzle. The nozzle is 5 ft above the hydrant and 35 ft above the pump. Assuming frictional losses of 10 ft from the pump to the hydrant, 7 ft in the hydrant, and 40 ft from the hydrant to the base of the nozzle, and a loss in the nozzle of 6 per cent of the velocity head in the jet, to what vertical height can the jet be thrown if the gage pressure at the pump is 80 lb per sq in.?

8. The pipe line shown in Fig. A takes water from a reservoir and terminates in a nozzle having a jet diameter of 2 in. With the pipe losses shown in the figure, and assuming a nozzle loss of 8 per cent of the velocity head in the jet, compute the discharge and make a table showing (to the nearest foot) the elevation head, velocity head, pressure head, and total head at each of the six stations.

9. The pump in Fig. B draws water from a reservoir and discharges through $2\frac{1}{2}$-in. hose which terminates in a nozzle having a jet diameter of 1 in. The head losses are: 1 to 2, 5 ft; 2 to 3, negligible; 3 to 4, 30 ft; 4 to 5, 10 ft. Compute the required horsepower output of the pump and make a table of the heads at the five points. $Q = 250$ gpm.

10. The pressure head difference between inlet and throat of the 3-in. by 1-in. Venturi meter shown in Fig. C is measured by means of a mercury differential manometer equipped with a scale which reads in inches. With water flowing through the meter and filling the gage tubes

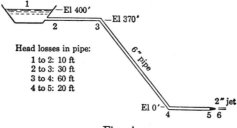

Head losses in pipe:
1 to 2: 10 ft
2 to 3: 30 ft
3 to 4: 60 ft
4 to 5: 20 ft

Fig. A

to the tops of the mercury columns: (a) Compute the numerical constant N in the equation for theoretical flow, $Q_t = N\sqrt{z_i}$, where z_i is the mercury difference in inches. (b) Compute the meter coefficient if the average mercury difference $z_i = 11.86$ in. during a test run in which 19.8 cu ft of water were discharged in 130.5 sec.

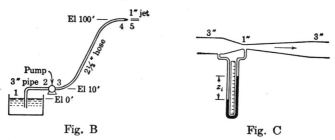

Fig. B   Fig. C

11. The Venturi meter in Fig. C is being used to measure the flow of a medium fuel oil at 60° F. The differential gage fluid is water. The meter coefficient is 0.975. Compute the discharge for the following values of z_i: (a) 7.2 in.; (b) 14.4 in.; (c) 28.8 in.; (d) 57.6 in.

12. A diverging tube discharges water from a reservoir, into the air, at a depth of 6 ft below the water surface. The diameter gradually increases from 6 in. at the throat to 9 in. at the outlet. Neglecting friction, determine (a) the rate of discharge in cubic feet per second; (b) the corresponding pressure at the throat.

13. A diverging tube discharges water from a reservoir at a depth of 36 ft below the water surface. The diameter gradually increases from 6 in. at the throat to 9 in. at the outlet. Neglecting friction, determine: (a) the maximum possible rate of discharge in cubic feet per second through this tube; (b) the corresponding pressure at the throat.

14. A diverging tube discharges water from a vessel at a point 10 ft below the surface on which the gage pressure is 8.5 lb per sq in. If the diameter of the throat is 4 in., at which point the absolute pressure is 10 lb per sq in., determine: (a) the discharge in cubic feet per second, neglecting friction; (b) the diameter of the tube at the discharge end.

15. The center of the intake end of a suction pipe is 5 ft below the water surface in the river. The pipe has a uniform rise of 1 ft per 100 ft to the pump. The velocity in the pipe is 8 ft per sec, the frictional loss at the pipe entrance is 0.5 ft, and the frictional loss in the pipe is 1 ft per 1000 ft. Determine the greatest length the pipe can have without

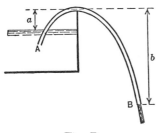

Fig. D

out causing the pressure at the pump to drop more than 6.0 lb per sq in. below atmospheric pressure.

16. Figure D shows a siphon discharging from a reservoir into the atmosphere. The pipe diameter is 6 in. The loss of head is $1.2 \, V^2/2g$ from A to the summit and $1.4 \, V^2/2g$ from the summit to B, V being the velocity of flow in the pipe. If $a = 6$ ft and $b = 20$ ft, compute the discharge and the absolute pressure at the summit if the fluid is: (a) water; (b) oil (sp gr 0.82); (c) brine (sp gr 1.15).

17. In problem 16, with water flowing, what length of vertical pipe added to end B will cause the siphon to flow at its maximum capacity? What is the flow then? Assume that the loss of head is $0.04 \, V^2/2g$ per foot of pipe. Neglect vapor pressure.

18. Water is delivered by a scoop from a track tank to a locomotive tender that has a speed of 20 miles per hour. If the entrance to the tender is 7 ft above the level of the track tank and 3 ft of head is lost in friction, at what velocity will the water enter the tender?

19. Water is delivered by a scoop from a track tank to a locomotive tender, the point of delivery being 7 ft above the level of the track tank. Neglecting friction, what is the lowest possible speed of the train at which water will be delivered to the tender?

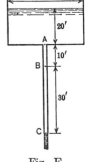

Fig. E

20. Figure E shows a vertical pipe discharging water from an elevated tank into the atmosphere. If the pipe is 6 in. in diameter and the loss of head is $0.04 \, V^2/2g$ feet per foot of pipe, compute the discharge and the pressure head in the pipe 1 ft below point A.

21. From A to B in Fig. E the pipe is 4 in. in diameter and the loss of head is $0.075 \, V_4^2/2g$ feet per foot of pipe. From B to C the pipe is 6 in. in diameter and the loss of head is 0.04 $V_6{}^2/2g$ feet per foot. Compute the discharge and the pressure head in the pipe 1 ft below point A.

22. In Fig. F, 3.0 cfs of water enters through the 5-in. diameter pipe at A and discharges radially in all directions between the two circular plates 24 in. in diameter and 1 in. apart, discharging into the air. Neglecting friction, determine the absolute pressure in pounds per square inch at B.

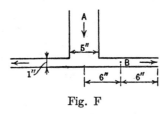

Fig. F

23. A Pitot tube in a pipe in which air is flowing is connected to a manometer containing water as in Fig. 50. If the difference in water levels in the manometer is 3.5 in. what is the velocity of flow in the pipe, assuming a tube coefficient, C_p, of 0.99?

Chapter VI

ORIFICES, TUBES, AND WEIRS

Orifice

56. Description. An orifice is an opening with a closed perimeter through which a fluid flows. The usual purpose of an orifice is the measurement or control of the flow.

The upstream edge of an orifice may be rounded or sharp. An orifice with prolonged sides, such as a piece of pipe two or three diameters in length, is called a tube. An orifice in a thin wall has the hydraulic properties of a tube. Longer tubes such as culverts under embankments are frequently treated as orifices although they may also be treated as short pipes.

Orifices used for measuring flow are usually circular, square, or rectangular in cross section. Because of simplicity of design and construction, sharp-edged circular orifices are most common for fluid measurement and have been most thoroughly investigated by experiment, although much remains to be learned about the laws governing their discharge, particularly for fluids other than water.

57. Velocity of Discharge. Figure 52 represents the general case of a liquid discharging through a vertical rounded orifice. There are two chambers, A and B, the gas pressures in these chambers being respectively p_A and p_B, and the relative amounts of the pressures being such that flow is from A to B. The liquid particles follow path lines of which mn is one, m being a point in chamber A and n a point in the jet. The pressure in the jet is p_B (Art. 54).

The path line mn passes through the orifice at a distance h below the surface of the liquid. The point m is a distance h_m below the surface and a distance z above n. Velocities at m and n are respectively v_m and v_n. The energy equation between these points, neglecting lost head, is:

$$\frac{v_m{}^2}{2g} + \left(h_m + \frac{p_A}{w}\right) + z = \frac{v_n{}^2}{2g} + \frac{p_B}{w} \tag{1}$$

and, since $h_m + z = h$,

$$v_n = \sqrt{2g\left[h + \frac{v_m{}^2}{2g} + \left(\frac{p_A}{w} - \frac{p_B}{w}\right)\right]}$$ (2)

expressing the general relation between velocity and head for any path line.

Since particles at different elevations discharge through a vertical orifice under different heads their velocities are not the same. In orifice flow, however, the mean velocity is ordinarily taken as the velocity due to the mean head. The mean velocity thus obtained is represented by the symbol V_t, while the mean velocity in the channel of approach, called the velocity of approach, is

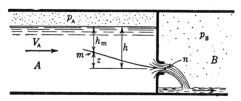

FIG. 52.　Discharge from orifice.

represented by V_A. If, then, the energy equation is written for the entire stream, neglecting unequal distribution of velocity in the cross section,

$$\frac{V_A{}^2}{2g} + h + \frac{p_A}{w} = \frac{V_t{}^2}{2g} + \frac{p_B}{w}$$ (3)

and

$$V_t = \sqrt{2g\left[h + \frac{V_A{}^2}{2g} + \left(\frac{p_A}{w} - \frac{p_B}{w}\right)\right]}$$ (4)

The condition most commonly encountered is that in which the surface of the liquid in chamber A and the jet in chamber B are each exposed to the atmosphere. Then $p_A = p_B$, and

$$V_t = \sqrt{2g\left[h + \frac{V_A{}^2}{2g}\right]}$$ (5)

If also the cross-sectional area of the reservoir or channel leading to the orifice is large in comparison with the area of the orifice the velocity of approach becomes negligible, and

$$V_t = \sqrt{2gh}$$ (6)

In equations 4 and 5, the quantities in brackets, and in equation 6 the quantity *h*, represent the *total head producing flow.* If this total head is represented by *H*, the theoretical velocity of discharge from an orifice, that is, the velocity which would exist if there were no loss of head, is given by the equation

$$V_t = \sqrt{2gH} \tag{7}$$

Equation 7 is also the formula for the velocity acquired by a body in falling from rest through a height *H*. The theoretical velocity of discharge from an orifice is therefore the velocity acquired by a body falling freely in a vacuum through a height equal to the total head on the orifice. This principle, discovered by Torricelli in 1644, is known as Torricelli's theorem.

58. Coefficient of Velocity. The actual velocity in the jet is less than the theoretical velocity because of the frictional resistance that occurs as the fluid enters and passes through the orifice. The ratio of the actual mean velocity *V* to the velocity V_t which would exist without friction is called the coefficient of velocity and is designated C_v. Thus $C_v = V/V_t$, and

$$V = C_v V_t = C_v \sqrt{2gH} \tag{8}$$

59. Coefficient of Contraction. Figure 53 represents a cross section of a vertical sharp-edged orifice discharging a liquid from a

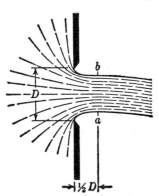

reservoir into the atmosphere. The particles of the liquid approach the orifice in converging paths from all directions. Because of the inertia of those particles with velocity components parallel to the plane of the orifice, they cannot make abrupt changes in their directions the instant they reach the orifice, and they therefore follow curvilinear paths, thus causing the jet to contract for a short distance beyond the orifice. This phenomenon is referred to as the contraction of the jet. The section *ab* where contraction caused by the orifice ceases is called the *vena contracta.* The vena contracta for a sharp-edged circular orifice of diameter *D* has been found to be at a distance of about $\frac{1}{2}D$ from the plane of the orifice.

Fig. 53. Vertical sharp-edged orifice.

The ratio of the cross-sectional area of the jet at the vena contracta to the area of the orifice is called the coefficient of contraction. Thus, if a and A are, respectively, the cross-sectional area of the jet at the vena contracta and the area of the orifice, and C_c is the coefficient of contraction,

$$C_c = \frac{a}{A} \quad \text{or} \quad a = C_cA$$

If V is the actual mean velocity in the vena contracta the discharge through the orifice is

$$Q = aV = C_cA \times C_v\sqrt{2gH} \tag{9}$$

Beyond the vena contracta the cross-sectional area of the jet, neglecting air resistance, does not undergo any change excepting

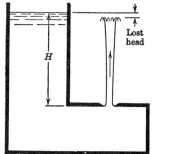

FIG. 54. Horizontal orifice discharging upward.

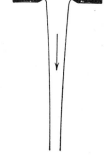

FIG. 55. Horizontal orifice discharging downward.

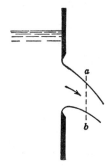

FIG. 56. Vertical orifice under low head.

so far as it is affected by gravity and surface tension. If the direction of the jet is vertically upward, as in Fig. 54, or if it has an upward component, gravity retards the velocity of the jet and thus increases its cross-sectional area, whereas in a jet discharging downward, as in Fig. 55, gravity increases its velocity and decreases its cross-sectional area.

Under very low heads the top elements of the jet from a vertical orifice do not have sufficient velocity to become horizontal at any point and the exact location of the vena contracta becomes more difficult. The form of jet from an orifice under a low head is shown in Fig. 56.

60. Coefficient of Discharge. It is usual to replace the product C_cC_v in equation 9 with a single coefficient C, called the coefficient of discharge.   The equation for the discharge of a fluid through an orifice thus becomes

$$Q = CA \sqrt{2gH} \qquad (10)$$

61. Velocity of Approach. The condition under which the velocity of approach is appreciable is encountered so frequently that further analysis is helpful.   The usual application is to the

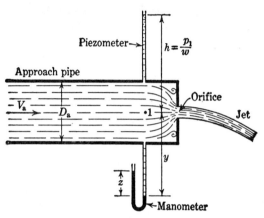

FIG. 57.   Orifice in end of pipe.

flow through an orifice in a flat plate either on the end of a pipe, as in Fig. 57, or inserted in a pipe line, as in Fig. 58.

The mean velocity in the approach pipe is V_a, and the velocity head, assuming $\alpha = 1$ (Art. 49), is $V_a^2/2g$.   The total head on the orifice in Fig. 57, with discharge into the atmosphere, is $p_1/w + V_a^2/2g$, where p_1/w is the pressure head at point 1.   In Fig. 58 the total head on the orifice is $(p_1 - p_2)/w + V_a^2/2g$ where p_1 and p_2 are respectively the pressures at points 1 and 2 upstream and downstream from

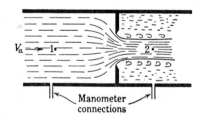

FIG. 58.   Pipe orifice.

the orifice plate.   From equation 10, for the orifice on the end of a pipe,

$$Q = CA \sqrt{2g\left(\frac{p_1}{w} + \frac{V_a^2}{2g}\right)} \qquad (11)$$

By definition, $V_a = Q/(\pi D_a{}^2/4)$, where D_a is the pipe diameter. Substituting this value of V_a in equation 11 and solving for Q,

$$Q = \frac{CA \sqrt{2g(p_1/w)}}{\sqrt{1 - C^2(D/D_a)^4}} \qquad (12)$$

Expanding the denominator by the binomial theorem gives a diminishing series, and dropping as negligible all terms except the first two, a closely approximate equation of discharge is obtained:

$$Q = CA \sqrt{2g\,(p_1/w)}\,[1 + \tfrac{1}{2}C^2(D/D_a)^4] \qquad (13)$$

D being the diameter of the orifice. The quantity in brackets is the corrective factor for velocity of approach. Its value approaches unity as the ratio of orifice diameter to pipe diameter becomes small.

Similarly for an orifice in a pipe:

$$Q = CA \sqrt{2g[(p_1 - p_2)/w]}\,[1 + \tfrac{1}{2}C^2(D/D_a)^4] \qquad (14)$$

62. Flow of Gases through Orifices. If a gas under pressure p_1 discharges through an orifice into a space in which the pressure p_2 is slightly less than p_1, variation of density can be neglected or its effect included in the value of the orifice coefficient. The laws applying to the flow of liquids through orifices then hold approximately. In this case w (equations 13 and 14) is the unit weight of the gas under a pressure of p_1 or p_2, or the average of p_1 and p_2. The value of p used to determine w affects the coefficient C.

As the pressure drop through the orifice increases, thermodynamic effects of compression or expansion of the gas become more important and the formulas for liquids give less accurate results.

63. Head Lost in an Orifice. Orifice flow is no exception to the general rule that fluid motion is always accompanied by an expenditure of energy. Loss of energy, or of head, in flow through an orifice is illustrated in Fig. 54. Even if air resistance could be completely eliminated, the jet from the horizontal orifice would not rise as high as the liquid level in the supply tank because of the loss of energy which occurs between points in the supply tank where the velocity is practically zero and the vena contracta.

For use in hydraulic engineering problems, the loss of head due to flow through an orifice is conveniently expressed in two ways: (1) as a function of the velocity head in the jet; (2) as a function of the original head.

1. Consider a fluid to be discharging from an orifice under a total head H. The velocity of discharge is $V = C_v\sqrt{2gH}$, from which the original head

$$H = \frac{1}{C_v{}^2} \cdot \frac{V^2}{2g} \tag{15}$$

The head remaining in the jet is velocity head, $V^2/2g$. The lost head, H_0 = original head minus remaining head, or

$$H_0 = \frac{1}{C_v{}^2} \cdot \frac{V^2}{2g} - \frac{V^2}{2g} = \left(\frac{1}{C_v{}^2} - 1\right)\frac{V^2}{2g} \tag{16}$$

2. From equation 15 the velocity head in the jet

$$\frac{V^2}{2g} = C_v{}^2 H$$

Hence the lost head

$$H_0 = H - C_v{}^2 H = (1 - C_v{}^2)H \tag{17}$$

Equations 16 and 17 are applicable to any orifice or tube for which the coefficient of velocity is known.

64. Inversion of the Jet. The form assumed by jets of liquid issuing from orifices of different shapes presents an interesting phenomenon. The cross section of the jet is similar to the shape of the orifice until the vena contracta is reached. Figure 59 shows cross sections of jets issuing respectively from square, triangular, and elliptical orifices. The left-hand diagram in each is a cross section of the jet near the vena contracta. The succeeding diagrams are cross sections at successively greater distances from the orifice. This change in form, which is common to all shapes of orifices, is known as the inversion of the jet. After passing through the fourth stage shown in the figure the jet reverts to its

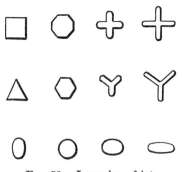

FIG. 59. Inversion of jet.

original form and continues to pass through the cycle of changes described above as long as it flows freely or is not broken up by wind or air friction or modified by surface tension.

PROBLEMS

1. Figure A represents two large tanks with an orifice in the dividing partition. The orifice has a diameter of 2 in. and is rounded so that $C_c = 1.00$ and $C_v = 0.97$. Pressures p_A and p_B are atmospheric. The liquid is oil with $h = 16$ ft. Determine the theoretical velocity in the jet, the actual velocity, and the discharge.

2. The orifice in Fig. A is 2 in. in diameter and is sharp-edged, with $C_c = 0.62$ and $C_v = 0.98$. Pressures p_A and p_B are atmospheric. The liquid is water with $h = 16$ ft. Determine the diameter of the jet, the actual velocity of the jet, and the discharge.

3. The liquid in Fig. A has a specific gravity of 3.00. The gage pressure $p_A = 10$ lb per sq in. and p_B is atmospheric. The orifice diameter is 3 in. and $C = C_v = 0.95$. With $h = 5$ ft, determine the discharge and the head lost in the orifice.

4. In problem 3, maintaining the liquid head h at 5 ft, to what pressure must p_A be raised in order to double the discharge?

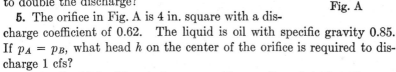

Fig. A

5. The orifice in Fig. A is 4 in. square with a discharge coefficient of 0.62. The liquid is oil with specific gravity 0.85. If $p_A = p_B$, what head h on the center of the orifice is required to discharge 1 cfs?

6. The liquid in Fig. A has a specific gravity of 1.50. The gas pressure p_A is $+5$ lb per sq in. and p_B is -2 lb per sq in. The orifice is 4 in. in diameter with $C = C_v = 0.95$. Determine the velocity in the jet and the discharge when $h = 4$ ft.

7. In problem 6, maintaining the liquid head h at 4 ft, and neglecting vapor pressure, what is the maximum percentage by which the discharge can be increased by decreasing p_B?

8. The horizontal orifice in Fig. 54 is 3 in. in diameter with $C_c = 0.63$, $C_v = 0.98$. When $H = 7.5$ ft, neglecting air resistance, compute the height to which the jet will rise above the plane of the orifice. What will be the diameter of the jet 3 ft above the vena contracta?

9. The orifice in Fig. 57 is 2 in. in diameter and is sharp-edged with $C_c = 0.63$ and $C_v = 0.97$. The diameter of the pipe is 6 in. The liquid is water and stands at a height h of 9.5 ft in the piezometer. Compute the discharge, the diameter of the jet, the mean velocity in the jet, and the lost head.

10. The liquid in Fig. 57 is oil (sp gr 0.82) discharging from a 10-in. pipe through a 4-in. rounded orifice with $C_c = 1.00$ and $C_v = 0.96$. The U-tube contains mercury with a difference z in mercury levels of 6.5 in. while height $y = 2.2$ ft. Compute Q.

11. A 1-in. sharp-edged orifice in the end of a 6-in. pipe (Fig. 57) discharges air into the atmosphere at a temperature of 80° F. A U-tube containing water shows a difference z in water levels of 8 in. while $y = 3$ ft. If C, based on unit weight of air at point 1 in the pipe, is 0.597, compute the discharge in cubic feet per minute.

12. The pipe orifice in Fig. 58 has a diameter of 4 in. and a coefficient of discharge of 0.650. The pipe diameter is 10 in. The manometer connections are attached to a differential gage partly filled with mercury, the remainder of the tube being filled with the same liquid as that flowing in the pipe, which is a heavy fuel oil at 70° F. Compute the discharge when the difference in mercury levels in the gage is: (a) 2.10 in.; (b) 13.5 in.; (c) 27.6 in.; (d) 52.2 in.

65. Experimental Determination of Orifice Coefficients. Since in practice it is usually the discharge from orifices that is required, it is the coefficient of discharge that is of greatest value to engineers. This coefficient, C, can be computed from equation 10 if the flow Q, the area of the orifice A, and the total head H are determined. C is not ordinarily constant for a given orifice, but varies with the head, with approach conditions, and with the viscosity of the fluid.

Experimental determination of C_c and C_v is more difficult. The diameter of the jet from a circular orifice can be measured at the vena contracta with calipers and the coefficient of contraction computed from the relation $C_c = a/A$. If the vena contracta is difficult to locate, as in Fig. 56, an arbitrary location for determining contraction must be chosen, as for instance plane ab which is $\frac{1}{2}D$ from the plane of the orifice.

The velocity in the jet can be computed from the area of the jet and the measured Q. If the total head is known, the coefficient of velocity can then be computed from the relation $C_v = V/\sqrt{2gH}$. The velocity in the jet can also be measured with fair accuracy with a Pitot tube or by the coordinate method outlined in the following article.

66. Coordinate Method of Determining Velocity of Jet. Let Fig. 60 represent a side view of a jet from a vertical orifice. The jet at the vena contracta is traveling horizontally with velocity V. The force of gravity causes the jet to curve downward. Let x and

y represent the coordinates of any other point in the jet. Neglecting air resistance, the horizontal component of the jet velocity is constant with the time t, from which

$$x = Vt$$

The jet has a downward acceleration which conforms to the law of falling bodies, and therefore

$$y = \tfrac{1}{2}gt^2$$

Eliminating t between the two equations

$$x^2 = \frac{2V^2}{g}\,y$$

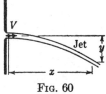

Fig. 60

which is the equation of a parabola with its vertex at the vena contracta. Solving for V,

$$V = \frac{4.01x}{\sqrt{y}} \tag{18}$$

This coordinate method is also of practical value in determining the flow from the open end of a horizontal pipe.

67. Standard Orifice Coefficients. The values of orifice coefficients cited on the following pages are typical, but it should be remembered that slight variations in design and installation of an orifice may cause appreciable variation in the coefficient. Hence an orifice to be used for flow measurements should, if possible, be calibrated in place.

Coefficients of discharge of sharp-edged circular orifices discharging water, as determined by Medaugh and Johnson, are given on page 128. The orifices tested were drilled in smooth ground brass plates, which were bolted to the side of a tank large enough to make the velocity of approach negligible. Special care was taken in the construction of the orifice holes to make the upstream edge square.

It is seen that C decreases as the head and the size of orifice increase. A possible reason is that the contraction of the jet then becomes more nearly perfect, the coefficient of contraction C_c approaching a minimum value of approximately 0.61. The coefficient of velocity meanwhile increases slowly to a maximum value of approximately 0.98, but the product C_cC_v decreases.

DISCHARGE COEFFICIENTS FOR VERTICAL SHARP-EDGED
CIRCULAR ORIFICES

For water at 60° F discharging into air at same temperature

Head in Feet	Orifice Diameter in Inches					
	0.25	0.50	0.75	1.00	2.00	4.00
0.8	0.647	0.627	0.616	0.609	0.603	0.601
1.4	.635	.619	.610	.605	.601	.600
2.0	.629	.615	.607	.603	.600	.599
4.0	.621	.609	.603	.600	.598	.597
6.0	.617	.607	.601	.599	.597	.596
8.0	.614	.605	.600	.598	.596	.595
10.0	.613	.604	.600	.597	.596	.595
12.0	.612	.603	.599	.597	.595	.595
14.0	.611	.603	.598	.596	.595	.594
16.0	.610	.602	.598	.596	.595	.594
20.0	.609	.602	.598	.596	.595	.594
25.0	.608	.601	.597	.596	.594	.594
30.0	.607	.600	.597	.595	.594	.594
40.0	.606	.600	.596	.595	.594	.593
50.0	.605	.599	.596	.595	.594	.593
60.0	.605	.599	.596	.594	.593	.593
80.0	.604	.598	.595	.594	.593	.593
100.0	.604	.598	.595	.594	.593	.593
120.0	.603	.598	.595	.594	.593	.592

Source: F. W. Medaugh and G. D. Johnson, *Civil Eng.*, July, 1940, p. 424.

The values of C given above agree closely with values obtained
many years ago by Hamilton Smith and long considered standard.
Other experimenters have obtained somewhat larger values, the
most probable cause of the discrepancies being slight differences
in the amount of rounding of the upstream edge of the orifices
tested. It is also possible that temperature differences cause such
discrepancies since Medaugh found that lowering the air temperature 20° F while keeping the water temperature approximately
constant increased the discharge about $\frac{1}{2}$ per cent.

68. Variation in Orifice Coefficients. A sharp-edged orifice is
usually not knife-edged but preferably of the form shown in Fig.
61a and b. This form is easier to machine and avoids the tendency
of a knife edge to wear dull. In order to ensure full and complete
contraction of the jet it is essential that the upstream corner be
square-cut so that the jet will spring clear as shown in Fig. 61a.

The orifice is then in effect sharp-edged, and the coefficients for sharp-edged orifices apply.

At low heads, however, or with viscous liquids there is a tendency for the jet to cling to the flat part of the orifice as shown in Fig. 61b, thereby reducing the contraction of the jet and increasing the value of C_c. A slight rounding or dullness of the upstream edge increases this tendency, and further rounding of the edge may cause the jet to fill the orifice, as shown in Fig. 60c. With this condition $C_c = 1$, whereas C_v may be little changed from its sharp-edged value of 0.98. For the bell-mouthed orifice shown in Fig. 61d, C and C_v may be reduced to 0.95 or less.

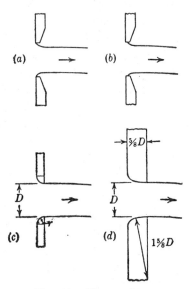

FIG. 61.   Forms of jet.

If r (Fig. 61c) is the radius of rounding of the upstream edge of an orifice of diameter D, Schoder and Dawson[1] state that the per cent increase in Q (and hence in C) due to rounding equals 3.1 times the per cent that r is of D. This finding was based on tests made at Cornell University and holds as a general rule for values of r/D up to 0.10.

By this rule, if a 1-in. circular sharp-edged orifice has its upstream edge rounded to a radius of 0.01 in., the discharge would be increased 3.1 per cent.   If the coefficient of discharge at a given head were originally 0.600, the new coefficient would be 0.619.   This effect shows the necessity of careful machining of orifices which are to be used for flow measurements with standard values of coefficients.

The increase in value of C (page 128) as the size of the orifice decreases can probably be partly accounted for by the fact that, with precisely the same sharpness of edge in two orifices of different size the relative rounding of the edge is greater for the smaller orifice.   No square corner can be microscopically perfect.   Slight

[1] E. W. Schoder and F. M. Dawson, *Hydraulics*, McGraw-Hill Book Co., 1934, p. 408.

imperfections of orifice edge which may allow almost perfect contraction in a 4-in. orifice may appreciably reduce the contraction and therefore increase the discharge coefficient of a $\frac{1}{4}$-in. orifice. A more rapid increase in C as the head decreases at the lower heads is shown in the table on page 128. At lower heads the viscosity of the fluid has greater effect on the flow, resulting in less contraction of the jet and hence larger values of C_c and C.

Roughening the upstream surface adjacent to the orifice causes the components of velocities in directions parallel to the plane of the orifice to be retarded and the contraction to be reduced.

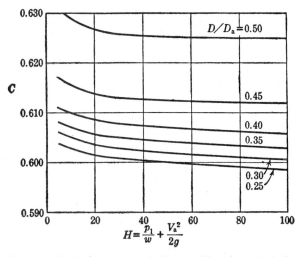

Fig. 62. Variation of C with H for orifices in end of pipe.

A reduction of the cross-sectional area of the channel through which the fluid approaches the orifice increases not only the effect of velocity of approach (Art. 61) but also the value of the coefficient of discharge. The reason is that as the sides of the approach channel are brought nearer to the edges of the orifice the contraction of the jet is decreased.

Figure 62 shows curves obtained from tests by Blackburn[1] on the flow of water through $1\frac{1}{4}$-in., 2-in., 3-in., and $3\frac{1}{2}$-in. orifices at the end of 4-in., 8-in., and 10-in. pipes. Each curve shows the variation of discharge coefficient C in equations 11 and 13 with total head on the orifice for the indicated ratio of diameter of orifice D

[1] G. B. Blackburn, "The End-cap Pipe Orifice as a Measuring Device," master of science thesis, University of Wisconsin, 1929.

to diameter of approach pipe D_a. The total head on the orifice included pressure head and velocity head in the approach pipe (Fig. 57). The increase in C with increase in the ratio D/D_a is evident. For values of D/D_a above 0.40 the rate of increase in C becomes much more rapid.

Values of the discharge coefficient C in equation 14 as obtained by Kowalke, Bain, and Moss[1] for liquids of different viscosities flowing through pipe orifices (Fig. 58) are shown in Fig. 63 for four different ratios of orifice diameter to pipe diameter. The quantity DV_0/ν is known as *Reynolds' number* (Art. 167), D being the diameter of orifice in feet, V_0 the mean velocity of flow at the orifice in feet per second, and ν the kinematic viscosity of the fluid

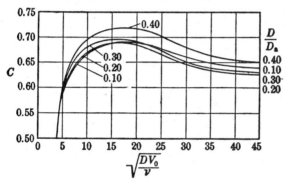

FIG. 63. Variation of C with Reynolds' number for pipe orifices.

in square feet per second. Pressures were measured one pipe diameter up-stream and one-half pipe diameter downstream from the orifice. The curves extend to much lower values of Reynolds number than the table for water on page 128 or the curves in Fig. 62. As with the orifice on the end of a pipe, C changes much more rapidly with changes in diameter ratio at ratios above 0.4. More reliable values of C can therefore be obtained with the lower diameter ratios.

Discharge coefficients of sharp-edged pipe orifices for measuring the flow of air as computed from values given by Bean, Buckingham, and Murphy[2] are shown in the accompanying table. The coefficients vary with the ratio of orifice diameter D to pipe diam-

[1] W. A. Bain and F. D. Moss, " Characteristics of Circular Orifices in the Measurement of Viscous Liquids," master of science thesis, University of Wisconsin, 1937.

[2] Bureau of Standards *Research Paper* 49, 1929, p. 591.

eter D_a and with the ratio of downstream absolute pressure p_2 to upstream absolute pressure p_1. The pressures were determined at taps located at distances D_a upstream and $\frac{1}{2}D_a$ downstream from the upstream face of the orifice plate. The coefficients given in the table apply in equation 14, w being the unit weight of air at the mean absolute pressure $(p_1 + p_2)/2$.

COEFFICIENT C FOR AIR DISCHARGING THROUGH SHARP-EDGED CIRCULAR ORIFICE IN PIPE

p_2/p_1(abs.) \\ D/D_a	0.2	0.3	0.4	0.5	0.55	0.6
1.00	0.597	0.600	0.604	0.616	0.624	0.636
.95	.599	.601	.606	.617	.625	.636
.90	.600	.602	.606	.617	.625	.636
.85	.600	.603	.607	.617	.625	.636
.80	.600	.603	.607	.616	.624	
.75	.600	.602	.606	.615		
.70	.599	.601	.604			
.65	.597	.599	.603			
.60	.595	.597				
.55	.593					
.50	.590					

PROBLEMS

1. Compute the discharge of water through a standard sharp-edged orifice $\frac{3}{4}$ in. in diameter under heads of 2, 28, and 72 ft.

2. Compute the discharge of water through a 2-in. circular sharp-edged orifice under heads of 0.8, 1.6, and 3.2 ft.

3. A standard sharp-edged orifice $2\frac{1}{2}$ in. in diameter discharges water under a head of 15 ft. Compute the discharge. How much will the discharge be increased if the upstream edge of the orifice is rounded on a radius of $\frac{1}{32}$ in.?

4. What diameter of standard, sharp-edged orifice will be required to produce a discharge of 1.25 cfs of water under a head of 11.5 ft.?

5. Under what head will a standard, sharp-edged orifice, 3 in. in diameter, discharge 0.250 cfs; 1.25 cfs; 2.50 cfs.?

6. If the orifice shown in Fig. 52, page 119, has a diameter of 2 in. with a coefficient of contraction of 1.00, determine the discharge if $h = 3.6$ ft, $p_A = 9.7$ lb per sq in., $p_B = 1.3$ lb per sq in. and the head lost is 0.8 ft. The liquid is oil (sp gr 0.85). Determine C.

7. A sharp-edged orifice, 3 in. in diameter, lies in a horizontal plane, the jet being directed upward. If the jet rises to a height of 26.5 ft and the coefficient of velocity is 0.98, determine the head under which the orifice is discharging, neglecting air friction.

8. A sharp-edged orifice, 4 in. in diameter, in the vertical wall of a tank, discharges under a constant head of 4 ft. The volume of water discharged in 2 minutes weighs 6350 lb. At a point 2.57 ft below the center of the orifice the center of the jet is 6.28 ft distant horizontally from the vena contracta. Determine C_c, C_v, and C.

9. An orifice 6 in. in diameter, having a coefficient of contraction of 0.62, discharges oil under a head of 24.5 ft. The average velocity at the orifice is 23.6 ft per sec. Determine C, C_v, and the frictional loss.

10. A vertical triangular orifice has a base 3 ft long, 2 ft below the vertex, and 4 ft below the water surface. Determine the theoretical discharge, neglecting velocity of approach.

11. A sharp-edged orifice in the vertical wall of a large tank has a diameter of 1 in. C_c is 0.62 and C_v is 0.98. If the jet drops 3.07 ft in a horizontal distance of 8.17 ft from the vena contracta, determine the head and the discharge.

12. A sharp-edged orifice 2 in. in diameter in the end of a 6-in. pipe as in Fig. A is used to measure water. The pressure in the pipe is indicated by a mercury-water manometer, the mercury levels being read on a vertical scale which at the level of the center of the orifice reads 6.035 ft. Determine the discharge for the following pairs of scale readings of mercury levels: (a) left 3.216 ft, right 2.452 ft; (b) left 3.679 ft, right 1.989 ft; (c) left 5.568 ft, right 0.100 ft.

13. A 6-in. circular sharp-edged orifice in a steel plate is placed on the end of a 12-in pipe. With water flowing a pressure gage

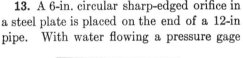

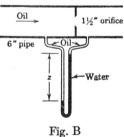

Fig. A   Fig. B

tapped into the side of the pipe 1 ft from the orifice shows a pressure of 35.0 lb per sq in. Compute the discharge.

14. The flow of a light dust-proofing oil at 60° F is measured by means of a 1½-in. circular sharp-edged orifice in a 6-in. pipe. The pressure drop across the orifice is determined by a differential manometer as shown in Fig. B. When $z = 4.50$ ft, what discharge can be expected?

15. A sharp-edged pipe orifice $\frac{3}{4}$ in. in diameter is installed in a 3-in. pipe. Determine the discharge of a heavy fuel oil at 40° F if the pressure drop across the orifice is 0.130 lb per sq in.

16. A sharp-edged pipe orifice 6 in. in diameter is installed in an 18-in. pipe with pressure taps 18 in. upstream and 9 in. downstream from the orifice. Air-water U-tube open manometers are connected to the taps. Determine the discharge of air at 32° F if the upstream manometer reads 15.6 in. of water and the downstream manometer reads 3.6 in. of water.

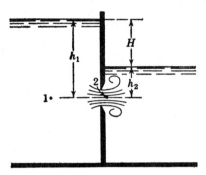

FIG. 64.   Submerged orifice.

69. Submerged Orifice. An orifice with submerged discharge is illustrated in Fig. 64. The greater depth on the center of the orifice is h_1 and the lesser depth is h_2. The assumption is usually made that every filament passing through the orifice is being acted upon by the head, $h_1 - h_2 = H$, the difference in elevation of the liquid surfaces. On this assumption, writing the energy theorem from point 1 to point 2 in the jet, neglecting lost head,

$$h_1 = h_2 + \frac{V_t^2}{2g}$$

from which

$$V_t = \sqrt{2g(h_1 - h_2)}$$

Applying the usual orifice coefficients

$$Q = CA \sqrt{2gH} \tag{10}$$

Coefficients of discharge for sharp-edged submerged orifices are about the same as for similar orifices discharging into air.

The assumption that h_2 is the pressure head on the center of the orifice at its lower side is not strictly true unless all the velocity head due to the velocity of the liquid leaving the orifice is lost in friction and turbulence as the velocity is reduced to zero. It has been shown experimentally that less than 90 per cent of this velocity head may be lost. With a loss of 90 per cent the pressure head at the center of the orifice is $h_2 - 0.10v^2/2g$.

70. Orifices under Low Heads.   The discharge of orifices under low heads can be determined by the formula

$$Q = CA \sqrt{2gH} \qquad (10)$$

if the value of C is known.   However, where the head on a vertical orifice is small in comparison with the height of the orifice there is

theoretically an appreciable difference between the discharge obtained by assuming the mean velocity to be that due to the mean head and the discharge obtained by taking into consideration the variation in head. Because it affords the simplest treatment, the rectangular orifice (Fig. 65) will be investigated.

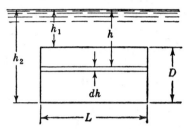

FIG. 65.   Rectangular orifice under low head.

Both the surface of the liquid and the jet are subjected to atmospheric pressure.   The width of the opening is L and the height is D.   The respective heads on top and bottom of the orifice are h_1 and h_2.   Velocity of approach will be neglected.   The theoretical discharge through any elementary strip of area $L\,dh$ at a distance h below the water surface is

$$dQ_t = L \sqrt{2gh}\, dh$$

which, integrated between the limits of h_2 and h_1 and with coefficient of discharge C' introduced, gives

$$Q = C'\tfrac{2}{3} \sqrt{2g}\, L(h_2^{3/2} - h_1^{3/2}) \qquad (19)$$

As h_1 approaches zero, the discharge approaches the value

$$Q = C'\tfrac{2}{3} \sqrt{2g}\, Lh_2^{3/2} \qquad (20)$$

which is the formula for discharge over a weir without velocity of approach correction.   (See Art. 80.)   Because of the surface drop-down curve, weir flow begins before h_1 decreases to zero.

Since the value of C' in equation 19 varies with head even more than C in equation 10, there is no practical advantage in using equation 19, but its derivation is given to show the close relation between orifice flow and weir flow.

71. Gates. As the term is commonly used in engineering practice, a gate is an opening in a dam or other hydraulic structure to control the passage of water. Gates have the hydraulic properties of orifices. Flow may be either free or submerged. The coefficient of discharge varies widely, however, with the design, the points at which the head is measured, and the conditions of flow. Calibration tests of a given installation are advisable if accurate flow measurements are to be obtained.

Gates ordinarily have the contraction partially or entirely suppressed on one or more sides. If a gate has any portion of its edge flush with the bottom or a side of the channel of approach, contraction for that part of the gate is entirely suppressed. If

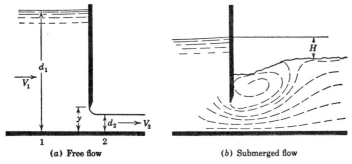

(a) Free flow   (b) Submerged flow

Fig. 66.   Flow through sluice gate.

one side of a rectangular gate is flush with the bottom or a side of the approach channel and there is opportunity for complete contraction on the other sides there will be larger velocity components parallel to the face of the gate on these sides and the coefficient of contraction will be very nearly the same as for a gate with complete contraction on all four sides.

A gate which has its lower edge in or near the bed of a channel is called a sluice gate. Typical profiles of flow through a sluice gate having a sharp top edge and no contraction at sides or bottom are shown in Fig. 66. The flow may be free, as in (a), or submerged, as in (b). With free flow, at fairly large ratios of upstream depth to height of gate opening, the surface of the stream issuing from the gate is quite smooth. With submerged flow the downstream surface may be extremely rough and turbulent.

Writing the energy theorem with respect to the stream bed as datum from point 1 to point 2 (Fig. 66a), assuming the pressure head equal to the depth, and omitting lost head,

$$\frac{V_1^2}{2g} + d_1 = \frac{V_2^2}{2g} + d_2$$

from which, introducing the velocity coefficient,

$$V_2 = C_v \sqrt{2g(d_1 - d_2) + V_1^2}$$

The coefficient of contraction $C_c = Bd_2/By = d_2/y$, where B is the width of flume.   Hence

$$Q = C_c ByC_v \sqrt{2g(d_1 - d_2) + V_1^2}$$
$$= CA \sqrt{2g(d_1 - d_2) + V_1^2} \qquad (21)$$

The cross section of a head gate such as is commonly used in diverting water from a river into a canal is shown in Fig. 67.   A

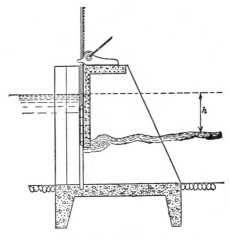

FIG. 67.   Head gate.

curtain wall extends between two piers, having grooves in which the gate slides.   The bottom of the opening is flush with the floor of the structure.   Such an opening has suppressed contraction at the bottom, nearly complete contraction at the top, and partially suppressed contractions at the sides.   Other equally complex conditions arise.   The selection of coefficients for gates is therefore a matter requiring mature judgment and an intelligent use of the few available experimental data.   Experiments have given coefficients of discharge for gates as low as 0.61 and as high as 0.91, with all intermediate values.

Gates are frequently placed on the top of a spillway dam to control the flow and head. Such gates may be either plane or curved surfaces structurally designed to withstand the water pressure and supported by piers on the crest of the dam or by the end abutments.

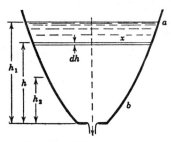

FIG. 68. Discharge under falling head.

Such gates act as orifices if the edge which forms the top of the opening makes contact with the water surface, otherwise the discharge follows the laws of weir flow.

72. Discharge under Falling Head. The vessel, Fig. 68, is shown to be discharging a liquid through an orifice under a head h_1. If there is no compensating inflow, the depth or head will gradually decrease. The time that will elapse while the head is being reduced from h_1 to h_2 is required.

At the instant when the head is h the discharge is

$$Q = CA \sqrt{2gh}$$

C being the coefficient of discharge and A the area of the orifice. In the infinitesimal time dt, the corresponding volume which flows out is

$$dV = CA \sqrt{2gh}\, dt \qquad (22)$$

In the same infinitesimal time the head will drop dh and the volume discharged will be

$$dV = A_s\, dh \qquad (23)$$

where A_s is the area of the liquid surface when the head is h. Equating the values of dV

$$A_s\, dh = CA \sqrt{2gh}\, dt$$

or

$$dt = \frac{A_s\, dh}{CA \sqrt{2gh}} \qquad (24)$$

By expressing A_s in terms of h and integrating between the limits h_1 and h_2, the time required to draw the reservoir down the desired amount can be determined. If the time required to empty the reservoir is desired, the lower limit of integration $h_2 = 0$. The discharge coefficient C is assumed constant.

The foregoing theory applies also to vertical or inclined orifices as long as they flow full. The heads h_1 and h_2 are then measured to the center of the orifice. The time required to empty a vessel completely can be determined only in the case of a horizontal orifice, and only approximately then, since, when the head becomes low, a vortex may be formed which materially alters the discharge coefficient.

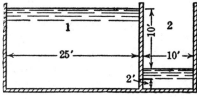

Fig. 69

EXAMPLE. Two chambers, 1 and 2 (Fig. 69), are separated by a partition in the bottom of which is an orifice 1 ft by 2 ft for which $C = 0.85$. Both chambers are 8 ft wide and have vertical sides. Chamber 1 is 25 ft long, and chamber 2 is 10 ft long. The chambers contain varying depths of water, but the orifice is at all times submerged. At a certain instant the water surface is 10 ft higher in chamber 1 than in chamber 2. After what interval of time will the water surfaces in the two chambers be at the same elevation?

Solution. Let h be the difference in elevation of water surfaces at any instant and dh the change in the difference in elevation of water surfaces in time dt. The amount of water flowing into chamber 2 in time dt will be $dV = CA \sqrt{2gh}\, dt = 0.85 \times 2 \times 8.02 \sqrt{h}\, dt = 13.6 \sqrt{h}\, dt$. Also in the same interval of time the head will drop $\frac{10}{35}\, dh$ in chamber 1 and rise $\frac{25}{35}\, dh$ in chamber 2. Then

$$dV = \frac{10 \times 25 \times 8}{35}\, dh = \frac{2000}{35}\, dh$$

Equating values of dV, solving for dt, and integrating,

$$t = \int dt = \int_0^{10} \frac{4.20\, dh}{\sqrt{h}} = 26.5 \text{ sec}$$

PROBLEMS

1. A vertical cylindrical tank discharges liquid through an orifice in the bottom. Show that the time required to lower the liquid in the tank from depth h_1 to depth h_2 is

$$t = \frac{2A_s}{CA \sqrt{2g}} (\sqrt{h_1} - \sqrt{h_2}) \tag{25}$$

where A_s is the cross-sectional area of the tank, A is the area of the orifice, and C is the coefficient of discharge (assumed constant).

2. Show from equation 25 that for a vertical cylindrical tank the time required to lower the liquid level a given amount is equal to the total volume of liquid discharged divided by the average of the initial and final rates of discharge. *Hint:* multiply numerator and denominator by $(\sqrt{h_1} + \sqrt{h_2})$.

3. A cylindrical vessel 4 ft in diameter and 6 ft high has a round-edged circular orifice 2 in. in diameter in the bottom. C for the orifice is 0.95. If the vessel is filled with water how long will it take to lower the water surface 4 ft?

4. A tank, which is the frustum of a cone having its bases horizontal and axis vertical, is 10 ft high and filled with water. It has a diameter of 8 ft at the top and 3 ft at the bottom. What is the time required to empty the tank through a sharp-edged orifice 3 in. square with a C of 0.61?

5. A tank 20 ft long and 10 ft deep is 8 ft wide at the bottom and 18 ft wide at the top. In the bottom is an orifice having an area of 24 sq in. and a coefficient of discharge of 0.60. If the tank is full at the beginning, how long will it take to lower the water surface 6 ft?

6. The tank in Fig. 68 between a where the depth is 20 ft and b where the depth is 4 ft is a frustum of a paraboloid with vertical axis. The side elements between a and b follow the curve $x^2 = 9h$, the origin being in the plane of the orifice. The orifice is sharp-edged, 4 in. in diameter, with $C = 0.607$. Compute the time required for the liquid level in the tank to drop from a to b.

7. A hemispherical shell, with base horizontal and uppermost, is filled with water. If the radius is 8 ft determine the time required to empty through a sharp-edged orifice 6 in. in diameter ($C = 0.60$) located at the lowest point.

8. Two vertical circular cylindrical tanks are connected near the bottom by a short tube having a cross-sectional area of 0.78 sq ft. The inside diameters of the tanks are 10 ft and 5 ft. The tanks contain oil (sp gr 0.80). With a valve in the connecting tube closed, the oil surface in the larger tank is 12 ft above the tube and in the smaller tank 2 ft above. Assuming a constant discharge coefficient for the tube of 0.75, find the time in which the oil surfaces in the two tanks will reach the same elevation following a quick opening of the valve.

Tubes

73. Standard Short Tube. A tube with square-cornered entrance and a length about $2\frac{1}{2}$ times its diameter is termed a standard short tube. If flow from such a tube is started suddenly at a relatively high head, the jet may spring clear of the walls (Fig. 70a) and the coefficients will then be the same as for a sharp-edged orifice in a thin plate.   By temporarily stopping the outlet and then allowing the liquid to escape, the tube can be made to flow full (Fig. 70b) if, as shown below, the head is less than about 41 ft of water. With the tube flowing full $C_c = 1.00$ and $C = C_v =$ about 0.82, the value varying slightly with the head and diameter of tube.   The discharge is thus about one-third greater than for a sharp-edged orifice of the same diameter, but the velocity of discharge is less.

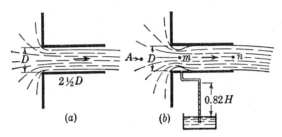

FIG. 70.   Standard short tube.

The lost head (Art. 63) in the entire length of a standard short tube flowing full is

$$H_0 = \left(\frac{1}{0.82^2} - 1\right)\frac{V^2}{2g} = 0.50\frac{V^2}{2g} \tag{26}$$

This case is important since the entrance to a pipe set flush with a vertical wall is usually considered to act as a standard short tube, and the head lost at entrance is taken as one-half the velocity head in the pipe.

The jet contracts on entering a sharp-edged short tube, as shown at m in Fig. 70b, and then expands to fill the tube.   The following theoretical analysis shows that the pressure head at the contracted section is about $-0.82H$, where H is the head producing flow through the tube.   This relation has been confirmed experimentally.

Assuming the coefficient of contraction at m to be 0.62, the same as for a sharp-edged standard orifice,

$$a_m = 0.62a_n, \quad V_m = 1.61V_n, \quad \text{and} \quad \frac{V_m{}^2}{2g} = 2.60\frac{V_n{}^2}{2g}$$

The velocity coefficient at point m is probably about 0.98.   The velocity V_m is, of course, considerably greater than $0.98\sqrt{2gH}$

since the head producing flow to point m is $H - p_m/w$, the latter term being negative. With $C_v = 0.98$, the head lost between the reservoir and m is $0.04V_m^2/2g = 0.10V_n^2/2g$.

Writing the energy theorem between point A, where the total head is H, and point m,

$$H = \frac{V_m^2}{2g} + \frac{p_m}{w} + \text{lost head}$$

Substituting values shown above,

$$H = 2.60\,\frac{V_n^2}{2g} + \frac{p_m}{w} + 0.10\,\frac{V_n^2}{2g} = 2.70\,\frac{V_n^2}{2g} + \frac{p_m}{w}$$

But

$$V_n = 0.82\,\sqrt{2gH} \quad \text{or} \quad \frac{V_n^2}{2g} = 0.82^2 H$$

Hence

$$H = 1.82H + \frac{p_m}{w}$$

and

$$\frac{p_m}{w} = -0.82H$$

Since p_m/w cannot be less than -34 ft of water, H cannot be greater than 41.5 ft of water if a short tube is to flow full.

74. Converging Tubes. Conical converging tubes having a circular cross section are frustums of cones with the larger end adjacent to the reservoir (Fig. 71). The jet contracts slightly beyond the end of the tube. The coefficient of contraction C_c, based on the area of the tip, decreases as the angle of convergence θ increases, becoming 0.62 for $\theta = 180°$ when the tube becomes a sharp-edged orifice. The coefficient of velocity C_v, on the other hand, decreases as θ decreases.

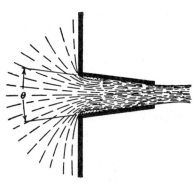

Fig. 71.   Converging tube.

The table gives coefficients for water discharging through converging, conical tubes with sharp-cornered entrances, interpolated from experiments by d'Aubuisson and Castel. These results are interesting in that they show the general laws of variation of coefficients, but, on account of the small models used in the experiments, they should not be taken as generally applicable to all tubes of this type.

COEFFICIENTS FOR CONICAL CONVERGING TUBES

Coeffi-	Angle of Convergence, θ (Fig. 71)								
cient	0°	5°	10°	15°	20°	25°	30°	40°	50°
C_v	0.829	0.911	0.947	0.965	0.971	0.973	0.976	0.981	0.984
C_c	1.000	.999	.992	.972	.952	.935	.918	.888	.859
C	0.829	.910	.939	.938	.924	.911	.896	.871	.845

The coefficient of velocity and therefore the coefficient of discharge are increased by rounding the entrance since this reduces the head lost in the tube. The coefficient of contraction is not materially changed.

75. Nozzles. Since a nozzle is a converging tube, the discharge through a nozzle can be determined from equation 10 as well as by the method of the energy theorem given in Art. 54. Velocity of approach must not be neglected.

The following mean values of coefficients for water discharging from smooth fire nozzles, similar to Fig. 44, having a diameter at the base of 1.55 in., and a C_c of 1.00, were determined from experiments by Freeman:[1]

TIP DIAMETER IN INCHES	$\frac{3}{4}$	$\frac{7}{8}$	1	$1\frac{1}{8}$	$1\frac{1}{4}$	$1\frac{3}{8}$
$C = C_v =$	0.983	0.982	0.980	0.976	0.971	0.959

The loss of head in a nozzle can be computed as for any other orifice or tube for which the coefficient of velocity is known. The nozzle loss is usually expressed as a factor K_n times the velocity head in the jet, where

$$K_n = \left(\frac{1}{C_v{}^2} - 1 \right)$$

A short converging tube, of the general design shown in Fig. 72, which can be inserted between flanges in a pipe line for the purpose

[1] John R. Freeman, "Experiments Relating to Hydraulics of Fire Streams," *Trans. Am. Soc. Civil Engrs.*, vol. 21, pp. 303–482, 1889.

of measuring the flow, is known as a flow nozzle. The coefficient
of discharge of a flow nozzle varies not only with the head on the
nozzle and the ratio of nozzle diameter to pipe diameter but also
with the design of the nozzle and the location of the pressure gage
connections. In order to secure accurate measurements with a
flow nozzle, therefore, it is necessary either to use a standard design
for which coefficients are known[1] or to calibrate the nozzle in place.

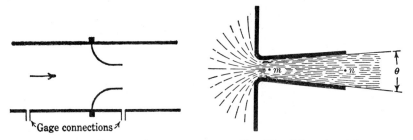

FIG. 72. Flow nozzle.   FIG. 73. Diverging tube.

76. Diverging Tubes. Figure 73 represents a conical diverging
tube, having rounded entrance corners, so that all changes in
velocity occur gradually. Such a tube will flow full provided that
the angle of flare is not too great nor the tube too long.

Experiments indicate that even under favorable conditions the
value of C_v, based on velocity at outlet, is small. Venturi and
Eytelwein, experimenting with the flow of water in a tube 8 in.
long, 1 in. in diameter at the throat, and 1.8 in. in diameter at the
outer end, obtained a value for C_v of about 0.46. The lost head
was, therefore, approximately $0.79H$. Even with this large loss of
head the discharge was about two and one-half times the discharge
from a sharp-edged orifice having the same diameter as the throat
of the tube.

The greater portion of the loss of head occurs between the throat
and outlet of the tube where the stream is expanding and thus has
a tendency to break up in eddies with a waste of energy. Experiments by Venturi indicate that an included angle θ of about 5°
and a length of tube about nine times its least diameter give the
highest coefficient of discharge. A diverging tube, such as that
shown in Fig. 73, is commonly called a Venturi tube.

The pressure head at the throat is less than atmospheric pres-

[1] American Society of Mechanical Engineers, *Flow Meters and Their Applications*, 1935

sure. This can be shown by writing Bernoulli's equation between m and n. When the ratio between the diameters at m and n is small, or when the head is great, the use of Bernoulli's equation may give a negative absolute pressure at m if the incorrect assumption is made that the tube flows full at n.

77. Re-entrant Tubes. Tubes having their ends projecting into a reservoir (Fig. 74) are called re-entrant or inward-projecting

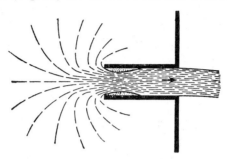

FIG. 74. Re-entrant tube.

tubes. Flow in such tubes is similar to that in standard short tubes (Art. 73), except that the contraction of the jet near the entrance is greater because some particles must change direction by as much as 180°.

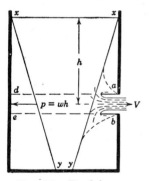

FIG. 75. Borda's mouthpiece.

A re-entrant tube $2\frac{1}{2}$ diameters in length, flowing full at the outlet ($C_c = 1$), has been found to have a C_v of about 0.75. The lost head is thus approximately $0.8V^2/2g$. The loss of head at entrance to a pipe which projects into a body of water is usually taken to be the same as the loss of head in such a re-entrant tube.

A special case of the re-entrant tube, consisting of a thin tube projecting into the reservoir about one diameter (Fig. 75), is called " Borda's mouthpiece " after the French scientist who applied the pressure-momentum theory to prove that the coefficient of contraction for such a tube under ideal conditions is 0.50.

The pressure-momentum theory (see Chapter IX) is based on the principle of mechanics that force equals the rate of change of momentum. The reservoir is assumed to be so large that the velocity of approach may be neglected. Also the pressure on the reservoir walls is assumed to increase uniformly with the depth, as indicated by the lines xy. Thus with such a tube the pressure in the corners a and b is full hydrostatic pressure corresponding to their depth below the free surface. The variation of pressure in the vicinity of an orifice without the inward-projecting tube is more nearly represented by the dotted lines, and in this case the pressure-momentum theory is less exact.

Excepting the pressure acting on the projection de of the mouth-piece on the opposite wall, the horizontal pressures on the walls balance each other. By the pressure-momentum theory, the unbalanced force on de must equal the resulting change per second in the momentum of the fluid.

If the area of the tube is A, the pressure on de is whA. If the area of the jet is a and its velocity is V, the mass of fluid per second passing any point is aVw/g. Since such a mass starts from rest every second and acquires a velocity V the change in momentum is aV^2w/g. Equating force to change in momentum

$$whA = \frac{aV^2w}{g}$$

from which

$$ghA = aV^2$$

Since

$$V = C_v \sqrt{2gh}$$

$$\frac{a}{A} = C_c = \frac{1}{2C_v{}^2}$$

With ideal flow, $C_v = 1.00$ and $C_c = 0.50$. When $C_v = 0.98$, $C_c = 0.52$, a value approximately verified by experiment.

78. Submerged Tubes. The discharge through a submerged tube, as through a submerged orifice (Art. 69), is $Q = CA\sqrt{2gH}$, C being the coefficient of discharge, A the area of the opening, and H the difference in elevation of the liquid surfaces. It is probable that as with orifices the coefficient is not materially changed by submergence. Loss of head in a submerged tube is affected by entrance conditions, being greatest for a re-entrant inlet and least for a bell mouth. Even moderate rounding of the entrance materially reduces the loss of head. Terminating the tube with a diverging end causes a gradual decrease in velocity and reduces loss of head at the outlet. To avoid vortex formation and loss of capacity through entrainment of air, the entrance must be submerged not less than the sum of the velocity head and the loss of head at entrance (Art. 73), the minimum submergence for a square-cornered entrance being thus about $1.5V^2/2g$, where V is the mean velocity in the tube.

Culverts to pass natural drainage water through embankments are an example of tubes which may flow submerged. Culverts usually have lengths of 10 or more diameters. During much of the time they may carry little or no flow, but they must be large enough to carry design floods without damage to the structures they safeguard. The discharge of culverts flowing partly full must be determined by use of the laws of non-uniform flow in open channels (Art. 138).

Under conditions of flood flow the outlet may be submerged as in Fig. 76 and the culvert may flow full. In this case it may be considered either as a short pipe (Art. 113) or as a submerged tube. A field examination frequently will be required to estimate the two water levels, the difference in these levels being the head H under which discharge occurs. More than 3000 experiments on various kinds of pipe and box culverts for determining C were performed at the University of Iowa by Yarnell, Nagler,

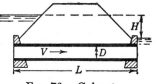

Fig. 76. Culvert.

and Woodward.[1] A summary of the results of these experiments, as they apply to concrete pipe culverts of length L up to 50 ft and diameter D from 1 to 6 ft, is given in the table on page 148.

A siphon spillway is illustrated in Fig. 77. Under normal operating conditions, both the intake and the outlet are submerged, sealing the ends of the tube and forming an inclosed chamber. When the upper water level gets high enough for flow through the siphon to begin, air in the chamber is carried out by the moving water and in a short time the passageway is flowing full. Flow continues until the seal is broken by lowering of the water surface below the lip of the intake, or below the top of the air vent, when such is provided. Such a siphon, if of airtight construction, has

[1] D. L. Yarnell, F. A. Nagler, and S. M. Woodward, " Flow of Water through Culverts," *Studies in Engineering,* University of Iowa, 1926.

COEFFICIENTS OF DISCHARGE FOR CONCRETE PIPE CULVERTS

	$\dfrac{D}{L}$	1	1.5	2	3	4	5	6
Beveled-lip entrance	10	0.86	0.89	0.91	0.92	0.93	0.94	0.94
	20	.79	.84	.87	.90	.91	.92	.93
	30	.73	.80	.83	.87	.89	.90	.91
	40	.68	.76	.80	.85	.88	.89	.90
	50	.65	.73	.77	.83	.86	.88	.89
Square-cornered entrance	10	.80	.81	.80	.79	.77	.76	.75
	20	.74	.77	.78	.77	.76	.75	.74
	30	.69	.73	.75	.76	.75	.74	.74
	40	.65	.70	.73	.74	.74	.74	.73
	50	.62	.68	.71	.73	.73	.73	.72

the hydraulic properties of a submerged tube, the head producing discharge being H, the difference in elevation of water surfaces.

The place at which the cross-sectional area, A, should be measured has been the subject of much discussion. Some experimental values of the discharge coefficient, C, have been based on the area of the throat, or highest point, of the siphon. It appears, however, that much more uniform values of C will be obtained if A is measured at the discharge end in a plane normal to the direction of flow. That this is true may be seen by considering the effect of reducing the cross-sectional area at the throat. For a given value of H (Fig. 77) a 50 per cent reduction of area in the region of the throat would (unless the pressure approached absolute zero) reduce the discharge but slightly, due to a slight increase in the loss of head, but would nearly double the value of C based on throat area.

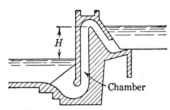

FIG. 77. Siphon spillway.

When a siphon spillway is operating, the pressure head at the throat is less than atmospheric. Pressure heads less than about -15 ft of water should be avoided in the hydraulic design of concrete-lined siphon spillways in order to avoid the danger of occurrence of the phenomenon known as cavitation, with resulting pitting of the concrete.

For purposes of design, the present tendency is to treat siphon spillways as short pipes in accordance with the principles of Chapter VII rather than as submerged tubes.

PROBLEMS

1. A standard short tube, 4 in. in diameter, discharges water under a head of 20 ft. What is the discharge: (a) in cubic feet per second; (b) in gallons per day?

2. A Venturi tube discharges water from a reservoir. The diameters at the throat and at the outlet are 3 in. and 4 in., respectively. Neglecting all losses, determine the maximum head under which this tube will flow full throughout its length.

3. A Borda's mouthpiece 6 in. in diameter discharges water under a head of 10 ft. Determine: (a) the discharge in cubic feet per second; (b) the diameter of the jet at the vena contracta.

4. A standard short tube 4 in. in diameter discharges water under a head of 16.5 ft. A small hole, tapped in the side of the tube 2 in. from the entrance, is connected with the upper end of a piezometer tube, the lower end of which is submerged in a pan of mercury. Neglecting vapor pressure, to what height will mercury rise in the tube? Also determine the absolute pressure in pounds per square inch at the upper end of the piezometer tube.

5. A nozzle similar to those tested by Freeman has a tip diameter of $\frac{7}{8}$ in. and is attached to a hose having a diameter of 1.55 in. If the pressure in the hose at the base of the nozzle is 40 lb per sq in., determine the discharge.

6. Determine the probable capacity of a concrete pipe culvert 4 ft in diameter and 40 ft long discharging as shown in Fig. 76 under a head H of 5.0 ft. The entrance is square-cornered.

7. What diameter of concrete pipe culvert 50 ft long with beveled-lip entrance should be installed to carry 500 cfs of water if the difference in water surface elevations at the two ends of the culvert is not to exceed 8.0 ft?

Weirs

79. Description and Definitions. A weir is an overflow structure built across an open channel for the purpose of measuring the flow. Weirs have been commonly used to measure the flow of water, but their use in measurement of other liquids is increasing. The same principles apply to all liquids, and the fundamental formulas based upon these principles are in all respects general.

Classified with reference to the shape of the opening through which the liquid flows, weirs may be rectangular, triangular, trapezoidal, circular, parabolic, or of any other regular form. The first three forms are most commonly used for measurement of water; the triangular weir is usually best adapted for measurement of other liquids.

The edge or top surface with which the flowing liquid comes in contact is termed the crest of the weir. Classified with reference to the form of the crest, weirs may be sharp-crested or broad-crested. The sharp-crested weir has a sharp upstream edge so formed that the liquid in passing touches only a line. The broad-crested weir has either a rounded upstream edge or a crest so broad that the liquid in passing comes in contact with a surface.

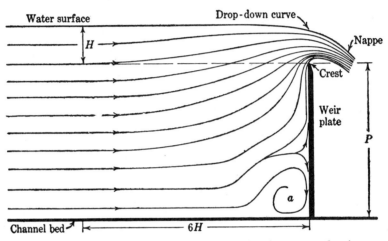

Fig. 78. Path lines of flow over rectangular sharp-crested weir.

The flow over a weir may be either free or submerged. If the water surface downstream from the weir is lower than the crest, the flow is free; if this downstream surface is higher than the crest, the weir is submerged (Art. 89).

The overfalling stream is termed the nappe. The nappe of a sharp-crested weir, Fig. 78, is contracted at its under side by the action of the vertical components of velocity just upstream from the weir. (See Art. 59.) This is called crest contraction. If the sides of the opening also have sharp upstream edges so that the nappe is contracted in width, the weir is said to have end contractions and is usually called a contracted weir. If a weir has a length L equal to the width of the channel, the nappe suffers no contraction in width and the weir has end contractions suppressed. Such a weir has commonly been called a suppressed weir although the name full-width weir is also used. If constructed in accordance with certain design specifications (Art. 82), a weir of this type has been called a standard weir.

There is a downward curvature of the surface of the liquid in the vicinity of the weir which is called the drop-down curve. The vertical distance H between the liquid surface and the crest of the weir, measured far enough upstream to be beyond the drop-down curve, is called the head. Surface curvature may be perceptible for a distance of several times the head upstream from the weir. The channel immediately upstream from a weir is termed the channel of approach, and the mean velocity in this channel is the velocity of approach, V. The height of weir P is the vertical distance of the crest above the bottom of the channel of approach.

Typical path lines of flow over a sharp-crested weir are shown in Fig. 78. These lines were determined by tracing on a glass-sided flume the paths followed by globules of carbon tetrachloride and benzine adjusted to have the same specific gravity as the water. The paths are approximately parallel until they reach a point about six times the head upstream from the weir. From this point they gradually curve upward to pass over the crest.

The path lines at about two-tenths the depth, where the velocity is usually close to the maximum, are the first to show upward curvature, and those near the bottom of the channel are affected last. At a is a dead-water region in which particles may remain for some time and from which they emerge in upward spirals or eddies in the corners between the weir plate and the sides of the flume.

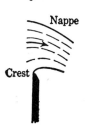

FIG. 79. Detail of crest.

The crest of a sharp-crested weir is not necessarily knife-edged but preferably of the form shown in Fig. 79. If the upstream corner is square-cut, the weir is still in effect sharp-crested, and excessive wear which would result from a thin knife edge is avoided.

80. Fundamental Theory. Development of formulas for weir discharge dates back into the early history of hydraulic theory. The base formulas are here developed for the rectangular full-width weir (Fig. 80) without end contractions.

The velocity head in the approach channel, corrected for unequal distribution of velocities (Art. 49), is denoted by αh_v, where $h_v = V^2/2g$. The depth of flow in the approach channel at the point where the head is measured is denoted by d.

By orifice theory, the theoretical discharge through the elementary strip of area $L\,dh$ under total head $h + \alpha h_v$ is

$$dQ_t = L\,dh\,\sqrt{2g(h + \alpha h_v)}$$

from which

$$Q_t = \int_0^H L\,\sqrt{2g}\,\sqrt{h + \alpha h_v}\,dh$$

$$= \tfrac{2}{3}\,\sqrt{2g}\,L[(H + \alpha h_v)^{3/2} - (\alpha h_v)^{3/2}] \qquad (27)$$

Introducing a coefficient C' to correct for vertical contraction of the nappe and for frictional loss,

$$Q = \tfrac{2}{3}\,\sqrt{2g}\,C'L[(H + \alpha h_v)^{3/2} - (\alpha h_v)^{3/2}] \qquad (28)$$

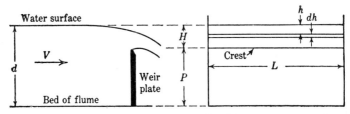

FIG. 80.   Rectangular sharp-crested weir without end contractions.

It is common practice to combine $\tfrac{2}{3}\,\sqrt{2g}\,C'$ into a single coefficient C and to assume $\alpha = 1$. The general formula for weir discharge including effect of velocity of approach thus becomes

$$Q = CL[(H + h_v)^{3/2} - (h_v)^{3/2}] \qquad (29)$$

If the ratio of head H to height of weir P is sufficiently small, the $h_v^{3/2}$ term in equation 29 may be neglected, in which case the weir discharge is given by the equation

$$Q = CL(H + h_v)^{3/2} \qquad (30)$$

If H/P is sufficiently small to make h_v negligible, the formula for weir discharge, without correction for velocity of approach, becomes

$$Q = CLH^{3/2} \qquad (31)$$

Since h_v is a function of Q, the solution of equation 29 or 30 for Q involves successive trials. A more convenient form of equation can be derived by the following transformation.

Multiplying and dividing the right side of (29) by $H^{3/2}$,

$$Q = CLH^{3/2}\left[\left(1 + \frac{h_v}{H}\right)^{3/2} - \left(\frac{h_v}{H}\right)^{3/2}\right] \tag{32}$$

By binomial expansion the quantity in brackets becomes

$$1 + \frac{3}{2}\frac{h_v}{H} + \frac{3}{8}\left(\frac{h_v}{H}\right)^2 \cdots - \left(\frac{h_v}{H}\right)^{3/2}$$

Dropping as negligible all terms except the first two, equation 32 becomes

$$Q = CLH^{3/2}\left[1 + \frac{3}{2}\frac{h_v}{H}\right]$$

Substituting for h_v its approximate value

$$h_v = \frac{V^2}{2g} = \frac{Q^2}{A^2 \cdot 2g} = \frac{(CLH^{3/2})^2}{(Ld)^2 \cdot 2g} = \frac{C^2}{2g} \cdot H \cdot \frac{H^2}{d^2}$$

the equation for discharge becomes

$$Q = CLH^{3/2}\left[1 + C_1\left(\frac{H}{d}\right)^2\right] \tag{33}$$

where

$$C_1 = \frac{3}{2}\frac{C^2}{2g} = 0.0233C^2$$

Thousands of experiments have been made with water flowing over rectangular sharp-crested weirs to determine the values of the coefficients in equation 33 or to provide a basis for modifying coefficients or exponents to fit better the actual conditions of flow. These experiments have covered a wide range of conditions, and although they substantiate the general form of the equation derived above they contain many inconsistencies. The experiments of an individual investigator are usually consistent in themselves, but experiments by different investigators sometimes give results differing from one another by several per cent.

81. Accuracy of Weir Measurements. In laboratory measurements of flowing water, as, for instance, in turbine or hydraulic model testing, an accuracy within $\frac{1}{2}$ per cent is usually desired. Such accuracy can be obtained in weir measurements only if: (1) a standard design is followed in the construction of the weir and approach flume; (2) a formula is used which is based on reliable tests of a weir of similar design with similar range of head; and (3) the necessary test measurements, particularly of the head on the weir, are conducted with the utmost care and precision.

82. Standard Weir. Out of many years of study of sharp-crested weirs have come certain fundamental principles of design which, as they apply to rectangular weirs without end contractions, are as follows:

1. The upstream face of the weir plate shall be vertical and smooth.

2. The crest edge shall be level, shall have a square upstream corner, and shall be so narrow that the water will not touch it again after passing the upstream corner.

3. The sides of the flume shall be vertical and smooth and shall extend a short distance downstream past the weir crest.

4. The pressure under the nappe shall be atmospheric.

5. The approach channel shall be of uniform cross section for a sufficient distance above the weir, or shall be so provided with baffles that a normal distribution of velocities exists in the flow approaching the weir, and the water surface is free of waves or surges.

A weir built in accordance with these specifications may be called a standard weir.

83. Standard Weir Formulas. Practically every set of experiments has been used as a basis for a weir formula with the result that many different formulas have been proposed. The following list of six formulas contains those which have been most commonly used in the measurement of flowing water by means of standard rectangular sharp-crested weirs. The formulas as stated have been reduced as nearly as possible to the general form of equation 33, and are expressed throughout in foot-pound-second units.

Francis:

$$Q = 3.33 \left[1 + 0.26 \left(\frac{H}{d} \right)^2 \right] L H^{3/2} \tag{34}$$

Bazin:

$$Q = \left(3.248 + \frac{0.0789}{H} \right) \left[1 + 0.55 \left(\frac{H}{d} \right)^2 \right] L H^{3/2} \tag{35}$$

King:

$$Q = 3.34 \left[1 + 0.56 \left(\frac{H}{d} \right)^2 \right] L H^{1.47} \tag{36}$$

Swiss:

$$Q = \left(3.288 + \frac{0.0108}{H + 0.0052}\right)\left[1 + 0.5\left(\frac{H}{d}\right)^2\right]LH^{3/2} \quad (37)$$

Rehbock:

$$Q = \left(3.228 + 0.435\frac{H_e}{P}\right)LH_e^{3/2} \quad (38)$$

where $H_e = H + 0.0036$

Harris:

$$Q = \left[3.27 + \frac{c}{H} + 1.5\left(\frac{H}{d}\right)^2\right]LH^{3/2} \quad (39)$$

where c varies with the water temperature from 0.023 at 39° F to 0.018 at 68° F.

In using any weir formula it is important to know the design conditions of the weir and the range of heads used in the tests on which the formula was based. Only brief notes regarding the basis for the foregoing formulas can be given here, but the original references are cited for each.[1]

The Francis formula[2] was based on a very few tests on a weir about 10 ft long and 5 ft high with a range of heads from about 0.7 to 1.0 ft. The head was measured 6 ft upstream from the weir. As originally stated, the Francis formula was

$$Q = 3.33L[(H + h_v)^{3/2} - h_v^{3/2}] \quad (40)$$

Substituting 3.33 for C in equation 33 leads to equation 34, which does not require trial solutions and gives results practically the same as 40. If velocity of approach is negligible the Francis formula reduces to the frequently used form

$$Q = 3.33LH^{3/2} \quad (41)$$

The Bazin formula[3] was based on several hundred tests on weirs from about 0.8 to 3.7 ft high, from 1.64 to 6.56 ft long, with a range

[1] For a more extensive summary see " Check List of Weir Formulas," *Engineering Bulletin* 53, State College of Washington, Pullman.

[2] J. B. Francis, *Lowell Hydraulic Experiments*, 4th ed., 1883. Also *Trans. Am. Soc. Civil Engrs.*, vol. 13, p. 303.

[3] H. Bazin, *Annales des ponts et chaussées*, October, 1888, translation by Marichal and Trautwine, *Proc. Eng. Club*, Philadelphia, January, 1890. Also, *Annales des ponts et chaussées*, 1894, 1er trimestre.

of heads from about 0.2 to 1.8 ft. The head was measured 16.4 ft upstream from the weir. The Bazin formula was verified by Nagler,[1] who reproduced Bazin's " standard " weir 3.72 ft high, and 6.56 ft long, and tested it under heads from about 0.4 to 4.0 ft.

The King formula[2] was based on an analysis of the experiments of earlier investigators including Francis, Bazin, and Nagler.

The Swiss formula[3] was proposed by the Swiss Society of Engineers and Architects and was substantiated by tests on a weir about 2.6 ft high and 9.8 ft long with a range of heads from about 0.35 to 2.6 ft.

The Rehbock formula[4] was based on laboratory tests covering a period of many years on weirs from about 0.4 to 1.6 ft high with a range of heads up to about 0.6 ft, as well as on tests by other experimenters on weirs from 0.5 to 4 ft high with heads ranging from about 0.03 to 2.7 ft.

Use of the "substitute head," H_e, as proposed by Rehbock results in a comparatively simple formula. If the measured head H is read only to the second decimal place, the addition of the 0.0036 term is superfluous and the measured head can be used directly in the formula, which in its approximate form then becomes

$$ Q = \left(3.23 + 0.43\,\frac{H}{P} \right) LH^{3/2} \qquad (42) $$

The Harris formula[5] was derived from a study of experiments and formulas of other investigators supplemented by tests on a weir 4 ft high and 2 ft long with heads up to 1 ft, the head being measured 4 ft upstream from the weir. The Harris formula is unique in including a correction for water temperature. The percentage variation in discharge indicated by the formula with ordinary variation in water temperature is appreciable only at low heads.

[1] F. A. Nagler, " Verification of Bazin Weir Formula by Hydro-Chemical Gaging," Trans. Am. Soc. Civil Engrs., vol. 83, 1919, p. 105.

[2] H. W. King, Handbook of Hydraulics, McGraw-Hill Book Co., 3rd ed., 1939, p. 87.

[3] " Contribution a l'étude des methodes de jaugeage," Bulletin 18, Swiss Bureau of Water Resources, Bern, 1926.

[4] Th. Rehbock, " Wassermessung mit scharfkantigen Ueberfallwehren," Zeitschrift des Vereines deutscher Ingenieure, June 15, 1929.

[5] C. W. Harris, Hydraulics, John Wiley & Sons, 1936.

If the standard design requirements of Art. 82 are met, and if the precautions noted in Art. 84 in regard to measurement of head are observed, it appears that any of the formulas 34 to 39 will give fairly accurate results. Under certain conditions, however, there may be differences of several per cent in the results obtained. The formulas in general are more reliable if the following limitations of flow conditions are observed:

1. The head is not smaller than 0.2 ft.

2. The head is not larger than one-half the height of the weir.

3. The head is not larger than one-half the length of the weir.

84. Measurement of Head. In using a weir to determine the rate of discharge the head is measured with some form of gage set in a fixed position. The elevation of the zero of the gage with reference to the crest of the weir must be accurately determined.

The head may be measured either in a stilling well connected to the channel by a small pipe or tube or directly in the channel itself. The stilling well provides a means of measuring the head in still water and reduces the effect of waves or surges which may be present in the channel of approach.

The pipe or tube leading to the stilling well should not project into the channel but should be flush with the side or bottom as with a piezometer tube (Art. 21). The stilling well may give incorrect readings if the temperature of the liquid in the connecting tube or in the stilling well itself is different from that of the liquid in the flume.

FIG. 81. Hook gage.

The water-surface elevation in the stilling well is most accurately measured with a hook gage or a point gage. The hook gage, Fig. 81, consists of a graduated metallic rod with a pointed hook at the bottom which slides vertically in fixed supports. By means of a vernier attached to one of the supports, readings to thousandths of a foot may be taken. For most precise work the gage can be constructed with a micrometer head by means of which readings may be taken to ten-thousandths of a foot.

The gage should be of such length and should be rigidly attached to a support at such an elevation that the movement of the hook covers the range of water-surface elevations to be read. In making a reading the point of the hook is lowered just below the surface and then raised by means of a slow-motion screw. Just before the point of the hook pierces the skin of the water a pimple is seen on the surface; the hook is then lowered slowly until the pimple just disappears and the vernier is read.

A point gage is like a hook gage except that it terminates in a point extending vertically downward. With a quiet water surface the instant of contact of the point with the water surface as the point is slowly lowered can easily be detected. The point gage can be used either in a stilling well or in the approach channel. If the head is measured in the approach channel, it is necessary, for precise results, that the water surface at the gage be free of waves and surges.

If accuracy is essential, the head used to compute discharge over a weir should be the mean of at least 10 and preferably 20 separate measurements taken at equal intervals. By this means the effect of small fluctuations in head may be largely eliminated.

Another device for determining head is a plummet attached to the end of a steel tape. This is used to measure the vertical distance from a fixed point above the channel of approach to the water surface. The head on a weir can also be read by holding a scale on a hub in the approach channel. A close approximation of the head can be made by holding a scale which is at least an inch wide on the crest of the weir with the flat side to the stream. The velocity of the nappe piles up the water on the scale to a height practically equal to the head on the weir.

The head always should be measured far enough upstream from the weir to be well above the effects of surface contraction. The distance used in the experiments on which the weir formulas in Art. 83 were based was in general from 4 to 6 ft. The distance selected and the method of head measurement should conform approximately to those of the experiments on which the formula to be used in computing discharges is based.

PROBLEMS

1. Compute the discharge over a standard rectangular sharp-crested weir for several combinations of the following formulas and data.

Formula	Length of Weir L (ft)		Height of Weir P (ft)		Head H (ft)		Water Temperature
(a) Francis	(g)	1.0	(l)	0.5	(r)	0.2	(y) 39° F
(b) Bazin	(h)	2.0	(m)	1.0	(s)	0.4	(z) 68° F
(c) King	(i)	3.0	(n)	1.5	(t)	0.7	
(d) Swiss	(j)	5.0	(o)	2.0	(u)	1.0	
(e) Rehbock	(k)	10.0	(p)	4.0	(v)	1.5	
(f) Harris			(q)	10.0	(w)	2.0	
					(x)	3.0	

2. Compute by the Francis, Bazin, and Harris formulas the value of C in the base formula $Q = CLH^{3/2}$ for a weir 4 ft high with the seven heads indicated in problem 1. Assume a water temperature of 60° F. For each formula plot a curve with C as abscissa and H as ordinate.

3. Compute by the Harris formula the discharge over a standard weir 4.0 ft high in a channel 3.0 ft wide under heads of 0.465 and 1.82 ft for water temperatures of 39°, 50° 60°, and 68° F.

4. Compute by the King and the Rehbock formulas the discharge per foot of length over a standard weir 3 ft high under heads of 0.250 and 1.250 ft.

5. A test measurement of flow over a standard weir 1.00 ft high in a flume 1.996 ft wide at a head of 0.5171 ft showed a total volume of 941.4 cu ft of water discharged in 362.3 sec. The water temperature was 42° F. Assuming the measurement of flow to be correct, compute the percentage of error in the discharge computed by each of the weir formulas 34 to 39.

6. A sharp-crested weir 4.0 ft high extends across a rectangular channel 10.0 ft wide. If the measured head is 1.22 ft, determine the discharge using the Francis and the Rehbock formulas. Compare with the discharge given by the approximate Francis and Rehbock formulas 41 and 42.

7. A sharp-crested weir 3.5 ft high extends across a rectangular channel 12 ft wide. If the measured head is 1.54 ft, determine the discharge, using the King and the Swiss formulas.

8. A sharp-crested weir 2 ft high extends across a rectangular channel 8 ft wide conducting irrigation water. If the measured head is $4\frac{3}{4}$ in., determine the discharge.

9. A rectangular channel 20 ft wide has a 3 ft depth of water flowing with a mean velocity of 2.45 ft per sec. Determine the height of standard sharp-crested weir that will increase the depth in the channel of approach to 5 ft.

10. A sharp-crested weir 3.0 ft high extends across a rectangular channel, 20 ft wide, in which there is 100 cfs flowing. Determine the depth of water upstream from the weir.

85. Rectangular Contracted Weirs. The full-width weir is not adapted for use in a flume or channel of other than rectangular cross section. In such a channel some form of notch in a bulkhead is preferred for measuring the flow.

The rectangular sharp-edged notch (Fig. 82) was one of the earliest forms of weir. Such a notch has end contractions the effect of which is to reduce the flow below that which would occur under the same head over a standard weir of the same length.

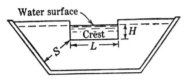

FIG. 82. Rectangular contracted weir.

Francis stated that the discharge through such a sharp-edged rectangular notch of length L under a head H can be computed by substituting in place of L in the standard weir formula the value

$$L' = L - 0.2H \qquad (43)$$

This correction for end contractions is approximate, so that great refinement in computation is not necessary. The use of the Francis correction leads to an absurdity when the length of weir becomes small in proportion to the head. For instance, for a weir 0.2 ft long under a head of 1 ft, $L - 0.2H = 0$, so that $Q = 0$, which is evidently not true. The use of the Francis correction is therefore ordinarily limited to weirs in which L is at least $3H$.

Although the Francis formula for computing correction for end contractions was originally recommended for use with the Francis formula (equation 40), it is equally applicable to other standard formulas.

A more precise formula based on tests by Cone[1] for the flow of water through rectangular sharp-crested weirs with complete end and bottom contractions is

$$Q = 3.247LH^{1.48} - \left(\frac{0.566L^{1.8}}{1 + 2L^{1.8}}\right)H^{1.9} \qquad (44)$$

It is recommended that L be at least equal to H for the use of equation 44.

[1] V. M. Cone, " Flow through Weir Notches with Thin Edges and Full Contractions," *Journal of Agricultural Research*, U. S. Department of Agriculture, March, 1916. See also " Measurement of Water in Irrigation Channels," *U. S. Department of Agriculture Farmers' Bulletin* 1683.

To obtain complete contraction, the minimum distance S (Fig. 82) from any point on the edges of the weir to the sides or bottom of the channel should be at least $2H$. The head should be measured at least $4H$ distant from the nearest point of the crest to avoid surface curvature.

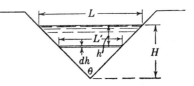

86. Triangular Weirs. The triangular or V-notch weir is preferable to the rectangular weir for the measurement of

Fig. 83. Triangular weir.

widely variable flows. Figure 83 represents a triangular weir over which a liquid is flowing. The measured head is H, and the distance between the sides of the weir in the plane of the liquid surface is L. The sides make equal angles with the vertical.

The area of an elementary horizontal strip dh in thickness is $L'\, dh$. Neglecting velocity of approach and friction loss, the velocity through this strip for a head h is $\sqrt{2gh}$, and the discharge is

$$dQ_t = L'\,\sqrt{2gh}\,dh \qquad (45)$$

From similar triangles $L' = (H - h)L/H$. Inserting this value of L' in equation 45,

$$dQ_t = L\,\sqrt{2g}\,\frac{(H-h)\,\sqrt{h}}{H}\,dh \qquad (46)$$

Integrating between the limits 0 and H and reducing,

$$Q_t = \tfrac{4}{15}\,\sqrt{2g}\,LH^{3/2} \qquad (47)$$

If θ is the notch angle, $L = 2H \tan \theta/2$. Substituting this value of L in equation 47, and introducing a discharge coefficient C,

$$Q = C\,\tfrac{8}{15}\,\sqrt{2g}\,\tan\frac{\theta}{2}\,H^{5/2} \qquad (48)$$

The most common angle of notch is 90°, for which, with a value of C of about 0.6, the approximate formula for discharge is

$$Q = 2.5H^{2.5} \qquad (49)$$

Experiments have shown that the coefficient and the exponent in equation 49 are not exactly 2.5, but that they do not vary far from this value. From tests by Barr[1] with water discharging over 90° weirs, Barnes[2] derived the formula

$$Q = 2.48H^{2.48} \qquad (50)$$

Tests by Lenz[3] with various angles of notch from 10° to 90° and with oil and water at various temperatures indicate that the value of C in equation 48 varies with the head and angle of notch as well as with the density, viscosity, and surface tension of the liquid. For cold water, the water temperature was found to have little effect on the coefficient, which in that case is a function only of head H and notch angle θ. Lenz derived the following formula for triangular weirs, for water only:

$$Q = \left(2.395 + \frac{N}{H^n}\right)\tan\frac{\theta}{2}H^{5/2} \qquad (51)$$

Values of N and n are given in the following table. The minimum value of the ratio N/H^n is 0.090.

θ	90°	60°	45°	30°
N	0.068	0.087	0.102	0.131
n	0.588	0.582	0.579	0.576

The most common angle of notch is 90°. For any liquid of density ρ, kinematic viscosity ν, and surface tension σ, Lenz derived the following formula for 90° V-notches only:

$$Q = \left[2.395 + \frac{1.247}{H^{0.59}(1/\nu)^{0.165}(\rho/\sigma)^{0.170}}\right]H^{5/2} \qquad (52)$$

the minimum value of the second term in the brackets being 0.090.

The effect of velocity of approach on triangular weirs is similar to the effect on rectangular weirs. From the nature of the triangular weir, however, the cross-sectional area of the nappe is usually much smaller than that of the channel of approach. The velocity of approach is therefore small, and the error introduced by neglecting it is usually inappreciable.

Sharpness of crest edge is as important with triangular weirs as with any sharp-edged orifices or weirs since a slight dullness or

[1] James Barr, " Flow of Water over Triangular Notches," *Engineering,* April 8 and 15, 1910.

[2] *Hydraulic Flow Reviewed*, E. and F. N. Spon, Ltd., 1916.

[3] *Trans. Am. Soc. Civil Engrs.,* 1943. p. 759.

rounding of the upstream edge results in appreciable increase in flow.

87. Trapezoidal Weirs. Figure 84 represents a trapezoidal weir having a horizontal crest of length L. The sides are equally inclined, making angles $\theta/2$ with the vertical.

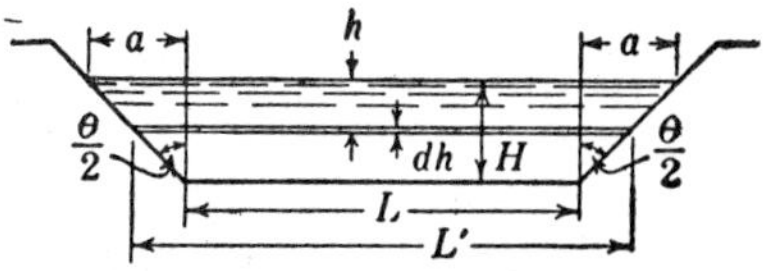

By writing the equation

$$dQ_t = L' \sqrt{2gh}\ dh$$

Fig. 84. Trapezoidal weir.

and expressing L' in terms of h and known quantities in a manner similar to that used in the preceding article for triangular weirs, and integrating and reducing, the formula for the discharge over trapezoidal weirs without velocity of approach correction becomes

$$Q = C'\tfrac{2}{3}\sqrt{2g}\ LH^{3/2} + C''\tfrac{8}{15}\sqrt{2g}\ \tan\frac{\theta}{2}H^{5/2} \qquad (53)$$

A trapezoidal weir, having a value of $\tan \theta/2$ of $\tfrac{1}{4}$, is called a Cipolletti[1] weir. This slope of the sides is approximately that required to secure a discharge through the triangular portion of the weir opening that equals the decrease in discharge resulting from end contractions. The advantage claimed for this type of weir is that it does not require a correction for end contractions. The method employed for arriving at the value of $\tan \theta/2$ is as follows:

The decrease in discharge resulting from end contractions, according to Francis (Art. 85), is

$$Q = C'\tfrac{2}{3}\sqrt{2g}\ 0.2H^{5/2}$$

The discharge through the triangular portion of the weir (Art. 86) is

$$Q = C''\tfrac{8}{15}\sqrt{2g}\ \tan\frac{\theta}{2}H^{5/2}$$

Equating the right-hand members of these equations, assuming $C' = C''$, and reducing,

$$\tan\frac{\theta}{2} = \frac{1}{4}$$

[1] C. Cipolletti, *Canal Villoresi*, 1887, a description of a trapezoidal weir.

The formula given by Cipolletti for determining discharge over a sharp-crested weir of this type is

$$Q = 3.367LH^{3/2} \tag{54}$$

Later experiments indicate that this formula gives too great discharges for the higher heads when the velocity of approach is low. Complete contractions are assumed for the use of the Cipolletti weir formula. The design requirements for obtaining complete contractions are approximately the same as for rectangular weirs.

The Cipolletti weir is used quite extensively in western United States for measuring irrigation water.

PROBLEMS

1. A contracted rectangular sharp-crested weir 6.0 ft long discharges water under a head of 1.55 ft. Compute the discharge by: (a) Francis correction applied to Francis formula; (b) Cone formula.

2. A contracted rectangular weir 10.0 ft long discharges water under a head of $9\frac{3}{16}$ in. Compute the discharge by: (a) Francis correction applied to Rehbock formula; (b) Cone formula.

3. In a river 60 ft wide, having an average depth of 3.4 ft and a mean velocity of 1.15 ft per sec, a contracted rectangular weir 30 ft long is to be constructed. Determine the head over the weir, assuming free overfall.

4. A contracted rectangular weir 10 ft long is built in the center of a rectangular channel 20 ft wide. How high is the weir if the depth of water upstream is 3.50 ft when the discharge is 40 cfs?

5. Determine the discharge of water over a 60° triangular weir if the measured head is (a) 0.623 ft; (b) 1.15 ft.

6. The discharge of water over a 45° triangular weir is 0.728 cfs. What is the head?

7. Compute the discharge of a heavy fuel oil at 80° F over a 90° triangular weir at a head of 0.542 ft.

8. The discharge of a light dust-proofing oil at 50° F over a 90° V-notch weir is 0.245 cfs. Compute the head.

9. The head of water on a 90° triangular weir is 0.725 ft. Compute the discharge by: (a) approximate formula; (b) Barnes formula; (c) Lenz formula.

10. Compute the discharge of water over a Cipolletti weir 5.0 ft long under a head of 0.85 ft. Compare with the discharge over a standard rectangular weir with the same length and head.

11. A stream 80 ft wide carries 65 cfs. Determine the resulting head over a Cipolletti weir 3 ft high and 12 ft long, assuming free overfall.

12. What length of Cipolletti weir should be constructed so that the measured head will not exceed 1.50 ft when the discharge is 120 cfs?

88. Weirs Not Sharp-crested. Weirs in which the water touches the surface of the crest rather than merely a line can have an infinite variety of form of cross section. Weirs of this type in which the crest is a flat surface are ordinarily called broad-crested weirs. Overflow dams are also in the class of weirs not sharp-crested.

The discharge over weirs not sharp-crested may be computed by any of the base formulas of Art. 80, provided the correct value of the coefficient C is known. These formulas are repeated here for convenience.

$$Q = CL[(H + h_v)^{3/2} - h_v^{3/2}] \qquad (28)$$

or its equivalent

$$Q = CLH^{3/2}\left[1 + 0.0233C^2\left(\frac{H}{d}\right)^2\right] \qquad (33)$$

$$Q = CL(H + h_v)^{3/2} \qquad (30)$$

$$Q = CLH^{3/2} \qquad (31)$$

Since experimental values of C may be based on any one of these formulas, it is important to use the one for which a given value of C was determined, especially if velocity of approach is appreciable.

Figures 85 and 86 show examples of flat-crested weirs with square upstream corners. If the crest breadth b in Fig. 85 is less than about 2/3 H, the nappe will ordinarily spring clear and the weir is in effect sharp-crested. Greater crest breadth, as in Fig. 86a, changes the form of the nappe, and sharp-crested weir coefficients no longer apply. As the breadth increases further (Fig. 86b), the weir becomes in effect a short flume.

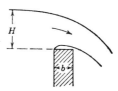

FIG. 85. Weir with rectangular cross section, nappe springing clear.

For flat-crested weirs 1 to 10 ft broad with square upstream corner and for values of b/H from 2 to 5, King[1] gives a value of about 2.7 for C in equation 31. As b/H decreases below 2, C in-

[1] H. W. King, *Handbook of Hydraulics*, McGraw-Hill Book Co., 1939, p. 164.

creases, reaching a value of about 3.3 when the nappe clears the crest.

The separation of nappe from crest which occurs at a sharp-cornered entrance can be avoided by sufficient rounding of the entrance, as shown in Fig. 87. Rounding reduces the amount of contraction and increases the coefficient of discharge.

The theoretical discharge of a broad-crested weir has been derived on the assumption that flow over the weir occurs at critical

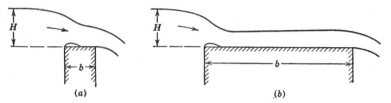

FIG. 86.   Profiles of flow over flat-crested weir with square upstream corner.

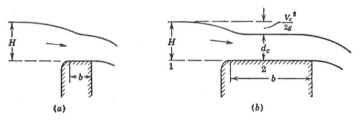

FIG. 87.   Profiles of flow over flat-crested weir with rounded upstream corner.

depth.   Under this condition, as will be shown in Chapter VIII, the velocity head is equal to half the depth, or one-third of the total head.   In Fig. 87b, assuming no loss of head between points 1 and 2,

$$\frac{V_c^2}{2g} = \frac{H}{3}$$

The discharge is

$$Q = AV = Ld_cV_c = L \cdot \tfrac{2}{3}H \cdot \sqrt{2g \cdot \frac{H}{3}}$$

from which

$$Q = 3.09LH^{3/2} \tag{55}$$

Actually, loss of head reduces the coefficient below the theoretical value of 3.09.

Profiles of flow at various depths over a broad-crested weir with
a level crest and rounded entrance are shown in Fig. 88.   At the
lower heads a succession of smooth waves appears on the crest.
The position of the waves at a given head remains fixed, and the
number of waves depends on the ratio of head H to breadth b of
the weir.   As the head increases, the spacing of the waves increases
and the waves disappear one by one over the downstream fall until
a head is reached at which the stream surface over the weir forms
a smooth double-reverse curve without waves.   The waves are
accounted for by the fact that the depth of flow over a weir of this
type is very close to critical depth, a condition which always
causes a disturbed stream surface because a slight change in energy
can result in a considerable change in depth.

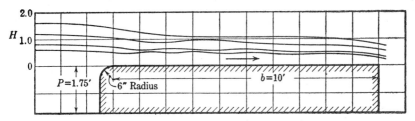

FIG. 88.   Profiles of flow over flat-crested weir.

Tests[1] on the weir shown in Fig. 88 in a flume 2 ft wide showed
a value of C in equation 30 increasing gradually from 2.78 at a
head of 0.5 ft to a constant value of 2.85 at heads of 1.2 ft and
higher.   The value of C increases as the ratio b/H decreases, that
is, as the weir becomes narrower in proportion to the head; giving
the weir surface a slight downstream slope from the entrance also
increases the value of C.   Such an increase in slope may so increase
the velocity of flow across the weir that the depth becomes less
than the critical and in that event no waves appear on the crest.

Overflow masonry dams should be so designed that the crest of
the dam follows approximately the lower curve of the nappe of a
sharp-crested weir discharging at some given head called the design
head, which is usually that of the maximum flood to be expected.
The theory is that at this discharge the under surface of the nappe
will follow the surface of the dam without exerting pressure on it,
and, what is more important, without tending to pull away from

[1] J. G. Woodburn, " Tests on Broad-Crested Weirs," with appendix by
A. R. Webb. *Trans. Am. Soc. Civil Engrs.*, 1932.

it and thus to create a region of partial vacuum which would increase the overturning moment on the dam. Laboratory tests[1]

FIG. 89. Detail of crest of dam.

of model dams have substantiated this theory. The discharge coefficient C in equation 31 for an overflow dam of this type with negligible velocity of approach can be shown to be about 4.0, the head H being measured from the high point of the crest. With reference to Fig. 89, the height z to which the under surface of a weir nappe rises above a sharp crest has been found to be from $0.11H$ to $0.13H$. When $z = 0.12H$, $H = 0.88H_s$, and, assuming the Francis coefficient of 3.33 for a sharp-crested weir, the value of C for the dam must be $3.33/0.88^{3/2} = 4.0$.

If the head on the dam increases above the design head, the nappe tends to spring free from the downstream face, thus decreasing the pressure under the nappe, drawing more water over the dam, and increasing the coefficient. This reduced pressure on the downstream face increases both the overturning and the sliding tendencies and reduces the safety of the structure. A reduction in head below the design value decreases the coefficient. It has been found[2] that as H varies from 0.5 to 1.5 times the design head, C varies from about 3.6 to 4.3.

It is often convenient to use an existing weir or overflow dam for measuring discharge. In such cases it is not likely that the shape of crest will conform exactly to one for which C is known. If there are no satisfactory experimental data and continuous discharge records are required it may be found desirable to obtain coefficients corresponding to different heads by means of current-meter measurements. Existing dams are sometimes used for estimating flood discharges of streams where direct measurements of discharge by other means are impracticable.

Horton[3] prepared tables and curves of C for use in equation 30, corresponding to different heads, for many shapes of weir sections for which experimental data were available. The table gives

[1] H. Rouse and L. Reid, Model Research on Spillway Crests, *Civil Engineering*, January, 1935, p. 9.

[2] *Ibid.*; also J. Hinds, W. P. Creager, and J. D. Justin, *Engineering for Dams*, John Wiley & Sons, 1945, Chapter 11.

[3] Robert E. Horton, " Weir Experiments, Coefficients and Formulas," *U. S. Geol. Survey Water Supply and Irrigation Paper* 200, 1907.

HORTON'S VALUES OF WEIR COEFFICIENT, C (EQUATION 30)
for weir sections shown in Fig. 90

Cross section	Head in Feet, H									
	0.5	1.0	1.5	2.0	2.5	3.0	3.5	4.0	4.5	5.0
A	2.70	2.64	2.64	2.70	2.80	2.89				
B	3.29	3.29	3.32	3.36	3.40	3.43	3.48	3.53	3.62	3.72
C	3.27	3.38	3.46	3.51	3.55	3.58	3.61	3.67	3.74	3.83
D		3.26	3.28	3.32	3.38	3.47	3.53	3.59	3.63	3.66
E		3.72	3.82	3.85	3.82	3.76	3.68	3.68	3.73	3.82
F			3.58	3.56	3.57	3.58	3.60	3.62	3.65	3.68

Horton's values of C for a few models of weir crests shown in Fig. 90. Sections resembling these or other models may be used for overflow dams or for spillways from reservoirs or canals or for other similar structures.

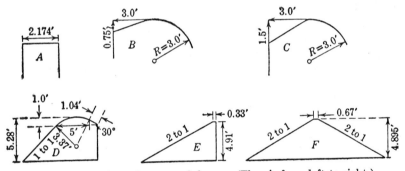

FIG. 90. Sections of weirs and dams. (Flow is from left to right.)

89. Submerged Weirs. If the elevation of the downstream water surface is higher than the crest of a weir, the weir is said to be submerged. The depth of submergence is the difference in elevation between the downstream surface and the crest.

Water flows over the crest at a velocity which is higher than the velocity of the water downstream, and a portion of this velocity is retained temporarily after leaving the weir. Where the slope of the channel is not sufficient to maintain this high velocity a piling-up effect is produced. This condition is illustrated in Fig. 91. The water has a higher velocity at a and a lower velocity at b than the normal velocity in the channel. This produces a standing wave, a being the trough and b the crest of the wave. Below the main wave a series of smaller waves form which gradually reduce in size and finally disappear.

Cox found[1] by an extensive series of tests of submerged sharp-crested weirs of various heights that the coefficient of discharge is a function of the submergence ratio $D/(H + h_v)$ but also depends on the behavior of the nappe. At small submergence ratios it is possible to have the nappe plunge below the surface and return to the surface at some distance downstream, from which point the surface velocity is directed both upstream toward the weir and downstream. A surface roller condition is thus created which is unfavorable for use of the weir as a measuring device.

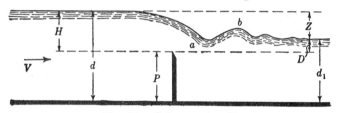

FIG. 91.   Submerged weir.

If the submergence ratio is increased sufficiently, however, the nappe no longer plunges below the surface but remains above. The surface velocity is then directed downstream at all points and the standing wave condition of Fig. 91 results. There is a range of submergence ratio in which either type of flow can take place.

With the nappe on the surface Cox found that the value of the weir coefficient in equation 30 is given by the equation

$$C = 4.3 \left[1 - \left(\frac{D}{(H + h_v)} + 0.002 \right) \right]^{0.25} - 0.82 \qquad (57)$$

A formula by King for the discharge over submerged sharp-crested weirs is

$$Q = 3.34 L Z^{1.47} \left[1 + 0.56 \left(\frac{H}{d} \right)^2 \right] \left(1 + 0.2 \sqrt{\frac{HD}{d_1 Z}} \right) \left(1 + 1.2 \frac{D}{Z} \right) \qquad (58)$$

[1] G. N. Cox, " The Submerged Weir as a Measuring Device," *Bulletin* **67**, Madison, University of Wisconsin Engineering Experiment Station.

GENERAL PROBLEMS

1. In Fig. A, the orifice in the side of the closed tank is 2 in. square, with $C = 0.60$ and $C_c = 0.62$. An open mercury manometer indicates the pressure in the air at the top of the tank. Compute the discharge when: (a) the liquid in the tank is water; (b) the upper 10 ft of liquid in the tank is oil (sp gr 0.82), and the remainder is water; (c) the liquid in the tank is oil (sp gr 0.82); (d) the liquid in the tank is molasses (sp gr 1.50); (e) the upper 8 ft of liquid in the tank is oil (sp gr 0.82) and the remainder is sea water ($w = 64$ lb per cu ft).

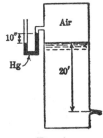

Fig. A

2. Compute the head lost in each case in problem 1.

3. A sharp-edged orifice 6 in. in diameter discharges oil under a head of 11.2 ft. The average velocity at the vena contracta is 26.0 ft per sec. How much head is lost in friction? Compute C_v.

4. An orifice in the side of a tank discharges water under a head of 9.0 ft. If the diameter of the orifice is 2 in., the coefficient of contraction is 0.630, and the head lost is 0.80 ft, determine the discharge. Compute C_v and C.

5. A circular orifice 1 in. in diameter, in the vertical side of a vessel, discharges water under a head of 7.60 ft. The jet strikes a horizontal plane 5.0 ft below the center of the orifice at a point 12.05 ft distant horizontally from the vena contracta. The weight of water discharged in 2 min 0 sec is 556 lb. Compute the three orifice coefficients.

6. Oil (sp gr 0.90) is discharging through a 4-in. circular sharp-edged orifice in the end of a 6-in. pipe. The pressure head on the center line of the pipe just upstream from the orifice is 30 ft of oil, and discharge takes place into a closed tank in which the vapor pressure is -3.0 lb per sq in. The discharge is not submerged. The orifice coefficients for this condition are: $C = 0.68$; $C_v = 0.97$. Compute the discharge.

7. A calibration test of a $\frac{3}{4}$-in. circular sharp-edged orifice in the end of a 2-in. pipe showed a discharge of 0.0275 cfs of water when the pressure head on the center line of the pipe just upstream from the orifice was 3.34 ft. The diameter of the jet was found to be 0.591 in. Compute the three orifice coefficients.

8. A calibration test of a $\frac{1}{2}$-in. circular sharp-edged orifice in the vertical side of a large tank showed a discharge of 132 lb of water in 81 sec at a head of 15.5 ft. Measurement of the jet showed that it traveled 7.70 ft horizontally while dropping 12 in. Compute the three orifice coefficients.

9. A nozzle having a tip diameter of 1 in. is attached to the end of a 2-in. hose. A calibration test showed a total discharge of 129 cu ft of water in 5 min. The average pressure in the hose at the base of the nozzle was 41.5 lb per sq in. If $C_c = 1.00$, compute C_v and the loss of head in the nozzle.

10. A sharp-edged orifice, 2 in. in diameter, in the vertical side of a large tank, discharges water under a head of 10 ft. If C_c is 0.62 and C_v is 0.98, how far horizontally from the vena contracta will the jet strike a horizontal plane which is 6 ft below the center of the orifice? Compute the discharge in gallons per minute.

11. A pump raises water from a well and discharges it into a pond through a horizontal 3-in. diameter pipe which is 4.5 ft higher than the pond level. The horizontal distance from the open end of the pipe to the point where the jet strikes the pond is 15.7 ft. Compute the discharge in gallons per minute.

12. A sharp-edged orifice 4 in. in diameter, in the side of a tank having a horizontal cross section 6 ft square, discharges water under a constant head. The rate of inflow by which the head is kept constant is suddenly changed from 0.80 cfs to 1.20 cfs. How long will it be, after this change occurs, until the head on the orifice becomes 7 ft? The coefficient of discharge may be considered constant and equal to 0.60.

13. A standard short tube 4 in. in diameter, in the side of a cylindrical tank 6 ft in diameter, and having its axis vertical, discharges under a constant head. The rate of inflow by which the head is kept constant is suddenly changed from 1.00 cfs to 1.35 cfs. How long will it be, after this change occurs, until the head becomes 5 ft?

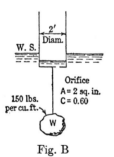

W. S.

2'
Diam.

Orifice
A = 2 sq. in.
C = 0.60

150 lbs.
per cu. ft.

W

Fig. B

14. In Fig. B, the cylinder weighs 175 lb and is empty at the time the orifice is opened. Neglecting the thickness of the cylinder walls, determine the value of W if the cylinder sinks 3 ft in 2 minutes.

15. In Fig. B, the cylinder weighs 175 lb and $W = 600$ lb. If the cylinder is empty at the time the orifice is opened, determine the time required for the cylinder to sink 2 ft, neglecting the thickness of the cylinder walls.

16. A sharp-edged sluice gate provides a rectangular opening 1 ft high in a channel 6 ft wide. Side and bottom contractions are entirely suppressed. When the water immediately upstream from the gate is 7.5 ft deep, and the downstream flow is free, compute the probable depth of flow just downstream from the gate and the probable discharge.

17. A cylindrical tank with its axis vertical has a diameter of 4 ft and a depth of 16 ft. A standard sharp-edged orifice 4 in. in diameter is located in the side 1 ft above the bottom. When at rest, water stands to a depth of 10 ft. Assuming that the volume of water in the tank is kept constant, determine the rate of discharge from the orifice: (a) when the tank has a constant velocity of 10 ft per sec in a direction opposite to that of the jet; (b) when it has a constant acceleration of 20 ft per sec per sec in a direction opposite to that of the jet.

18. A tank 20 ft long and 10 ft deep is 10 ft wide at the top and 20 ft wide at the bottom. In the bottom is an orifice having an area of 30 sq in. and a coefficient of discharge of 0.60. If the tank is full of oil (sp gr 0.90) at the beginning how long will it take to lower the oil surface 5 ft?

19. A tank 12 ft long has its ends vertical, top and bottom horizontal, and is 6 ft high. The top and the bottom are rectangular, having widths of 8 ft and 5 ft, respectively. A standard short tube 4 in. in diameter is located in one end 1 ft above the bottom. If at the beginning the tank is full of water, find the time necessary to lower the water surface 4 ft.

20. A rectangular channel 16 ft wide carries water at a depth of 2.5 ft and a mean velocity of 2.85 ft per sec. If a standard sharp-crested weir 3.0 ft high is built across this channel, what will be the depth of water upstream?

21. A trapezoidal canal, 20 ft wide on the bottom and having side slopes of 2 horizontal to 1 vertical, carries water at a depth of 1.2 ft and a mean velocity of 2.0 ft per sec. What length of contracted rectangular weir 2.75 ft high should be placed in the middle of the canal if the depth of water upstream is to be 4.0 ft?

22. A channel is carrying 10 cfs of water. Assuming that an error of 0.005 ft is made in measuring the head, determine the resulting percentage of error in discharge: (a) if a 90° triangular weir is used, and (b) if a Cipolletti weir 10 ft long is used.

23. If there is a measured head of water of 2.05 ft on a 90° triangular weir, what length of Cipolletti weir could be substituted so that the length of weir would be four times the head on it?

24. A contracted rectangular weir is to be constructed in a stream of water in which the discharge varies from 2 to 50 cfs. Determine a length of weir, such that the measured head will never be less than 0.2 ft or greater than one-third of the length of weir.

25. A contracted rectangular weir 18 ft long has a head of 1.82 ft of water over it. What length of dam similar to C in Fig. 90 would be required in the same stream in order to have the same head?

26. A submerged sharp-crested weir 2.5 ft high extends across a rectangular channel 10 ft wide. The depth of water in the channel of

approach is 4.0 ft, and 35 ft downstream from the weir the depth of water is 3.0 ft. Determine the discharge, assuming that the nappe remains on top.

27. A standard sharp-crested weir 2.5 ft high is built across a rectangular flume 30 ft wide. With water flowing, the measured head is 1.25 ft. Some distance upstream in the flume is another sharp-crested weir having a height of 3.5 ft, the middle of the weir being on the center line of the flume. If the measured head on the second weir is 1.62 ft, what is the length of crest? Assume free overfall.

28. The measured discharge over a dam 100 ft long is 520 cfs when the head is 1.28 ft. Determine the weir coefficient for this head.

29. An overflow masonry dam is to be constructed across a stream. The stream is estimated to have a maximum flood discharge of 30,000 cfs when the elevation of water surface at the dam site is 1132.0. Six sluice gates each 8 ft high and 6 ft wide ($C = 0.85$) are to be constructed in the dam with their sills at elevation 1122.5. The main overflow weir for which $C = 2.63$ will be 200 ft long with a crest elevation of 1184.0. An auxiliary weir 600 ft long with a crest elevation of 1185.3 will operate during floods. For this weir $C = 3.40$. With all sluice gates open, what will be the elevation of the water surface upstream from the weir when the discharge is 30,000 cfs? Neglect velocity of approach.

Chapter VII

PIPES

90. Description. A pipe may be defined as a closed conduit through which liquids or gases flow. In hydraulics, pipes are commonly understood to be conduits of circular cross section which flow full. Conduits flowing partially full are considered to be open channels. (See Chapter VIII.)

City water and gas mains in which flow occurs under pressure are examples of pipes. Sewers and drainage tile, which normally do not flow full, are classed as open channels. Since frictional losses in general are independent of pressure, the same laws apply to flow in both pipes and open channels, and the formulas for each take the same general form.

91. Critical Velocities in Pipes. The limiting conditions (see Art. 43) which determine whether flow will occur with laminar or turbulent motion were first investigated experimentally by Reynolds[1] by the method illustrated in Fig. 92. Water was drawn through a small glass tube a from a large tank b with glass sides. A cock c regulated the outflow. By means of the arrangement shown in the figure a fine stream of colored water

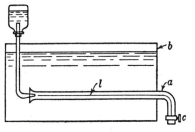

FIG. 92.   Reynolds' experiment.

was admitted into the entrance of the glass tube. The experiments were extended to include tubes of different diameters and water at various temperatures.

With the water in the tank very quiet, and when the velocity in tube a was low enough, the colored water appeared as a straight line extending through the tube, showing the flow to be laminar. As the rate of flow was gradually increased, a velocity was finally obtained at which the thread of color suddenly broke up and

[1] Osborne Reynolds, *Trans. Roy. Soc. (London)*, 1882 and 1895.

mixed with the surrounding water, showing the flow to be turbulent. If, then, the flow was gradually decreased, a velocity was finally obtained at which the flow changed back from turbulent to laminar and the thread of color formed as before.

The velocity at which the change from laminar to turbulent flow occurred was found to be higher than that which caused the change from turbulent to laminar flow. Reynolds called these velocities respectively the higher and the lower critical velocities.

As a result of extending his experiments to include the flow of water at different temperatures through tubes of different diameters, Reynolds established a criterion applicable to all fluids for determining the type of flow occurring under stated conditions. The numerical value of the expression

$$\frac{DV\rho}{\mu}$$

where D is the diameter of pipe, and V, ρ, and μ are respectively velocity, density, and viscosity of the fluid, is commonly called the Reynolds number and designated by N_R. In this volume these symbols are expressed in foot-pound-second units, but since the Reynolds number is dimensionless its value is independent of the system of units employed.

Since $\mu/\rho = \nu$, the kinematic viscosity, the value of Reynolds' number can also be written

$$N_R = \frac{DV}{\nu} \tag{1}$$

It has been found and verified by many careful experiments that for commerical pipes of circular cross section when Reynolds' number is less than about 2100 the flow will be laminar and when greater than about 3000 the flow will in practically all instances be turbulent. By exercising extreme care, laminar flow has been produced in laboratories when the Reynolds' number was far greater than 3000, but it is unlikely that such a condition will be encountered in practice.

92. Analysis of Velocities. In laminar flow, although the fluid particles while moving forward occupy successively the same relative transverse positions, the fluid near the axis advances a given distance in a shorter period than that nearer the conduit walls. (See Art. 43). In turbulent flow, however, notwithstanding the irregular paths traversed by the fluid particles, the average longitudinal speed of each particle is approximately the same. This can be shown by suddenly injecting a charge of colored liquid into a pipe in which water is flowing and observing the water at the outlet. The coloring matter will be found to remain in a comparatively short prism having a length equal to about one-tenth of the distance traveled. This principle is sometimes utilized in measuring the velocity of flow in pipes.

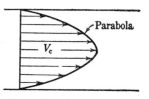

FIG. 93.   Laminar flow.

The flow of a fluid with laminar motion between two parallel plates of indefinite width is illustrated in Fig. 93. The velocity varies as the ordinates to a parabola, from zero at each plate to a maximum velocity at mid-distance, the average velocity being two-thirds of the maximum.

When laminar flow occurs in a conduit of circular cross-section, the movement of the mass of fluid can be compared to the telescoping of a large number of extremely thin concentric tubes. If the liquid wets the conduit wall, the outer tube adheres to the wall while the next one moves with extremely low velocity. The velocity of each successive tube increases gradually until the maximum velocity is reached at the center. In this case the velocity varies as the ordinates to a paraboloid of revolution, and the average velocity is one-half the maximum velocity.

In turbulent flow, there are transverse as well as longitudinal components of velocities of fluid particles, but it is only the latter that have any effect in producing motion in the fluid as a whole. In speaking of velocities, therefore, we always refer to the components in the direction of flow.

The distribution of velocity in the cross section of a circular pipe with turbulent flow has been found to vary with Reynolds' number. The velocity is again practically zero at the side walls but increases more rapidly for a short distance from the walls than in laminar flow. Throughout the central core, however, the mixing resulting from turbulence tends to equalize the velocities of the particles. Turbulence increases with Reynolds' number; hence the velocity distribution becomes more uniform as the Reynolds number increases.

Velocity distribution curves for a circular pipe are shown in

Fig. 94.[1] In laminar flow the velocities along any diameter vary as shown by curve A, which is a parabola, the maximum velocity V_c at the center of the pipe being twice the average velocity V.

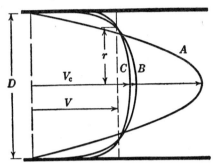

FIG. 94. Velocity distribution in straight pipe.

With turbulent flow the velocity distribution curves are much flatter, as indicated by curves B and C. Tests[2] have shown that the ratio of average to maximum velocity in a pipe of circular cross section varies with the Reynolds number approximately as shown in the following table:

$N_R = \dfrac{DV}{\nu}$	$\dfrac{V}{V_c}$
1700 and under	0.50
2000	0.55
3000	0.71
5000	0.76
10000	0.78
30000	0.80
100000 and over	0.81

Since the velocity of flow of water in pipe lines is almost always such that the value of Reynolds' number is larger than 10,000, it can be stated in general that the average velocity of water in a pipe is about 0.80 of the maximum velocity. The circle of mean velocity has a radius r (Fig. 94) of approximately $\frac{3}{8}D$.

With either laminar or turbulent flow, any irregularity or obstruction or any condition which causes a change in direction of flow will change the regular distribution of velocities. A bend in

[1] H. Rouse, " Modern Conceptions of the Mechanics of Fluid Turbulence," *Trans. Am. Soc. Civil Engrs.*, 1937.

[2] T. E. Stanton and J. R. Pannell, " Similarity of Motion in Relation to the Surface Friction of Fluids," *Trans. Roy. Soc. (London)*, A214, 1914.

a pipe, for example, causes the line of maximum velocity to move from the axis of the pipe toward the concave side. Figure 95 shows the actual distribution of velocities of water with turbulent flow in a curved pipe from measurements by Saph and Schoder.

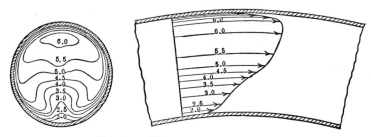

FIG. 95. Velocities in curved pipe.

93. Velocity Head in a Pipe. The velocity head at any cross section of a stream of fluid is $\alpha V^2/2g$, where V is the average velocity at the cross section and α (Art. 49) is a coefficient greater than unity which corrects for unequal distribution of velocities. Experiments by Bazin and others indicate that for water flowing with turbulent motion in a straight pipe α has a mean value of about 1.06. It can be shown that when laminar flow occurs in a pipe α has a value of 2.

In problems involving flow in pipes it is common to assume that the velocities at all points of a cross section are equal, or that α equals unity and that, therefore, the kinetic energy contained in 1 lb of fluid is equal to $V^2/2g$. The energy equation, when written between two points in a stream, then applies to the entire cross section in which the points lie. The error introduced by assuming α equal to unity is not usually of serious consequence.

94. Continuity of Flow in Pipes. In any pipe carrying liquid and flowing full, within the limits of error resulting from the assumptions that the liquid is incompressible and the pipe inelastic, at any given instant the same quantity of liquid is passing every cross section of the pipe. This statement implies continuity of flow (see Art. 47) and holds true even when the flow is unsteady, a condition which exists when the head producing discharge is variable. With gases, which are highly compressible, continuity of flow in a pipe can exist only when flow is steady.

95. Loss of Head. Loss of head in feet of fluid, meaning loss of energy expressed in foot-pounds per pound of fluid (Art. 51), occurs in any flow of fluid through a pipe. The loss is caused by: (1) " pipe friction " along the straight sections of pipe of uniform diameter and uniform roughness; and (2) changes in velocity or direction of flow. Losses of these two types are ordinarily referred to respectively as major losses and minor losses.

Major Loss. This is a continuous loss of head, h_f, assumed to occur at a uniform rate along the pipe as long as the size and quality of pipe remain constant, and is commonly referred to as the loss of head due to pipe friction.

Minor Losses. These consist of:

1. A loss of head, h_c, due to contraction of cross section. This loss is caused by a reduction in the cross-sectional area of the stream and the resulting increase in velocity. The contraction may be sudden or it may be tapered. The loss of head at the entrance to a pipe from a reservoir is a special case of loss due to contraction.

2. A loss of head, h_e, due to enlargement of cross section. This loss is caused by an increase in the cross-sectional area of the stream with resulting decrease in velocity. The enlargement may be either sudden or gradual. The loss of head at the outlet end of a pipe where it discharges into a reservoir is a special case of loss of head due to enlargement.

3. A loss of head, h_g, caused by obstructions such as gates or valves which produce a change in cross-sectional area in the pipe or in the direction of flow. The result is usually a sudden increase or decrease in velocity followed by a more gradual return to the original velocity.

4. A loss of head, h_b, caused by bends or curves in pipes, in addition to the loss which occurs in an equal length of straight pipe. Such bends may be of any total deflection angle as well as any radius of curvature. Occasionally, as in a reducing elbow, the loss due to the bend is superimposed on a loss due to change in velocity.

If the symbol H_L is used to designate all losses of head in a pipe line in which there is steady, continuous flow

$$H_L = h_f + h_c + h_e + h_g + h_b \qquad (2)$$

96. Loss of Head Due to Pipe Friction. The following discussion applies to all liquids and approximately to gases when the pressure drop is not more than 10 per cent of the initial pressure.

Changes in density of gases which result from larger drops in pressure introduce factors which will not be considered. The discussion, so far as it applies to fluids in general, is therefore subject to this limitation.

Consider a straight pipe of internal diameter D in which fluid is flowing at a mean velocity V. Let the loss of head in length L be denoted by h_f.

Certain general laws based upon observation and experiment appear to govern fluid friction in pipes and are expressed in all the generally accepted pipe formulas. These laws briefly stated are:

1. Frictional loss in turbulent flow generally increases with the roughness of the pipe. As will be shown later (Art. 98), when the flow is laminar the frictional loss is independent of the roughness.

2. Frictional loss is directly proportional to the area of the wetted surface, or to πDL.

3. Frictional loss varies inversely as some power of the pipe diameter, or as $1/D^x$.

4. Frictional loss varies as some power of the velocity, or as V^n.

5. Frictional loss varies as some power of the ratio of viscosity to density of the fluid, or as $(\mu/\rho)^r$.

Combining these factors, a rational equation for loss of head due to pipe friction for any fluid can be written in the form

$$h_f = K' \times \pi DL \times \frac{1}{D^x} \times V^n \times \left(\frac{\mu}{\rho}\right)^r \tag{3}$$

in which K' is a combined roughness coefficient and proportionality factor.

If $m + 1$ is substituted for x, equation 3 can be written in the form

$$h_f = \left[K'\pi\left(\frac{\mu}{\rho}\right)^r\right] \times \frac{L}{D^m} \times V^n \tag{4}$$

The historic development of hydraulic-flow formulas was related almost entirely to water at natural temperatures. The effect of viscosity and density of water on loss of head at usual flow velocities is so small that it was long neglected. What little effect there was could be easily included in a general coefficient.

K being substituted for the quantity in brackets in equation 4, the base formula for loss of head in pipe flow was thus stated as:

$$h_f = K \frac{L}{D^m} V^n \tag{5}$$

A determination of K, m, and n is necessary for practical application of equation 5 to flow problems. Chezy (1775) pointed out that the loss of head in the flow of water in conduits varied approximately as the square of the velocity. About the middle of the nineteenth century, Darcy, Weisbach, and others, accepting Chezy's value of 2 for n, further modified equation 5 by proposing a value of 1 for m, and divided and multiplied by $2g$, so that

$$h_f = (K'' \times 2g) \times \frac{L}{D} \times \frac{V^2}{2g} \tag{6}$$

By substituting a so-called " friction factor " f for $K'' \times 2g$, the well-known pipe formula, called the Darcy-Weisbach formula, was obtained:

$$h_f = f \frac{L}{D} \frac{V^2}{2g} \tag{7}$$

This formula is of convenient form since it expresses the loss of head in terms of the velocity head in the pipe. Moreover, it is dimensionally correct since f is a numerical factor, L/D is a ratio of lengths, and h_f and $V^2/2g$ are both expressed in units of length.

The defects of the Darcy-Weisbach formula are:

1. The loss of head with turbulent flow varies not as the square of the velocity but as some power varying from 1.7 to 2 or more. This discrepancy must be taken care of by varying the value of f. It will be shown in Art. 98 that with laminar flow the loss of head varies as the first power of the velocity.

2. Since $V = Q/A = Q/\left(\frac{\pi}{4} D^2\right)$, for a given Q, f and L, the loss of head by the Darcy-Weisbach formula varies inversely as the fifth power of the diameter. Tests have shown, however, that the actual variation is closer to the 5.25 power and that the exponent of D in the formula should therefore be in the neighborhood of 1.25. Again the discrepancy is taken care of by varying the value of f. It will be shown in Art. 98 that for a given Q with laminar flow the loss of head varies inversely as the fourth power of the diameter.

3. The friction factor f must therefore be a function of velocity and diameter as well as of the pipe roughness and of the viscosity and density of the fluid. Much research has been directed, without complete success, toward the discovery of a comprehensive formula for f. Reliance must therefore be placed on tables and diagrams, which are usually limited in scope to the fluids and test conditions on which they were based. With modern research, however, advance is continually being made toward a complete mathematical evaluation of friction loss. Meanwhile empirical methods of determination of f have long been used successfully by engineers.

97. Values of f for Water. Since pipes are most frequently designed to carry water, that liquid will be considered first. The table on page 184 shows average values of f as given by Fanning[1] for the turbulent flow of water at natural " cold-water " temperatures in straight smooth pipes. This description probably represents the conditions of new cast-iron pipe, welded-steel pipe, wood pipe made of planed staves, concrete pressure pipe of best quality, and cement-lined steel pipe. Brass and copper pipe, glass tubing, and asbestos-cement pipe may be expected to have slightly lower values of f. The table also includes values of f for fire hose as computed from test data by Underwriters' Laboratories, Inc.[2]

For any given velocity of flow, the value of f is seen to decrease as the diameter of the pipe increases. This decrease in f is largely accounted for by the decrease in " relative roughness " of the material in the pipe wall. By relative roughness is meant the ratio of the magnitude of surface irregularities to the pipe diameter. Of two pipes constructed of the same kind of material, a 6-in. pipe is thus relatively twice as rough as a 12-in. pipe. The relationship of f to relative roughness and to Reynolds' number is discussed in Art. 99.

Some kinds of pipe become rougher with age with resulting increase in f. This possibility is usually taken care of in design by increasing the value of f for new pipe by a certain percentage. The increase in f for cast-iron or steel pipe may be 50 to 100 per cent after some years of service, due to corrosion or tuberculation of the surface. On the other hand, wood pipe and asbestos-cement pipe have shown little or no increase in f after many years of service.

[1] H. W. King, *Handbook of Hydraulics*, McGraw-Hill Book Co., 1939, p. 205.
[2] *Bulletin of Research* 12, 1939.

VALUES OF f IN THE DARCY–WEISBACH FORMULA, $h_f = f \dfrac{L}{D} \dfrac{V^2}{2g}$

For water flowing in straight smooth pipe

Diameter of Pipe in Inches[1]	Mean Velocity (V) in Feet per Second								
	0.5	1.0	2.0	3.0	4.0	5.0	10.0	15.0	20.0
$\frac{1}{2}$	0.042	0.038	0.034	0.032	0.030	0.029	0.025	0.024	0.023
$\frac{3}{4}$	.041	.037	.033	.031	.029	.028	.025	.024	.023
1	.040	.035	.032	.030	.028	.027	.024	.023	.023
1$\frac{1}{2}$	.038	.034	.031	.029	.028	.027	.024	.023	.023
2	.036	.033	.030	.028	.027	.026	.024	.023	.022
3	.035	.032	.029	.027	.026	.025	.023	.022	.022
4	.034	.031	.028	.026	.026	.025	.023	.022	.021
5	.033	.030	.027	.026	.025	.024	.022	.022	.021
6	.032	.029	.026	.025	.024	.024	.022	.021	.021
8	.030	.028	.025	.024	.023	.023	.021	.021	.020
10	.028	.026	.024	.023	.022	.022	.021	.020	.020
12	.027	.025	.023	.022	.022	.021	.020	.020	.019
14	.026	.024	.022	.022	.021	.021	.020	.019	.019
16	.024	.023	.022	.021	.020	.020	.019	.019	.018
18	.024	.022	.021	.020	.020	.020	.019	.018	.018
20	.023	.022	.020	.020	.019	.019	.018	.018	.018
24	.021	.020	.019	.019	.018	.018	.018	.017	.017
30	.019	.019	.018	.018	.017	.017	.017	.016	.016
36	.018	.017	.017	.016	.016	.016	.016	.015	.015
42	.016	.016	.016	.015	.015	.015	.015	.015	.014
48	.015	.015	.015	.015	.014	.014	.014	.014	.014
54	.014	.014	.014	.014	.014	.014	.013	.013	.013
60	.014	.013	.013	.013	.013	.013	.013	.013	.012
72	.013	.012	.012	.012	.012	.012	.012	.012	.012
84	.012	.012	.011	.011	.011	.011	.011	.011	.011

For water flowing in cotton rubber-lined fire hose

Nominal Diameter	Velocity in feet per second					
	4	6	10	15	20	25
1$\frac{1}{2}$ in.	0.024	0.023	0.023	0.022	0.021	. . . . .
2$\frac{1}{2}$ in.	. . . . .	0.020	0.019	0.018	0.018	0.018

[1] Values given are nominal diameters. The actual diameters of commercial pipe are slightly different.

Methods of cleaning iron and steel pipe have been developed which can restore the pipe practically to its original smoothness.

Even under the descriptions of pipe shown in the table, slight differences in manufacture or in laying the pipe may change the **values of f** by several per cent. Interpolation closer than the second significant figure is therefore seldom warranted. More-over, an answer based on these values of f can seldom be considered correct to more than two significant figures.

A few examples are given to show the use of the table on page 184 in connection with the Darcy-Weisbach pipe formula.

EXAMPLE 1. Determine the loss of head in 200 ft of 6-in. new cast-iron pipe carrying 250 gpm of water.

Solution. Since 1 cu ft = 7.48 gal, 1 cfs = 449 gpm; hence $Q = 0.557$ cfs, $V = Q/A = 2.83$ ft per sec, $V^2/2g = 0.124$ ft. From page 184, $f = 0.025$. Thus

$$h_f = 0.025 \times \frac{200}{\frac{1}{2}} \times 0.124 = 1.2 \text{ ft of water}$$

EXAMPLE 2. Determine the capacity of a 30-in. wood-stave pipe carrying water with a loss of head of 10 ft per mile.

Solution. Since f depends on the unknown velocity of flow as well as on the known diameter, an assumed value of f is used in a trial solution. Tabular values for 30-in. pipe vary only from 0.019 to 0.016. Using an intermediate value of 0.018,

$$10 = 0.018 \times \frac{5280}{2.5} \times \frac{V^2}{2g}$$

from which $V^2/2g = 0.263$ and $V = 4.1$. For this trial velocity, $f = 0.017$. Correcting the solution, $V^2/2g = 0.279$, $V = 4.23$, and $Q = 21$ cfs.

EXAMPLE 3. What size of best-quality concrete pipe will carry 10 cfs of water with a loss of head of 2.0 ft per 1000 ft?

Solution. Again a trial solution can be made by assuming a value of f although with neither the velocity nor diameter known the error may be greater than in Example 2. Assuming $f = 0.020$, and since $V = \dfrac{Q}{(\pi/4)D^2}$,

$$2.0 = 0.020 \times \frac{1000}{D} \times \frac{100}{0.785^2 \times D^4 \times 2g}$$

Thus $D^5 = 25.2$, $D = 1.91$ ft (= 23 in.), $A = 2.86$ sq ft, $V = 3.5$ ft per sec. Revising f to 0.019, $D^5 = 23.9$ and $D = 1.89$ ft, indicating no change from the 23-in. size.

Checking h_f: $A = 2.88$ sq ft, $V = 3.47$ ft per sec, $V^2/2g = 0.187$ ft.

$$h_f = 0.019 \times \frac{1000}{23/12} \times 0.187 = 1.85 \text{ ft}$$

which is slightly less than the allowed 2 ft, and therefore satisfactory. The next larger commercial size of pipe (24-in.) would probably be selected.

PROBLEMS

1. A new cast-iron pipe 1200 ft long and 6 in. in diameter carries 1.5 cfs of water. Determine the frictional loss of head.

2. A 60-in. wood-stave pipe discharges 100 cfs of water. Determine the frictional loss of head per 1000 ft. of pipe.

3. A city-water-supply pipe line consists of new 24-in. cast-iron pipe. Compute the frictional loss of head per mile of pipe when the discharge if 8 mgd (millions of gallons per day).

4. Determine the discharge of water through a new 12-in. cast-iron pipe if the loss of head in a 3000-ft length is 30 ft.

5. Determine the discharge of water through a 36-in. wood-stave pipe if the loss of head is 15 ft in a length of 5000 ft.

6. Determine the discharge in gallons per minute of water through a 2-in. wrought-iron pipe if the frictional loss is 12 lb per sq in per 100 ft of pipe.

7. What diameter (to the nearest inch) of new cast-iron pipe 1 mile long is required to discharge 4.4 cfs of water with a loss of head of not more than 55 ft?

8. If the frictional loss remains the same, what will be the capacity of the pipe of problem 7 after ten years of service if the friction factor f is doubled in that length of time?

9. What diameter of smooth concrete pipe 8000 ft long is required to discharge 40 cfs of water with a loss of head of 8 ft?

10. What diameter of wood-stave pipe should be installed to carry 50 cfs of water 5 miles with a loss of head of 5 ft?

11. A 48-in. wood-stave pipe is laid on a downgrade of 3 ft per mile. The pressure at A in the pipe is 5.5 lb per sq in. If the discharge is 45 cfs of water, determine the pressure at point B, the distance from A to B being 1 mile.

12. Points A and B are 3000 ft apart along a 10-in. new steel pipe. B is 220 ft higher than A. With a flow of 3.2 cfs of water from A to B, what pressure must be maintained at A if the pressure at B is to be 50 lb per sq in.?

13. Water is pumped through a vertical 2-in. new galvanized-iron pipe to an elevated tank on the roof of a building. The pressure on the discharge side of the pump is 200 lb per sq in. What pressure can be expected at a point in the pipe 250 ft above the pump when the flow is 150 gpm?

14. Points A and B are 3 miles apart along a 24-in. new cast-iron pipe carrying water. A is 30 ft higher than B. If the pressure at B is 20 lb per sq in. greater than at A, determine the direction and amount of flow.

98. Frictional Loss with Laminar Flow. Figure 96 represents a longitudinal section and a cross section of a straight pipe in which a fluid of uniform unit weight w is moving from left to right with steady laminar motion. The velocity distribution curve is pictured in the longitudinal section.

Consider a circular cylinder of fluid, $abcd$, of length L extending from section 1 where the unit pressure is $p_1 = wh_1$ to section 2

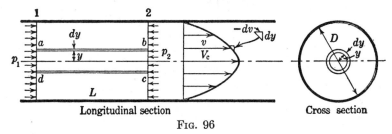

Longitudinal section   Cross section

FIG. 96

where the unit pressure has decreased to $p_2 = wh_2$. The difference in total pressure on the two ends of the cylinder is thus

$$w(h_1 - h_2)\,\pi y^2 = wh_f \pi y^2 \tag{8}$$

It is considered that the cylinder is in equilibrium between this pressure difference and the shear resistance exerted by the surrounding fluid on the curved surface of the cylinder.

From the definition of viscosity (Art. 4), the unit shearing stress on this surface is

$$\tau = \mu\left(\frac{-dv}{dy}\right)$$

since, as indicated in Fig. 96, for each increment dy in distance from the pipe axis, there is a decrement dv in velocity. The total shear stress on the surface of the cylinder is thus

$$-2\pi y L \mu \frac{dv}{dy} \tag{9}$$

Equating (8) and (9) leads to a simple differential equation

$$dv = -\frac{wh_f}{2L\mu}\,y\,dy \tag{10}$$

Integrating,

$$v = -\frac{wh_f y^2}{4L\mu} + C_1 \tag{11}$$

The constant of integration C_1 is evaluated from the conditions at the pipe wall where it is assumed that the velocity is zero. Thus, when $y = D/2$, $v = 0$, and

$$C_1 = \frac{wh_f D^2}{16L\mu}$$

Substituting this value in (11),

$$v = \frac{wh_f}{4L\mu}\left(\frac{D^2}{4} - y^2\right) \tag{12}$$

This equation gives the velocity v at any distance y from the pipe axis. The discharge through the ring of width dy is

$$dQ = v \times 2\pi y\, dy$$

Substituting the value of v from equation 12,

$$dQ = 2\pi \frac{wh_f}{4L\mu}\left(\frac{D^2}{4}y - y^3\right)dy \tag{13}$$

Integrating, the limits of y being 0 and $D/2$,

$$Q = \frac{\pi wh_f D^4}{128L\mu} \tag{14}$$

or the loss of head, substituting ρg for w, and ν for μ/ρ,

$$h_f = \frac{128L\nu Q}{\pi D^4 g} \tag{15}$$

Since $Q = (\pi/4)D^2 V$,

$$h_f = \frac{32L\nu V}{gD^2} \tag{16}$$

which is a mathematical statement of what is known as the Hagen-Poiseuille law for loss of head with laminar flow.

Equation 16 can be put into the Darcy-Weisbach form by multiplying numerator and denominator by $2V$ and replacing DV/ν, the Reynolds number, by N_R. Thus

$$h_f = \frac{64}{N_R}\frac{L}{D}\frac{V^2}{2g} \tag{17}$$

from which it is evident that, for laminar flow,

$$f = \frac{64}{N_R} \qquad (18)$$

The relation between the maximum velocity V_c and the average velocity V can now be shown. Equation 12 shows that the velocity distribution curve along any diameter is a parabola, the maximum velocity being at the center of the pipe ($y = 0$) and having the value

$$V_c = \frac{h_f g D^2}{16 L \nu} \qquad (19)$$

Transposing equation 16,

$$V = \frac{h_f g D^2}{32 L \nu} \qquad (20)$$

From the last two equations,

$$V_c = 2V \qquad (21)$$

99. General Method of Determining f in Darcy-Weisbach Formula. The values of f on page 184 are intended to apply only to the turbulent flow of water at temperatures less than about 75° F in pipes of a certain smoothness. For rougher or smoother pipes different values of f must be used. Moreover, modern engineering practice frequently requires the determination of the loss of energy in pipe lines carrying fluids other than water. This article describes methods of determining f for any kind of liquid flowing in pipes of various degrees of roughness.

If the Reynolds number for a fluid flowing in a circular pipe is less than about 2100, the flow is almost certain to be laminar. In this case the Hagen-Poiseuille law applies, as developed in the preceding article, and

$$f = \frac{64}{N_R} \qquad (18)$$

The loss of head with laminar flow is seen to be independent of the degree of roughness of the conduit surface.

When the Reynolds number is greater than about 3000, the flow is practically always turbulent and the value of f then may be dependent not only on the Reynolds number but also on the relative roughness of the pipe. Relative roughness is defined as the ratio of the average height k of the protuberances on the pipe surface (Fig. 97) to the diameter D of the pipe.

It has been determined that even with turbulent flow there exists next to the wall of the pipe a very thin layer of fluid in which the flow is laminar. The thickness of this boundary layer decreases with increase in the Reynolds number.

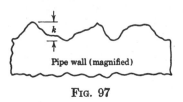

Pipe wall (magnified)

FIG. 97

A pipe is spoken of as "hydraulically smooth" if the height of the protuberances on the pipe wall is less than the thickness of the boundary layer. In such a pipe, variations in relative roughness do not affect the value of f.

If, on the other hand, the protuberances are greater in height than the thickness of the boundary layer, their presence affects the amount of turbulence and hence the value of f. The pipe has then ceased to be smooth. As the height of the protuberances increases, or as the thickness of the boundary layer decreases with increasing Reynolds' number, the turbulence increases to a maximum value at which it is said to be "fully developed."

Studies by Prandtl and von Karman led to the following equations for determining f in the Darcy-Weisbach formula for the two extreme conditions of flow in pipes:

For smooth pipes:

$$\frac{1}{\sqrt{f}} = 2 \log \left(\frac{N_R \sqrt{f}}{2.51} \right) \tag{22}$$

For pipes in which turbulence is fully developed:

$$\frac{1}{\sqrt{f}} = 2 \log \left(3.7 \frac{D}{k} \right) \tag{23}$$

These equations, which have been substantiated by experiment, show that for turbulent flow in pipes which are hydraulically smooth, f is independent of the relative roughness and is a function only of the Reynolds number, and that when turbulence is fully developed, f is independent of the Reynolds number and depends only on the relative roughness.

Between the two limiting conditions of flow is a transition region for which Colebrook and White[1] developed the following

[1] Journal, Institution of Civil Engineers, Vol. 11, 1939.

equation for f for use with commercial pipes:

$$\frac{1}{\sqrt{f}} = -2 \log \left(\frac{k}{3.7D} + \frac{2.51}{N_R \sqrt{f}} \right) \tag{24}$$

This transition equation merges at one end into the smooth-pipe law, and at the other into the law for fully developed turbulence. It is seen that for perfectly smooth pipes, when $k/D = 0$, equation 24 becomes equation 22, whereas at the other limit of the transition region, when N_R becomes large, equation 24 becomes equation 23.

Recommended values of k for common pipe materials are given in the table. These values, however, require further substantiation.[1]

KIND OF PIPE (new)	k (in feet)
Wrought iron and steel	0.00015
Asphalted cast iron	.0004
Galvanized iron	.0005
Cast iron	.00085
Wood stave	0.0006 to 0.0030
Concrete	.001 to .010
Riveted steel	.003 to .030

The application of equations 22, 23, and 24 is facilitated by the use of Fig. 98, which shows the variation of the Darcy-Weisbach friction factor f with Reynolds' number N_R as given by the Colebrooke-White equation for the transition region for a number of values of D/k, the reciprocal of the relative roughness. The diagram also shows the curve for smooth pipe and the minimum value of f for each value of D/k, in accordance with the Prandtl-von Karman equations. The relationship of f to N_R for laminar flow plots as a straight line on logarithmic paper and is shown at the left. The dashed line showing the practical upper limit of the transition region was suggested by Rouse[2] and is defined by the equation

$$N_R = 400 \frac{D}{k} \log \left(3.7 \frac{D}{k} \right) \tag{25}$$

[1] For a more extended discussion of the subject of pipe friction formulas see Julian Hinds, "Comparison of Formulas for Pipe Flow," *Journal Am. Water Works Assoc.*, Nov., 1946.

[2] Proceedings of the Second Hydraulics Conference, University of Iowa, 1942.

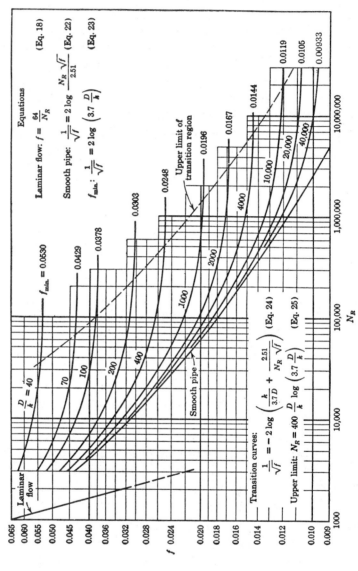

FIG. 98.   Variation of f with Reynolds' number and relative roughness, for flow of liquids in circular pipes.

When the value of N_R given by DV/ν is larger than that given by equation 25, the value of f_{min} determined by equation 23 can be used.

In the *critical region*, where Reynolds' number is approximately 2100 to 3000, the flow may be either laminar or turbulent and there is no method of predetermining which type will occur. In order to be on the side of safety it is usually considered advisable to assume that the type of flow giving the larger value of f is the one that prevails. This means that the flow should be assumed to be turbulent whenever the Reynolds number is greater than 2100. It is probably safe, for design in the critical region, to use values of f corresponding to an N_R of 3000.

EXAMPLE 1. Compute the pressure drop in pounds per square inch per 1000 ft of pipe and the horsepower lost in pipe friction when 0.5 cfs of crude oil (sp gr 0.86, $\nu = 0.0001$ sq ft per sec) is pumped through level 4-in. new cast-iron pipe.

Solution. From page 191, $k = 0.00085$. Hence $D/k = 392$. $V = 5.73$ ft per sec. $N_R = 19{,}100$. Interpolating between D/k curves in Fig. 98, $f = 0.031$. Hence

$$h_f = 0.031 \times 3000 \times 0.51 = 47.4 \text{ ft of oil} = 17.7 \text{ lb per sq in.}$$

$$hp = 0.5 \times 53.7 \times 47.4/550 = 2.3$$

EXAMPLE 2. What is the discharge capacity of 108-in. wood-stave pipe of best quality carrying water at 60° F with a loss of head of 8 ft per mile?

Solution. $D/k = 9/0.0006 = 15{,}000$. N_R is evidently large, hence from Fig. 98, estimated $f = 0.012$. $L/D = 587$. Hence,

$$8 = 0.012 \times 587 \times V^2/2g$$

$$V^2/2g = 1.135, \ V = 8.56, \text{ and } N_R = 6{,}300{,}000.$$

A closer value of f of 0.0115 could be used. Recomputing,

$$V^2/2g = 1.183, \ V = 8.72, \ A = 63.6, \text{ and } Q = 554 \text{ cfs.}$$

EXAMPLE 3. What size of steel pipe should be installed to carry 500,000 gpd of a heavy fuel oil at 75° F a distance of 1 mile with an allowable pipe friction loss of 50 lb per sq in.?

Solution. From page 10, sp gr = 0.906, $\nu = 0.00135$ sq ft per sec. $Q = 0.774$ cfs, $h_f = 127$ ft of oil.

To obtain an idea of whether the flow is turbulent or laminar, assume a normal economical velocity, such as 5 ft per sec. Then $A = Q/V =$

0.155 sq ft, $D = 0.445$ ft, and $N_R = 1650$, indicating laminar flow. From equation 15, $D^4 = 0.0550$, and $D = 0.484$ ft. Use 6-in. pipe.

Check: $A = 0.196$ sq ft, $V = 3.95$ ft per sec, $V^2/2g = 0.242$ ft, $N_R = 1460$, and $f = 64/N_R = 0.044$.

$$h_f = 0.044 \times 10,560 \times 0.242 = 112 \text{ ft of oil} = 44 \text{ lb per sq in.}$$

PROBLEMS

1. A 4-in. new steel pipe carries 250 gpm of a heavy fuel oil at a temperature of 80° F. Determine the pipe friction loss in pounds per square inch per mile of pipe.

2. With the same oil and the same friction loss as in problem 1, what would be the capacity of 8-in. new steel pipe, in gallons per minute?

3. A horizontal 6-in. new steel pipe carries 600,000 gpd of a medium fuel oil at a temperature of 40° F. Determine the pressure drop in pounds per square inch per 1000 ft of pipe.

4. If the temperature of the oil in problem 3 increased to 90° F, determine in gallons per day the discharge capacity of the 6-in. pipe with the same pressure drop.

5. Determine the discharge in gallons per minute of a new 16-in. cast-iron pipe if the friction loss is 5 lb per sq in. per 1000 ft and the liquid is: (a) water at 60° F; (b) a heavy fuel oil at 60° F.

6. A new 4-in. cast-iron pipe carries water at a velocity of 3 ft per sec. Determine the loss of head in feet per 1000 ft of pipe when the temperature of the water is (a) 40° F; (b) 100° F; (c) 200° F.

7. Compute the difference in horsepower lost in pipe friction in parts (a) and (c) of problem 6. What horsepower applied in the form of heat would be required to raise the temperature of the stream continuously from 40° to 200° F? (Note: 778 ft-lb of energy is the mechanical equivalent of 1 Btu, which is the average amount of heat required to raise 1 lb of water 1° F.)

8. What size of brass pipe would be required to carry standard gasoline at 80° F at a velocity of 10 ft per sec with a loss of head of 15 ft per 100 ft?

9. In pumping 1400 gpm of a heavy fuel oil through new steel pipe at a temperature of 90° F, a pipe friction loss of 30 lb per sq in. per 1000 ft is not to be exceeded. What diameter of pipe (to the nearest inch) should be used? With that diameter, what horsepower per 1000 ft will be required to overcome friction?

10. Determine the pumping horsepower required to overcome pipe friction in 50 miles of 20-in. new steel pipe through which 9,000,000 gpd of gasoline at 50° F are being pumped.

11. Points A and B are 3500 ft apart along a 6-in. new steel pipe. B is 60 ft higher than A. With a flow from A to B of 700 gpm of a medium fuel oil at 60° F, what pressure in pounds per square inch must be maintained at A if the pressure at B is to be 50 lb per sq in.?

12. If the roughness of the pipe in problem 11 increases 10 per cent of its original value each year, what per cent of increase in the necessary pressure at A can be expected after ten years of service?

13. What is the maximum velocity of flow in feet per second of: (a) water, (b) heavy fuel oil, (c) air, at a temperature of 68° F in a 12-in. pipe if the flow is to be laminar? Assume the air to be flowing at standard atmospheric pressure.

14. What diameter of glass tube 5 ft long will convey 1.0 gpm of castor oil at 50° F with a pressure-head drop of 10 in. of oil? ($\nu =$ 0.027 sq ft per sec.)

15. An oil with a specific gravity of 0.902 flows through a 4.0-ft length of $\frac{1}{4}$-in. glass tubing with a head loss of 6.5 in. of oil. The measured discharge is 0.405 lb in 5.0 min. Determine the viscosity of the oil in poises.

16. Water at 180° F is piped through 2-in. galvanized iron pipe. What will be the frictional loss in pounds per square inch per 100 ft of pipe when the flow is 150 gpm?

17. Points A and B are 4 miles apart along a 120-in. riveted-steel pipe of the best grade of smoothness. A is 350 ft higher than B. With water at 60° F flowing from A to B, the pressure head in the pipe at A is 20 ft, and at B is 340 ft. Compute the capacity of the pipe.

18. If the roughness of the pipe in problem 17 became the worst grade for riveted steel instead of the best, what per cent reduction in capacity could be expected for the same pressure-head conditions?

100. Wetted Perimeter and Hydraulic Radius. The wetted perimeter P of any conduit is the line of intersection of its wetted surface with a cross-sectional plane. Thus for a circular pipe flowing full, D being the diameter, the wetted perimeter is equal to the circumference, or πD; if flowing half full it is $\frac{1}{2}\pi D$.

The hydraulic radius R of a conduit is the area of cross section of the stream which it carries divided by the wetted perimeter of the section. For a circular conduit flowing either full or half full the hydraulic radius is $D/4$. All formulas for loss of head due to friction involve either the diameter or the hydraulic radius. It has usually been found more convenient to write D directly into pipe formulas than to use the hydraulic radius.

101. Hydraulic Gradient and Energy Gradient. Loss of head in straight pipe flow is illustrated graphically in Fig. 99, in which are shown two lines designated respectively the hydraulic gradient and the energy gradient. The former is defined by the locus of elevations to which liquid rises in successive piezometer tubes, and is thus a graphical representation, with respect to any selected datum, of the potential (pressure + elevation) head or energy which the liquid possesses at all sections of the pipe.

The energy gradient is above the hydraulic gradient a distance equal to the velocity head at each section, and is thus a graphical representation, with respect to the selected datum, of the total head or energy possessed by the liquid.

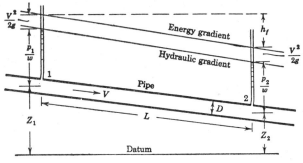

FIG. 99. Frictional loss in pipe.

102. Alternate Forms of Pipe Formulas. The base formula (equation 5) for loss of head in pipe flow

$$h_f = K \frac{L}{D^m} V^n \qquad (5)$$

gives the loss of head h_f which occurs in length L of the pipe. With reference to Fig. 99 it is seen that the ratio h_f/L is the slope of the energy gradient, denoted by S, which thus represents the loss of head in feet per foot of pipe. By substituting S for h_f/L, equation 5 may be transposed to the form

$$V = \left(\frac{1}{K}\right)^{1/n} D^{m/n} S^{1/n} \qquad (26)$$

or, substituting C' for $(1/K)^{1/n}$, y for m/n, and z for $1/n$,

$$V = C' D^y S^z \qquad (27)$$

This equation expresses the velocity of flow in terms of the diameter D of the pipe and slope S of the energy gradient. Introducing

the hydraulic radius in place of the diameter from the relation
$D = 4R$

$$V = (C' \times 4^y) R^y S^z$$

Or, substituting C'' for $(C' \times 4^y)$, the base formula for velocity of flow in pipes is obtained

$$V = C'' R^y S^z \qquad (28)$$

Formulas for flow in pipes are commonly written in any one of the three forms expressed by equations 5, 27, and 28. They are generally applicable to all fluids. The numerical values of coefficients and exponents must be determined from experimental data.

103. The Chezy Formula. The Darcy-Weisbach formula can be put in the form of equation 28 by substituting S for h_f/L and $4R$ for D. Thus, from equation 7,

$$V = \sqrt{\frac{2g}{f}} \times \sqrt{4R} \times \sqrt{S} = \sqrt{\frac{8g}{f}} \sqrt{RS}$$

Substituting a coefficient C for $\sqrt{8g/f}$,

$$V = C \sqrt{RS} \qquad (29)$$

This formula for velocity of flow in terms of the hydraulic radius of a conduit and slope of the energy gradient is called the Chezy (pronounced Shay-zee) formula. The Chezy coefficient C is a function of the same variables as the Darcy-Weisbach coefficient f, and the Chezy formula is therefore subject to the same defects as noted in Art. 96 for the Darcy-Weisbach formula. Tables of empirical values of C or f are necessary for its use.

104. Other Pipe Formulas. Many experiments are available on the flow of water in pipes and open channels which cover a wide range of conditions and which form the basis of a large number of empirical formulas. In each investigation of these experiments it has been the aim to secure a formula in which the coefficient would have a minimum range of variation and particularly one in which the coefficient would, so far as practicable, be a function of the degree of roughness of the conduit surfaces and not a function of R and S. It has not been found possible, however, to secure any formula that more than roughly approximates these conditions. The most satisfactory formulas have been of the form of equations 5, 27, or 28 with values of exponents which in the opinion of the investigators have appeared to correspond best with the available experiments.

The formulas given in the following paragraphs are intended for use only with water at temperatures less than about 75° F. For problems involving the flow of warmer water and of other fluids, the Darcy-Weisbach formula with coefficients obtained by the method described in Art. 99 should be employed. In the ordinary problems in hydraulic engineering which deal with natural waters the choice of formula is to a large extent a matter of personal preference. Considerations of comparative simplicity or of convenience or expediency may determine the choice.

The *Manning formula* is one of the best-known open-channel formulas (Art. 127), and it is quite commonly used for pipes. In the form of equation 28 the Manning formula is

$$V = \frac{1.486}{n} R^{2/3} S^{1/2} \tag{30}$$

in which n is a roughness coefficient. It is more conveniently applied to pipes in the form of equation 27:

$$V = \frac{0.59}{n} D^{2/3} S^{1/2} \tag{31}$$

obtained by substituting $D/4$ for R. Transposed to the form of equation 5, the Manning formula is

$$h_f = 2.87n^2 \frac{LV^2}{D^{4/3}} \tag{32}$$

The coefficient n increases with the degree of roughness of the conduit. The table on page 199 contains typical values of n recommended for water flowing in pipes.

The *Hazen-Williams formula* has been used extensively for designing water-supply systems in the United States. Written in the form of equation 28, the Hazen-Williams formula is

$$V = 1.318C_1 R^{0.63} S^{0.54} \tag{33}$$

This formula was designed for the flow of water in both pipes and open channels but is used more commonly for pipes. The following is written by the authors[1] of the formula:

[1] G. S. Williams and A. Hazen, *Hydraulic Tables*, 3d ed., John Wiley & Sons, 1933.

If exponents could be selected agreeing perfectly with the facts, the value of C_1 would depend upon the roughness only, and for any given degree of roughness C_1 would then be a constant. It is not possible to reach this actually, because the values of the exponents vary with different surfaces, and also their values may not be exactly the same for large diameters and for small ones, nor for steep slopes and for flat ones. Exponents can be selected, however, representing approximately average conditions, so that the value of C_1 for a given condition of surface will vary so little as to be practically constant. . . .

VALUES OF n TO BE USED WITH THE MANNING FORMULA

Kind of Pipe	Variation		Use in Designing	
	From	To	From	To
Brass and glass pipe..............	0.009	0.013	0.009	0.011
Asbestos-cement pipe..............	. . . .	. . . .	.010	.012
Wrought-iron and welded-steel pipe..	.010	.014	.011	.013
Wood-stave pipe..................	.010	.014	.011	.013
Clean cast-iron pipe..............	.010	.015	.011	.013
Concrete pipe....................	.010	.017	. . . .	. . . .
very smooth.....................	. . . .	. . . .	.011	.012
" wet mix," steel forms...........	. . . .	. . . .	.012	.014
" dry mix," rough forms..........	. . . .	. . . .	.015	.016
with rough joints................	. . . .	. . . .	.016	.017
Common-clay drainage tile.........	.011	.017	.012	.014
Vitrified sewer pipe..............	.010	.017	.013	.015
Riveted-steel pipe................	.013	.017	.015	.017
Dirty or tuberculated cast-iron pipe..	.015	.035	. . . .	. . . .
Corrugated-iron pipe..............	. . . .	. . . .	.020	.022

The exponents in the formula used were selected as representing as nearly as possible average conditions, as deduced from the best available records of experiments upon the flow of water in such pipes and channels as most frequently occur in waterworks practice.

The following table contains values of C_1 recommended by the authors of the formula for flow of water in pipes.

DESCRIPTION OF PIPE	VALUE OF C_1
Extremely smooth and straight	140
Very smooth	130
Smooth wooden or wood stave	120
New riveted steel	110
Vitrified	110

For estimating discharges of pipe lines where the carrying capacity after a series of years is the controlling factor, values of $C_1 =$ 100 for cast-iron pipe and $C_1 = 95$ for riveted steel are recommended. For the smaller sizes of pipes a somewhat lower value of C_1 should be used. For old iron pipes in bad condition, $C_1 = 80$ to 60; and for small pipes badly tuberculated, C_1 may be as low as 40. On the other hand it is claimed that asbestos-cement pipe has retained a value of C_1 of about 140 after many years of service.

Since the Manning and Hazen-Williams formulas have exponents differing but little in their respective values, the range in variation of their coefficients is practically the same. In both formulas the extent to which the coefficient varies with velocity and diameter is comparatively small, and in this respect they possess a marked advantage over the Darcy-Weisbach formula.

105. Pipe Diagrams. The solution of problems on flow of water in pipes is facilitated by the use of graphs called pipe diagrams, two types of which are shown in Figs. 100 and 101.

Figure 100 is a plotting on logarithmic paper of the variation of loss of head with discharge for a range of pipe diameters from 3 to 60 in. as given by the Manning formula with a value of n of 0.011. Similar diagrams could be drawn for other values of n.

The abscissas represent h_1, loss of head in feet per 1000 ft, and the ordinates represent discharge in cubic feet per second. The lines sloping up to the right represent pipe diameters in inches; those sloping down to the right represent velocities in feet per second. Thus, if any two of these quantities are known, the other two can be read directly from the diagram.

Figure 101 shows a pipe diagram of the straight-line type based on the Hazen-Williams formula with $C_1 = 120$. The four quantities — discharge, loss of head, diameter, and velocity — are represented by lines so spaced and graduated that the corresponding values of the four quantities lie in a straight transverse line. Thus if any two of the quantities are known, a straight edge joining these two points will intersect the other two lines at the values given by the formula.

EXAMPLE 1. What is the loss of head in 5000 ft of 12-in. pipe carrying 3 cfs?

Solution by Manning Diagram. Follow along the horizontal line repre-

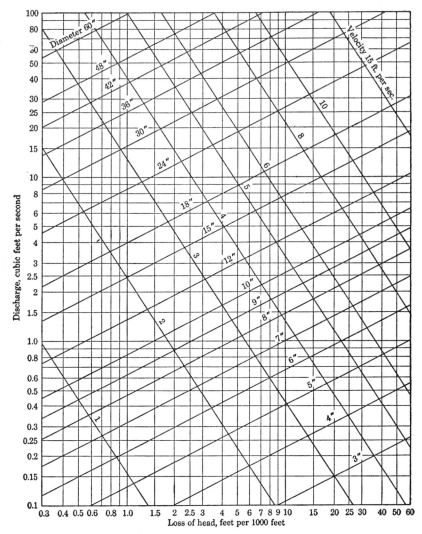

FIG. 100. Flow of water in pipes by Manning formula with $n = 0.011$.

senting 3 cfs until it intersects the sloping 12-in.-diameter line. Follow from this point vertically to the bottom of the diagram and read $h_1 = 4.9$ ft per 1000 ft. Multiply 4.9 by 5, giving 24.5 ft loss in 5000 ft.

Solution by Hazen-Williams Diagram. Lay a straight edge through $Q = 3$ and $D = 12$. Read $h_1 = 5.1$ ft per 1000 ft. Thus $h_f = 25.5$ ft.

EXAMPLE 2. What is the capacity of 30-in. new welded steel pipe if the allowable loss of head is 10 ft per mile?

Solution by Manning Diagram. The loss of head h_1 in feet per 1000 ft

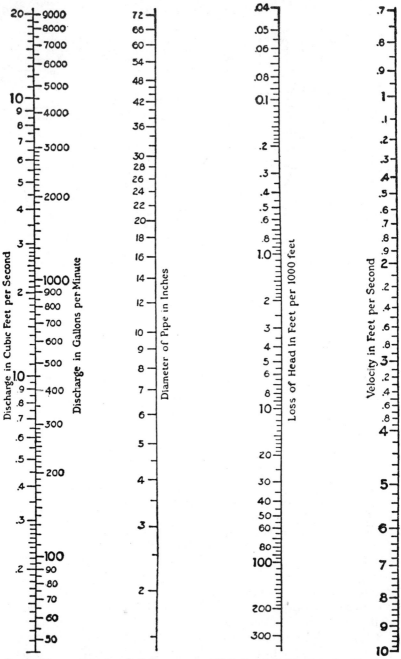

Reprinted by permission from F. E. Turneaure and H. L. Russell, Public Water Supplies, Fourth Edition, John Wiley & Sons, 1940.

FIG. 101. Flow of water in pipes by Hazen-Williams formula with $C_1 = 120$.

202

is 1.9. Locate at the bottom a loss of 1.9 ft per 1000 ft. Follow vertically to intersect the 30-in. diameter line. Then follow horizontally to left reading a value of 22 cfs.

By Hazen-Williams Diagram. $Q = 19$ cfs.

Note: The student should solve by means of pipe diagrams a number of problems in the list on page 186.

106. Minor Losses. In discussing loss of head due to pipe friction (Art. 97) it was shown that, other things being equal, loss of head varies as the velocity raised to some power which, for turbulent flow, usually is less than 2, but does not depart greatly from this value. In formulas of the Darcy-Weisbach type, therefore, where frictional loss is expressed as a function of the velocity head, the value of coefficient to be applied to the formula changes with the velocity, but the range of this variation is comparatively small.

In a similar manner it has been found that minor losses vary roughly as the square of the velocity, and they are commonly expressed by applying variable coefficients to the velocity head, or, using nomenclature given in Art. 95,

$$h_c = K_c \frac{V^2}{2g}, \quad h_e = K_e \frac{V^2}{2g}, \quad h_g = K_g \frac{V^2}{2g}, \quad h_b = K_b \frac{V^2}{2g}$$

where the subscripts c, e, g, and b denote contraction, enlargement, gate, or bend, and refer to the source of the change in amount or direction of velocity which gives rise to the loss of energy.

In general, the loss occurs in the eddying which is set up at the source of the loss and which is superimposed on the normal turbulence of flow. This eddying takes place mostly downstream trom its source and decreases gradually until, at a distance of 20 to 50 diameters, the normal turbulence of straight pipe flow is again established.

The loss therefore occurs not at but in the vicinity of the source, mostly downstream. In measuring minor losses it is necessary to trace the hydraulic and energy gradients for a short distance upstream from the source and for a considerable distance downstream. Otherwise, the entire loss may not be included.

107. Loss of Head Due to Contraction. The behavior of the energy and hydraulic gradients in the vicinity of a sudden contraction in a pipe is indicated in Fig. 102. The rate of loss of head in the larger pipe is indicated by the slope of the gradients between piezometers a and b. From b to f the location of the hydraulic gradient is different for different path lines. The gradient shown represents approximately the variation of pressure along the pipe axis. No piezometers are shown between b and f for the reason that the ordinary piezometer tube, which is set flush with the pipe wall, measures the pressure at the wall but does not necessarily measure the pressure at points in the same cross section at some distance from the wall. In smooth straight pipes the difference

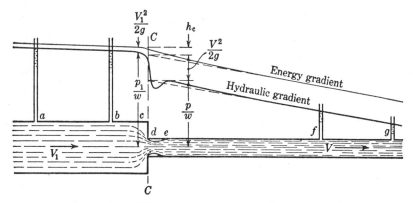

FIG. 102. Sudden contraction in pipe.

between pressures at the wall and interior points is probably not great, but it may be quite large near sections where changes in diameter occur.

Immediately following the contraction there is a region at d of reduced pressure similar to that described in the standard short tube (Art. 73). As the fluid expands to fill the smaller pipe at e the pressure rises slightly. From e to f the drop in hydraulic and energy gradients is caused by both pipe friction and the superimposed loss due to eddies set up by the contraction and subsequent enlargement. Beyond f where normal turbulence is restored the slope of hydraulic and energy gradients indicates the rate of loss due to pipe friction only.

If the energy gradient between a and b is extended downstream to the line $C-C$ while the energy gradient between f and g is extended upstream to $C-C$, the loss in head caused by the contraction is seen to be h_c. If the entire change in pressure, velocity, and total heads is pictured as occurring at $C-C$, the energy equation before and after contraction becomes

$$\frac{p_1}{w} + \frac{V_1{}^2}{2g} = \frac{p}{w} + \frac{V^2}{2g} + h_c \tag{34}$$

The loss of head due to contraction expressed as a function of the velocity head is

$$h_c = K_c \frac{V^2}{2g} \tag{35}$$

in which K_c is an empirical coefficient, and V is the velocity in the smaller pipe. The table gives experimental values of K_c for sudden contraction.

VALUES OF THE COEFFICIENT K_c FOR SUDDEN CONTRACTION

Velocity in Smaller Pipe, V	Ratio of Smaller to Larger Diameter									
	0.0	0.1	0.2	0.3	0.4	0.5	0.6	0.7	0.8	0.9
2	0.49	0.49	0.48	0.45	0.42	0.38	0.28	0.18	0.07	0.03
5	.48	.48	.47	.44	.41	.37	.28	.18	.09	.04
10	.47	.46	.45	.43	.40	.36	.28	.18	.10	.04
20	.44	.43	.42	.40	.37	.33	.27	.19	.11	.05
40	.38	.36	.35	.33	.31	.29	.25	.20	.13	.06

Problem. Determine the loss of head due to sudden contraction if a pipe carrying 2.0 cfs suddenly changes from a diameter of: (a) 8 in. to 6 in., (b) 12 in. to 6 in., (c) 18 in. to 6 in. Also determine the difference in pressure resulting from these changes.

If the change to a smaller diameter takes place gradually, as in a uniformly tapering section, or if the corners of the smaller pipe are rounded so as to reduce contractions, values of K_c will be much smaller than those given for a sudden reduction of diameter. If the change is made as gradually as in a Venturi meter or if a bell-mouth connection between the two pipes is used, K_c may become so small as to be practically negligible.

The loss of head at entrance to pipes is a special case of loss of head due to contraction. If the body of water is large the conditions conform approximately to a ratio of diameters of zero, and for a square-cornered entrance, where the end of the pipe is flush with a wall having a plane surface, the values of K_c are comparable with the values for 0.0 ratio in the preceding table.

Since the first two or three diameters of a pipe are similar to a short tube, entrance losses for pipes are usually considered to be the same as for short tubes. The general formula for loss of head at entrance to a pipe is, by equation 16, page 124,

$$h_c = \left(\frac{1}{C_v^2} - 1\right)\frac{V^2}{2g} = K_c \frac{V^2}{2g} \tag{36}$$

in which the coefficient of velocity, C_v, depends for its value upon the conditions at entrance, and $K_c = (1/C_v^2) - 1$. For convenience of reference, values of C_v and K_c, given in Chapter VI, are repeated in the following table.

VALUES OF THE COEFFICIENT K_c FOR DETERMINING LOSS
OF HEAD AT ENTRANCE TO PIPES

Entrance to Pipe	Reference	C_v	K_c
Inward projecting.............	Art. 77	0.75	0.8
Square cornered...............	Art. 73	.82	0.5
Slightly rounded..............	.....	.90	0.2
Bell mouth	Art. 68	.95	0.1

Since the effect of entrance conditions cannot be determined accurately the selection of a proper value of K_c is to some extent a matter of judgment. Unless the entrance is known to be other than square-cornered, a value of 0.5 is commonly used.

108. Loss of Head Due to Enlargement. The behavior of the energy and hydraulic gradients in the vicinity of a sudden enlargement in a pipe is indicated in Fig. 103. Again the slope of hydraulic and energy gradients between piezometers a and b represents the rate of loss of head in the approach pipe. The hydraulic gradient immediately preceding and following the enlargement follows approximately the line shown. As the stream expands to fill the larger pipe the velocity head decreases and there is an increase in the pressure head, the amount of the increase depending on the loss of head as well as on the decrease in velocity head.

In general, with an enlarging cross section and a reduction of velocity the eddying which is set up is much more extensive than with a contraction of cross section and an increase of velocity. It is therefore even more important with an enlargement to extend the study of the gradients a considerable distance downstream, probably closer to the larger value of the 20 to 50 diameters mentioned in Art. 106.

If normal turbulence is established by the time the stream reaches piezometer f, and the energy gradient from f to g is extended backward to the line E–E, the loss in head caused by the enlargement

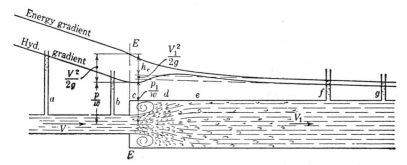

FIG. 103.   Sudden enlargement in pipe.

is h_e.   Again picturing the entire change as occurring at section E–E the energy equation before and after the enlargement becomes

$$\frac{p}{w} + \frac{V^2}{2g} = \frac{p_1}{w} + \frac{V_1^2}{2g} + h_e \tag{37}$$

The loss of head due to sudden enlargement, illustrated in the figure by the drop h_e in the energy gradient, expressed as a function of the velocity head, again in the smaller pipe, is

$$h_e = K_e \frac{V^2}{2g} \tag{38}$$

Archer[1] has shown from an investigation of his own experiments and the experiments of others that for water h_e is quite accurately represented by the equation

$$h_e = 1.10 \frac{(V - V_1)^{1.92}}{2g} \tag{39}$$

From equations 38 and 39,

$$K_e = \frac{1.10}{V^{0.081}} \left[1 - \left(\frac{D}{D_1}\right)^2 \right]^{1.92} \tag{40}$$

[1] W. H. Archer, "Loss of Head Due to Enlargements in Pipes," *Trans. Am. Soc. Civil Engrs.*, vol. 76, 1913.

D/D_1 being the ratio of the smaller to the larger diameter. The following values of K_e are computed from equation 40.

VALUES OF THE COEFFICIENT K_e FOR SUDDEN ENLARGEMENT

Velocity in Smaller pipe, V	Ratio of Smaller to Larger Diameter									
	0.0	0.1	0.2	0.3	0.4	0.5	0.6	0.7	0.8	0.9
2	1.0	1.0	0.96	0.86	0.74	0.60	0.44	0.29	0.15	0.04
5	0.96	0.95	.89	.80	.69	.55	.41	.27	.14	.04
10	.91	.89	.84	.76	.65	.52	.39	.26	.13	.04
20	.86	.84	.80	.72	.62	.50	.37	.24	.12	.04
40	.81	.80	.75	.68	.58	.47	.35	.22	.11	.03

Problem. Determine the loss of head due to the sudden enlargement if a pipe carrying 2.0 cfs suddenly changes from a diameter of: (a) 6 in. to 8 in., (b) 6 in. to 12 in., and (c) 6 in. to 18 in. Also determine the difference in pressure resulting from these changes.

The loss at submerged discharge from a pipe into a reservoir is a special case of loss of head due to enlargement in which the ratio of smaller to larger diameter is practically zero. Experiments at the University of Michigan indicate that Archer's formula holds quite accurately in the limit where V_1 is zero. Values of K_e for square-cornered exit may therefore be taken from the 0.0 ratio column of the preceding table. Since these values are nearly unity for the ordinary velocities encountered in pipes, it is commonly considered that the entire velocity head is lost.

VALUES OF THE COEFFICIENT K_e FOR GRADUAL ENLARGEMENT

Angle of Cone	Ratio of Smaller to Larger Diameter								
	0.1	0.2	0.3	0.4	0.5	0.6	0.7	0.8	0.9
5°	0.04	0.04	0.04	0.04	0.04	0.04	0.03	0.02	0.01
15°	.16	.16	.16	.16	.16	.15	.13	.10	.06
30°	.49	.49	.48	.48	.46	.43	.37	.27	.16
45°	.64	.63	.63	.62	.60	.55	.49	.38	.20
60°	.72	.72	.71	.70	.67	.62	.54	.43	.24

The loss of head due to enlargement may be reduced by changing the diameter gradually. For conical transitions the loss of head decreases with the rate of divergence, and it is practically negligible for very small angles. Experimental values of K_e have not been well determined for gradual enlargements, but those given in the table on page 208 are the approximate mean of such data as are available. There are not sufficient experiments to determine the variation with the velocity.

109. Loss of Head Due to Obstructions. Gates or valves when partially closed obstruct the flow and cause a loss of head in addition to the loss due to friction. If a pipe has the same diameter on both sides of the obstruction, the hydraulic gradient and the energy gradient drop the same amount, and each therefore indicates the loss of head. Following the form used for other losses, the loss of head in pipes due to gates, valves, or other obstructions is written

$$h_g = K_g \frac{V^2}{2g} \qquad (41)$$

V being the mean velocity in the pipe. Values of K_g as determined by Corp and Ruble for various heights of opening of gate valves with nominal diameters from $\frac{1}{2}$ to 12 in. are given in the table below. The value of K_g decreases with increase in ratio d/D of opening and in size of valve. Slightly open gate valves showed

LOSS OF HEAD DUE TO GATE VALVES

Values of K_g in $h_g = K_g \dfrac{V^2}{2g}$

Nominal Diameter of Valve, Inches	Ratio of Height d of Valve Opening to Diameter D of Full Valve Opening					
	$\frac{1}{8}$	$\frac{1}{4}$	$\frac{3}{8}$	$\frac{1}{2}$	$\frac{3}{4}$	1
$\frac{1}{2}$	450	60	22	11	2.2	1.0
$\frac{3}{4}$	310	40	12	5.4	1.1	0.29
1	230	32	9.0	4.1	0.90	0.23
$1\frac{1}{2}$	170	23	7.2	3.3	0.75	0.18
2	140	20	6.5	3.0	0.68	0.16
4	92	16	5.5	2.6	0.55	0.14
6	73	14	5.3	2.4	0.49	0.12
8	66	13	5.2	2.3	0.46	0.10
12	56	12	5.1	2.2	0.42	0.07

Source: Corp and Ruble, *University of Wisconsin Engineering Bulletin,* vol. 9, No. 1, 1922.

very high values of K_g. Even when the valve is wide open
($d/D = 1$), the valve seat obstructs the flow slightly and causes
some loss of head.

110. Loss of Head Due to Bends. Bends or curves in pipes
cause a loss of head in excess of that which would occur in an equal
length of straight pipe. The velocity at the center of the pipe
approaching the bend being greater than that near the pipe walls
results in a spiraling of the flow in the bend and the formation of
eddies which may persist as far as 50 diameters downstream before
normal distribution of velocities is restored. As with contraction
and enlargement, therefore, a large part of the loss of head occurs
downstream from the bend itself.

The loss of head due to a bend, in excess of that which would
occur in a straight pipe of equal length, is usually expressed as a
function of the velocity head in the pipe, that is,

$$h_b = K_b \frac{V^2}{2g} \tag{42}$$

Although test data on bend losses are fragmentary and conflicting, the most reliable tests indicate that the value of the coefficient
K_b varies with the ratio of radius of curvature of pipe axis r, to
pipe diameter D, with the roughness of the surface in the bend, and
with the Reynolds number. If the flow is turbulent, the effect
of variation in Reynolds' number is thought not to be of practical
importance, and in that case K_b is a function of the ratio r/D and
of the roughness of the bend.

Values of K_b for 90° smooth pipe bends, as determined by Beij[1]
for various values of r/D, are as follows:

r/D	K_b
1	0.35
2	.19
4	.16
6	.22
10	.32
15	.38
20	.42

Values of K_b for rough bends may be double the above values.

[1] *Pressure Losses for Fluid Flow in 90° Pipe Bends*, Bureau of Standards
Research Paper RP1110, 1938.

Losses of head in 45° bends are usually about 50 per cent less and in 180° bends about 25 per cent greater than in 90° bends.

It has also been shown[1] that for a given bend and a given rate of flow the loss of head may be large or small depending on the distribution of velocities in the approach pipe.   When the velocity in the approach pipe was high toward the inner side of the bend and low toward the outer side, the loss of head was found to be two to four times as much as for the same bend with normal distribution of velocity in the approach pipe.   With high velocity toward the outer side of the bend and low velocity toward the inner side the loss may be even less than with normal distribution.

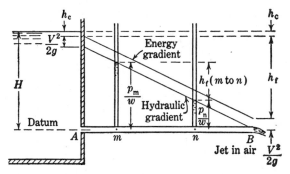

Fig. 104.   Pipe discharging from reservoir.

111. Pipe Discharging from Reservoir.   Figure 104 shows conditions of flow in a pipe of uniform diameter leading from a reservoir and discharging into air.   If there were no frictional losses the velocity of discharge would be $V = \sqrt{2gH}$, the same as the theoretical velocity from an orifice.   In any long pipe or system of pipes, however, by far the greater portion of the total head H is used in overcoming friction.

If there is no change in the diameter of a pipe, the difference in height of columns in piezometer tubes at any two sections measures the loss of head due to pipe friction between those sections.   In Fig. 104 the loss of head between sections at m and n is $h_m - h_n$. In the entire length of the pipe the loss of head due to pipe friction is h_f.   There is also a loss of head h_c due to contraction at the entrance and indicated by a drop in the energy gradient at A.   The

[1] B. L. Yarnell and F. A. Nagler, " Flow of Water around Bends in Pipes," *Trans. Am. Soc. Civil Engrs.*, 1935.

hydraulic gradient drops to a distance $V^2/2g$ below the energy gradient. From the figure,

$$H = \frac{V^2}{2g} + h_c + h_f \tag{43}$$

This equation is in effect a statement of the energy theorem considering the horizontal center line of the pipe as the datum. (See Art. 52.)

112. Part of Pipe above Hydraulic Gradient. Figure 105 shows a pipe of uniform diameter leading from a reservoir and discharging under the head H. The summit, M, is a distance y above the

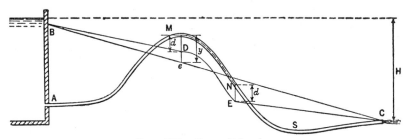

Fig. 105. Summit in pipe.

straight line BeC but at a lower elevation than the water surface in the reservoir. Two conditions will be considered: first, where $y < \dfrac{p_a - p_v}{w}$, and second, where $y > \dfrac{p_a - p_v}{w}$, p_a being atmospheric pressure and p_v being the vapor pressure corresponding to the temperature of the water in the pipe.

Assume the pipe $AMSC$ to be empty when water is turned into it at A. Water will first rise to the summit M and begin to flow toward the depression S, where it will collect and seal the pipe, entrapping air between M and S. Eventually water will discharge from the outlet C. If the velocity is high enough the air entrapped between M and S will be removed by the flowing water; otherwise it will remain there and obstruct the flow. Under such circumstances the air can be removed by a suction pump at the summit. If there is no air in the pipe and $y < \dfrac{p_a - p_v}{w}$, assuming the loss of head to be uniform, the hydraulic gradient will be the straight line BeC and the flow will be the same as though all the pipe were below the hydraulic gradient.

If $y > \dfrac{p_a - p_v}{w}$, the flow of water will be restricted, even though all air is exhausted from the pipe, and the hydraulic gradient will no longer be a straight line.   It will be straight to a point D, which is a distance $d = \dfrac{p_a - p_v}{w}$ below the summit M.   At M (assuming no air in the pipe) the absolute pressure in the pipe is the vapor pressure corresponding to the temperature within the pipe, and this pressure continues on down to N, the pipe flowing partially full between M and N.   Wherever the pipe is flowing full the velocity must necessarily be the same since the discharge is constant, and therefore, assuming a uniform degree of roughness, the slope of hydraulic gradient in such portions must be uniform.   In other words, the slope of EC must be the same as the slope of BD. Throughout the distance where the pipe is not flowing full, the hydraulic gradient, represented by the line DE, is the same vertical distance, d, below the water surface in the pipe.   The point E is the intersection of the line CE, parallel to BD, and the line DE. The section at N where the pipe begins to flow full is vertically above E.

The conditions of flow, especially at low velocities, are not usually as favorable as those described above, because of the tendency of air to collect at a summit.   Water flowing at low velocities will not remove air and may even liberate it, and cause air to collect at high places such as M.   The condition will be worse at summits above the hydraulic gradient if the pipe leaks, since the movement of air will be inward.   The occasional operation of an air pump at the summit will then be necessary to remove the air.   At a summit below the hydraulic gradient, where the pressure within the pipe is greater than atmospheric, the air which collects can be removed through a valve.

Air at a summit which is at a lower elevation than the water surface will not stop the flow of water entirely but will cause a portion of the pipe to flow partially full.   Summits in pipe lines are always objectionable, and especially so when they are above the hydraulic gradient.   Where summits cannot be avoided special provision should be made for removing the air which collects.

113. Pipe Connecting Two Reservoirs.   Figure 106 illustrates flow conditions for a pipe of uniform diameter conveying liquid from one reservoir to another reservoir at a lower elevation.   The liquid starts with zero velocity in the upper reservoir, has a velocity V in the pipe, and comes to rest in the lower reservoir. All the energy represented by the difference in elevation of liquid surfaces is utilized in overcoming resistance. The hydraulic gradient and the energy gradient are shown in the figure.

A distinguishing characteristic of the energy gradient as contrasted with the hydraulic gradient is that, except where energy is supplied from an outside source, as by a pump (Art. 118), the energy gradient always slopes downward in the direction of flow, whereas the hydraulic gradient may alternately rise and fall to

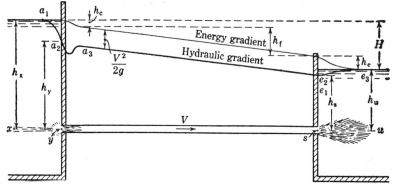

FIG. 106. Pipe connecting two reservoirs.

accord with velocity and pressure changes. The difference in elevation between the free surface in the supply reservoir and the energy gradient above any section represents the total of all losses in head that have occurred down to that section.

In changing from zero velocity in the reservoir to velocity V at the pipe entrance, pressure head is changed to velocity head, equal to $V^2/2g$. This loss in pressure is illustrated by the drop from a_1 to a_2 in the hydraulic gradient. It is assumed that there is no loss of head to point y and therefore no drop in the energy gradient. The line a_1a_2 must be considered as the hydraulic gradient of some particular path line, such as xy, since points in the other path lines which are the same horizontal distance from the entrance to the pipe may have different velocities and therefore different hydraulic gradients.

It may appear that the pressure at any point y in the stream should be that caused by the weight of the liquid column above it. This would be true if the laws of hydrostatics applied. The laws of hydrostatics do not, however, apply to liquid in motion, the pressure being less than it would be at the same depth for liquid at rest. That this is true has been proved experimentally. It also follows from writing the energy equation between a point x where the velocity is practically zero and a point y at the entrance to the pipe where the velocity is V. Assuming the points to be of the same elevation the equation becomes

$$h_x + \frac{V_x^2}{2g} = h_y + \frac{V^2}{2g}$$

or since V_x is practically zero,

$$h_y = h_x - \frac{V^2}{2g} \tag{44}$$

The behavior of energy and hydraulic gradients immediately downstream from the pipe entrance is similar to their behavior at a sudden contraction, as described in Art. 107. The loss of head in the pipe is represented by the drop h_f in the energy gradient.

Conditions at the outlet of a pipe with submerged discharge are illustrated to the right in Fig. 106. If there were no loss of head where the liquid enters the reservoir the hydraulic gradient would connect a_3 to e_1, the latter point being $V^2/2g$ below the water surface. The distance e_1e_2 represents the portion of the velocity head lost through shock and turbulence. It is also shown in the figure as h_e, the drop in the energy gradient.

Conditions in the reservoir can be illustrated by writing the energy equation between a point s at the outlet of the pipe where the velocity is V, and a point u where the velocity V_u is practically zero. If the two points are at the same elevation,

$$h_s + \frac{V^2}{2g} = h_u + \frac{V_u^2}{2g} + h_e \tag{45}$$

or, since $V_u = 0$,

$$h_s = h_u - \frac{V^2}{2g} + h_e \tag{46}$$

$h_u - h_s$ represents the portion of the velocity head which is not lost but which is reconverted into pressure head, as represented in the figure by the portion of the hydraulic gradient, e_2e_3.

PROBLEMS

(The hydraulic and energy gradients should be drawn for each problem.)

1. A new cast-iron pipe 8 in. in diameter and 100 ft long having a sharp-cornered entrance draws water from a reservoir and discharges into the air. What is the difference in elevation between the water surface in the reservoir and the discharge end of the pipe if the discharge is 5.0 cfs.

2. A new cast-iron pipe 12 in. in diameter and 1 mile long carries water from a reservoir and discharges into the air. If the entrance to the pipe is 10 ft below the water level in the reservoir and the pipe is laid on a downgrade of 2 ft per 1000 ft, determine the discharge.

3. A new cast-iron pipe 12 in. in diameter and 100 ft long connects two reservoirs, both ends being sharp-cornered and submerged. Determine the difference in elevation between the water surfaces in the two reservoirs if the discharge is 16.0 cfs.

4. What diameter of smooth concrete pipe 300 ft long will be required to carry 50 cfs between two reservoirs under a head of 3 ft, both ends of the pipe being sharp-cornered and submerged?

5. A wood-stave pipe 500 ft long is to carry 100 cfs across a ravine. Water enters one end of the pipe from an open flume, and discharges at the other end into another flume. If the difference in elevation between the water surfaces in the flumes is to be 3 ft, determine the necessary diameter of pipe, assuming well-designed transitions ($K_c = 0.1$, $K_e = 0.2$) and neglecting the effect of velocity in the flumes.

6. A horizontal corrugated-iron culvert is built through a road embankment, both ends of the pipe being inward-projecting and submerged. What head will be required to produce a discharge of 60 cfs if the pipe is 80 ft long and 3.0 ft in diameter?

114. Pipes of Different Diameters Connected in Series. Figure 107 represents a system of pipes conveying liquid from one reservoir to another. The diameter of pipe BC is less than either AB or CD. Assuming the liquid is at rest in both reservoirs, the difference H in elevation of free surfaces is the total head producing discharge. The losses of head, as indicated by the drops in the energy gradient, are successively: h_{c_1}, due to contraction at entrance at A; h_{f_1} due to friction in pipe 1; h_{c_2}, due to contraction to smaller pipe at B; h_{f_2}, due to friction in pipe 2; h_{e_1}, due to enlargement to larger pipe at C; h_{f_3}, due to friction in pipe 3; and h_{e_2}, due to enlargement at discharge at D. Therefore

$$H = h_{c_1} + h_{f_1} + h_{c_2} + h_{f_2} + h_{e_1} + h_{f_3} + h_{e_2} \qquad (47)$$

The hydraulic gradient is at all points in the three pipes a distance $V^2/2g$ below the energy gradient. Note that, since pipe 2 is smaller than pipe 1, the velocity head is greater and the hydraulic gradient is farther below the energy gradient. With the enlargement at C, however, the velocity head becomes less, resulting in a rise in the hydraulic gradient at that point.

In most hydraulic problems the major pipe friction losses (h_{f_1}, h_{f_2}, and h_{f_3} in equation 47) constitute most of the total head H, and the minor losses are frequently so small as to be negligible. If the pipe length in any problem is about 500 diameters, the error resulting from neglecting minor losses will ordinarily not exceed 5 per cent, and if the pipe length is 1000 diameters or more the

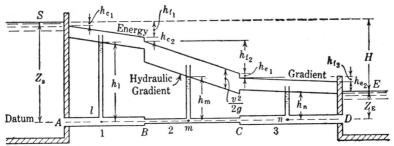

FIG. 107.   System of pipes connecting two reservoirs.

effect of minor losses can usually be considered negligible. If, however, it is desired to include these losses, a solution may be made first neglecting them and then correcting the results to include them.

Figure 108 shows a simplified diagram of flow through a pipe line of different diameters in series connecting two reservoirs. Minor losses are neglected and only the hydraulic gradient is shown. The flow is assumed to be continuous and steady. Two common problems of this type arise.

1. Sizes and lengths of pipes, and Q, given; to find total loss of head.

The loss of head, h_{f_1}, h_{f_2}, and h_{f_3} in each successive size of pipe can be determined by formula or diagram. The total head lost is then $H = h_{f_1} + h_{f_2} + h_{f_3}$. The minor losses, at entrance, enlargement, contraction, and discharge, can be computed and included if appreciable.

2. Allowable loss of head given, lengths and sizes of pipe given;

to find Q.   Four different methods of solution of this problem are outlined.

Method 1: Trial Solution.   Assume a Q.   Compute loss of head in each size of pipe by formula or diagram and add the losses. Compare with total allowable loss and revise Q in proper direction. Repeat until satisfactory check is obtained.

Method 2: Algebraic Solution.   Write $f\dfrac{L}{D}\dfrac{V^2}{2g}$ for each pipe, assuming values of f and equating sum of the terms to the allowable head loss.   Express all velocity heads in terms of velocity head in

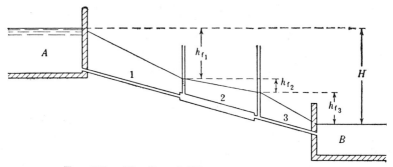

Fig. 108.   Pipe line of different diameters in series.

one of the given sizes of pipe.   Solve for that velocity head and velocity.   Compute the velocities in the other pipes.   Look up proper f's and check total loss of head.   Revise solution if necessary.

This method is adapted to the condition in which minor losses are appreciable since they can also be expressed in terms of the velocity head and included in the equation.   The minor loss coefficients selected for the first solution may also need revision.

Method 3: Equivalent-length Solution.   Reduce the over-all length of compound pipe to an equivalent length of some selected diameter as suggested in the following example.   With this selected diameter and computed equivalent length determine Q for the given loss of head.   Using this Q check summation of losses in the line.   A pipe diagram is of much assistance in the application of this method.   An example will illustrate the use of a pipe diagram in reducing a given length and size of pipe to an equivalent length of some other diameter.

EXAMPLE. Reduce 600 ft of 8-in. pipe to an equivalent length of 12-in., meaning thereby the length of 12-in. which gives the same loss of head as 600 ft of 8-in.

Solution. Assume any reasonable Q, such as 1 cfs. By the Manning diagram ($n = 0.011$) the loss of head in 8-in. pipe is 4.6 ft per 1000 ft while in 12-in. pipe the loss is 0.55 ft per 1000 ft. The equivalent length of 12-in. is, therefore:

$$\frac{4.6}{0.55} \times 600 = 5020 \text{ ft}$$

Method 4: Equivalent-diameter Solution. Reduce the different sizes of pipe in series to an equivalent diameter of the given overall length. With this diameter and length determine Q from the allowable overall loss of head. Using this Q check summation of losses in the line. An example will show the method of using a pipe diagram to determine equivalent diameter.

EXAMPLE. A pipe line consists of 1000 ft of 6-in., 500 ft of 4-in., and 800 ft of 8-in. in series. Determine the equivalent diameter, meaning thereby the diameter of pipe which if 2300 ft long would give the same loss of head as the compound pipe.

Solution. Assume $Q = 0.4$ cfs. Then by the Manning diagram ($n = 0.011$)

$$
\begin{aligned}
h_f \text{ in } 1000 \text{ ft of 6-in.} &= 3.7 \text{ ft} \\
h_f \text{ in } 500 \text{ ft of 4-in.} &= 16.0 \text{ ft} \\
h_f \text{ in } 800 \text{ ft of 8-in.} &= \underline{0.6 \text{ ft}} \\
\text{Total } h_f \text{ in } 2300 \text{ ft} &= 20.3 \text{ ft} \\
\text{Loss per } 1000 \text{ ft} &= 8.8 \text{ ft}
\end{aligned}
$$

With $Q = 0.4$ cfs, equivalent diameter by diagram $= 5.1$ in.

PROBLEMS

1. Three new cast-iron pipes are connected in series as shown in Fig. 108. The first has a diameter of 12 in. and a length of 1200 ft; the second has a diameter of 24 in. and a length of 2000 ft; and the third has a diameter of 18 in. and a length of 1500 ft. If the discharge is 8.0 cfs, determine the lost head, neglecting the minor losses.

2. Three new cast-iron pipes connected in series have diameters of 12 in., 10 in., and 8 in., each being 100 ft long. The largest pipe leads from a reservoir and the smallest discharges into the air, all changes in section being sharp-cornered. Determine the total lost head when the discharge is 6.8 cfs. What must be the elevation of the outlet end of the pipe with respect to the water-surface elevation in the reservoir?

3. Determine the discharge through three new cast-iron pipes connected in series, having diameters of 6 in., 8 in., and 10 in., and lengths of 900 ft, 1200 ft, and 2000 ft, respectively, when the total frictional loss, not including the minor losses is 15.5 ft.

4. A pipe line between points A and B consists of 1000 ft of 18-in., followed by 800 ft of 12-in., followed in turn by 500 ft of 8-in. If $n = 0.011$ and the total loss of head from A to B is 12.0 ft, determine Q: (a) by computing the equivalent length of 12-in. pipe, and (b) by computing the equivalent diameter with a length of 2300 ft.

5. Two pipes with $C_1 = 120$, connected in series, discharge 4.75 cfs with a loss of head of 27.0 ft. Each pipe has a length of 1000 ft. If one pipe has a diameter of 18 in., determine the diameter of the other, neglecting the minor losses.

6. What will be the discharge through three new cast-iron pipes connected in series, having diameters of 36 in., 24 in., and 30 in., respectively, each pipe being 500 ft long? The 36-in. pipe leads from a reservoir and the 30-in. pipe discharges into the air. The difference in elevation between the water surface in the reservoir and the open end of discharge pipe is 9.0 ft. All changes in section are abrupt.

7. In Fig. 107, assume that pipe line $ABCD$ is level, pipe AB is 12 in. in diameter and 100 ft long, pipe BC is 10 in. in diameter and 150 ft long, and pipe CD is 15 in. in diameter and 125 ft long. All pipes are smooth and all changes in cross section are square-cornered. When $Z_E = 8.0$ ft, what must be the value of Z_S to produce a discharge of 8.0 cfs?

115. Pipe System with Branches in Parallel. The reservoirs A and B are connected by a system of pipes, as shown in Fig. 109. Pipe 1 leading from reservoir A divides at S into pipes 2 and 3 which join again at T. Pipe 4 leads from the junction P to a reservoir B. Let L_1, D_1, and V_1 be respectively the length, diameter, and mean velocity for pipe 1, and let the same symbols with subscripts 2, 3, and 4 be the corresponding quantities for pipes 2, 3, and 4. Q_2 and Q_3 are the respective discharges for pipes 2 and 3, the sum of which discharges equals Q, the total discharge through pipes 1 and 4. Assuming piezometer tubes at S and T, H is the total head lost in the system of pipes, h_{f_1} is the head lost in pipe 1, $h_{f_2} = h_{f_3}$ is the head lost in pipe 2 or 3, and h_{f_4} is the head lost in pipe 4.

Questions. Is the loss of head from S to T equal to the sum of the losses in the branches, or is it equal to the loss in either one of the branches? Why must the loss of head be the same in both branches?

The solution of a problem involving a compound pipe in parallel requires a determination of the division of flow in the two or more branches.   A quick approximation of the division can be made by assuming a loss of head from S to T and computing the flow in each branch.   Within the normal range of economical velocities

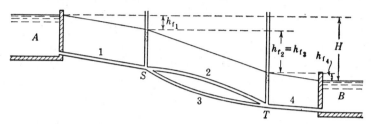

FIG. 109.   Pipe line with branches in parallel.

the *percentage* of the total flow passing through each branch will be fairly constant even with considerable variation in the amount of head lost.

A mathematical determination of the division of flow·can be made by use of the Darcy-Weisbach formula.   Substituting $\dfrac{Q}{(\pi/4)D^2}$ for V inequation 7 and letting M and N represent numerical coefficients,

$$h_f = M \cdot f \frac{LQ^2}{D^5}$$

from which

$$Q = N \cdot \sqrt{h_f} \cdot \frac{D^{5/2}}{\sqrt{f}\,\sqrt{L}} \qquad (48)$$

Equation 48 shows that, for a given h_f,

Q varies as the 5/2 power of D.
Q varies inversely as the square root of f.
Q varies inversely as the square root of L.

With two pipe branches in parallel, as in Fig. 109, it follows that

$$\frac{Q_2}{Q_3} = \left(\frac{D_2}{D_3}\right)^{5/2} \times \sqrt{\frac{f_3}{f_2}} \times \sqrt{\frac{L_3}{L_2}} \qquad (49)$$

If the diameters and the lengths of the branches are known and values of f are known or assumed, equation 49 reduces to

$$Q_2 = F \cdot Q_3 \qquad (50)$$

where F is a numerical factor. Moreover,

$$Q_2 + Q_3 = Q \tag{51}$$

With Q known or assumed, solution of 50 and 51 simultaneously gives Q_2 and Q_3. Then, using these discharges, the head lost in pipes 2 and 3 can be computed. These should be equal. If the computations do not show them equal, the discharges should be adjusted by trial until reasonable agreement is reached.

This method can be extended to any number of branches in parallel. For instance, with four branches in parallel, equation 49 can be used to give the relations

$$Q_2 = F'Q_1, \quad Q_3 = F''Q_1, \quad \text{and} \quad Q_4 = F'''Q_1$$

These equations can then be combined with the equation

$$Q = Q_1 + Q_2 + Q_3 + Q_4$$

to give the flow in each branch.

Three types of problems are explained with reference to Fig. 109.

1. Having given the discharge, and the diameters and lengths of all pipes; to determine the total lost head.

Determine the division of flow in the branches in parallel by one of the methods outlined above, and adjust by trial until the loss of head in the two branches is the same. This loss of head plus the loss of head in pipes 1 and 4 gives H, the total lost head.

2. Having given the discharge, the total lost head, the lengths of all pipes, and the diameters of three pipes; to determine the other diameter.

Assume that the diameter of pipe 2 is to be determined. Compute the head lost in pipes 1 and 4. Deduct from the total lost head the sum of these computed losses. With this difference, which is the head lost in each of pipes 2 and 3, determine Q_3. Then, $Q_2 = Q - Q_3$. With Q_2 and the lost head known, compute the diameter of pipe 2.

If the diameter of one of the single pipes, as for example, pipe 4, is to be determined, compute the head lost in pipe 1, and also the head lost in the two branches as described above. The difference between the total lost head and the sum of the above losses is the head lost in pipe 4, from which the diameter of this pipe can be computed.

3. Having given the lengths and diameters of all pipes and the total lost head; to determine Q.

This problem can best be solved by trial; but it will save time in trial solutions to determine first the portion of the total flow that passes through one of the branching pipes. Then successive values of Q can be assumed and the lost head in each pipe computed until the sum of the losses in the three pipes equals the total lost head. A final check should be made to see that Q equals approximately $Q_2 + Q_3$.

PROBLEMS

1. With a total flow of 14 cfs in Fig. A, determine the division of flow and the loss of head from A to B.

2. If the loss of head from A to B in Fig. A is 12 ft, determine the total flow.

3. The discharge of the pipe system shown in Fig. B is 10,000,000 gpd. Determine the head loss from A to D.

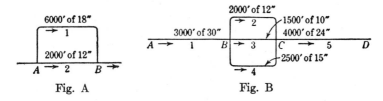

Fig. A   Fig. B

4. With the same loss of head as in problem 3, by what percentage would the capacity of the system shown in Fig. B be increased by the addition of another 10-in. pipe 1500 ft long between B and C?

5. Pipes 2, 3, and 4 in Fig. B are to be replaced with a single pipe 1500 ft long from B to C. What size should be installed if the loss of head from B to C is to be the same as in problem 3?

6. Determine h_f from A to D in the system shown in Fig. C if the discharge is 60 cfs.

3000' of 36″   2000' of 18″   1500' of 30″

$A \rightarrow$ B $C \rightarrow D$ 2400' of 24″

$\rightarrow 3$

Fig. C

7. Determine the discharge of the system shown in Fig. C if the loss of head between A and D is 18 ft.

8. Pipes 1, 2, 3, and 4 have lengths of 1000 ft, 800 ft, 700 ft, and 1500 ft, all being new cast iron. Pipes 2 and 3 are laid in parallel, both drawing from 1 and discharging into 4. If 1, 2, and 4 have diameters of 8 in., 6 in., and 10 in., respectively, determine the diameter of 3 when the discharge through the system is 1.6 cfs under a total h_f of 27 ft.

116. Flow in Pipe Networks. City water supply distribution systems are constructed in the form of many loops and branches, more or less complicated in arrangement. Such a system is called a network. Computation of the probable flow in each pipe of such a network may be quite laborious.

An important advance in the solution of pipe network problems was made by Cross,[1] who developed a method of successive approximations by which the distribution of flow can be determined. A brief summary of this method will be given here.

Consider an elementary loop A in a general network of pipes, as illustrated in Fig. 110. The arrowheads indicate direction of flow. Two conditions control the flow in the pipes of this and

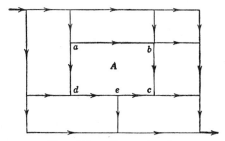

FIG. 110. Pipe network.

every other loop: (1) At any junction the total inflow is equal to the total outflow. (2) The loss of head due to flow in a clockwise direction around the loop in pipes ab and bc is equal to the loss of head in counterclockwise flow in pipes ad, de, and ec.

A flow is assumed in each pipe of the loop which will meet condition (1). Such assumed flow would meet condition (2) only by good luck. The computed loss of head in the assumed clockwise flow will thus ordinarily not be equal to the loss of head in counterclockwise flow. Cross developed the following mathematical method of computing a correction to the assumed flow that would tend to equalize the lost heads.

The standard formulas for flow of water in pipes show that for a given size, length, and roughness of pipe the loss of head varies as some power of the discharge, or

$$h_f = KQ^n \tag{52}$$

[1] *University of Illinois Experiment Station Bulletin* **286**, 1936.

where K is a proportionality factor and n has a numerical value which depends on the formula used. Thus with the Darcy-Weisbach formula (equation 7) and the Manning formula (equation 32), since $V = Q/(\pi D^2/4)$, $n = 2$. With the Hazen-Williams formula (equation 33), transposed, since $S = h_f/L$, $n = 1/0.54 = 1.85$.

In the following discussion, the symbols Σ_c and Σ_{cc} are used to denote the summation of quantities in, respectively, the clockwise and counterclockwise directions.

In any elementary loop A, the loss of head in clockwise flow is the sum of the losses in all pipes in which flow is clockwise around the loop, and can be expressed as

$$\Sigma_c h_f = \Sigma_c K Q_c{}^n \tag{53}$$

Likewise the loss of head in counterclockwise flow can be expressed as

$$\Sigma_{cc} h_f = \Sigma_{cc} K Q_{cc}{}^n \tag{54}$$

As pointed out above, the first assumed division of flow will ordinarily not result in equality of $\Sigma_c h_f$ and $\Sigma_{cc} h_f$. Assuming $\Sigma_c h_f$ to be the larger, the positive quantity given by the expression

$$\Sigma_c K Q_c{}^n - \Sigma_{cc} K Q_{cc}{}^n$$

represents the " error of closure " of the lost head. It is desired to determine the amount of the flow correction ΔQ which, when subtracted from Q_c and added to Q_{cc}, will equalize the head losses in the two directions, and thus satisfy the equation

$$\Sigma_c K (Q_c - \Delta Q)^n = \Sigma_{cc} K (Q_{cc} + \Delta Q)^n \tag{55}$$

Expanding the quantities in parentheses by the binomial theorem and retaining only the first two terms of the expansion,

$$\Sigma_c K (Q_c{}^n - n Q_c{}^{n-1} \Delta Q) = \Sigma_{cc} K (Q_{cc}{}^n + n Q_{cc}{}^{n-1} \Delta Q) \tag{56}$$

Solving for ΔQ,

$$\Delta Q = \frac{\Sigma_c K Q_c{}^n - \Sigma_{cc} K Q_{cc}{}^n}{n (\Sigma_c K Q_c{}^{n-1} + \Sigma_{cc} K Q_{cc}{}^{n-1})} \tag{57}$$

From equation 52, dividing by Q,

$$K Q^{n-1} = \frac{h_f}{Q} \tag{58}$$

Substituting terms from equations 53, 54, and 58 in equation 57,

$$\Delta Q = \frac{\Sigma_c h_f - \Sigma_{cc} h_f}{n \left[\Sigma_c \dfrac{h_f}{Q} + \Sigma_{cc} \dfrac{h_f}{Q} \right]} \tag{59}$$

EXAMPLE. Determine the flow in each pipe in the network shown in Fig. 111a.

Solution. An assumption is made of the flow in all pipes as shown in Fig. 111b. Care is taken to note that the total inflow equals the total outflow at each junction.

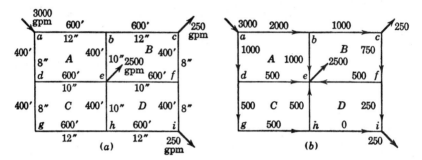

(a) Pipe network.

(b) First assumption of flow in gallons per minute.

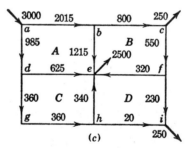

(c) Corrected flow after first approximation.

FIG. 111

Using the Hazen-Williams diagram (Fig. 101), the loss of head in successive pipes in a clockwise direction and in a counterclockwise direction is computed for each loop, A, B, C, and D, and the sum of the losses in each direction is determined. The quantity h_f/Q is also computed for each pipe. Equation 59 is then applied to determine

ΔQ for each loop. A convenient form for these computations is shown in the table, in which the clockwise losses are shown first for each loop.

First Approximation

Loop A				Loop B			
Pipe	h_1	h_f	h_f/Q	Pipe	h_1	h_f	h_f/Q
ab	11.0	6.6	0.0033	bc	3.0	1.8	0.0018
be	7.3	2.9	.0029	cf	12.8	5.1	.0068
		9.5		fe	2.0	1.2	.0024
						8.1	
ad	22.0	8.8	.0088				
de	2.0	1.2	.0024	be	7.3	2.9	.0029
		10.0	0.0174				0.0139

$$\Delta Q = \frac{9.5 - 10.0}{1.85 \times 0.0174} = -15 \qquad \Delta Q = \frac{8.1 - 2.9}{1.85 \times 0.0139} = +200$$

Loop C				Loop D			
de	2.0	1.2	0.0024	fi	1.6	0.64	0.0025
				he	2.0	0.8	.0016
dg	6.0	2.4	.0048			1.44	
gh	0.8	0.5	.0010				
he	2.0	0.8	.0016	fe	2.0	1.2	.0024
		3.7	0.0098				0.0065

$$\Delta Q = \frac{1.2 - 3.7}{1.85 \times 0.0098} = -140 \qquad \Delta Q = \frac{1.44 - 1.2}{1.85 \times 0.0065} = +20$$

The corrections are then applied to the flows of Fig. 111b. If in any loop the clockwise losses exceed the counterclockwise losses, the algebraic sign of their difference is positive, and the clockwise flow must be reduced by an amount ΔQ and the counterclockwise flow increased by the same amount. Pipes be, de, fe, and he are each common to two loops and each, therefore, requires a double correction. The result is shown in Fig. 111c. A second computation using the corrected flows is then made. The process is repeated until the corrections become negligible in amount.

PROBLEMS

1. Using the Manning formula, assuming $n = 0.011$, compute the flow in each pipe of the network shown in Fig. A.

2. Using the Hazen-Williams formula, assuming $C_1 = 120$, compute the flow in each pipe of the network shown in Fig. B.   If the pressure head at point a is 100 ft, what is the pressure head at point e?

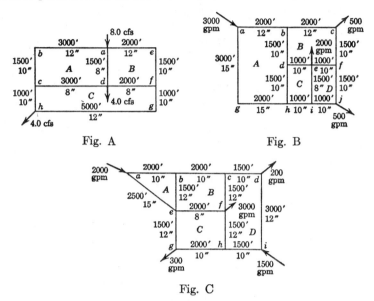

Fig. A   Fig. B

Fig. C

3. Using the Hazen-Williams formula, assuming $C_1 = 120$, compute the flow in each pipe of the network shown in Fig. C.   If the pressure at point a is 60 lb per sq in., what is the pressure at point f?

117. Branching Pipe Connecting Reservoirs at Different Elevations.   The "Three-reservoir" Problem.   In Fig. 112 A, B, and C are three reservoirs connected by pipes 1, 2, and 3. A condition of steady flow with constant reservoir level is assumed.

Let L_1, D_1, Q_1, and V_1 represent, respectively, the length, diameter, discharge, and mean velocity for pipe 1, and the same symbols, with subscripts 2 and 3, the corresponding terms for pipes 2 and 3.   If a piezometer is assumed to be at the junction P, the water surface in the tube will be a certain distance, h_{f_1}, below the surface in reservoir A.   The surface of reservoir B is a distance $H_B = h_{f_1} + h_{f_2}$ below that of reservoir A, and the surface of reservoir C is $H_C = h_{f_1} + h_{f_3}$ below the surface of reservoir A.   If $h_{f_1} < H_B$, reservoir A will supply reservoirs B and C.   If $h_{f_1} > H_B$, reservoirs A and B will supply reservoir C.   Many problems are suggested by this figure, in which certain quantities are given with others to be determined. Methods of solving three of these problems are given.

1. Having given the lengths and diameters of all pipes, and elevations of the three reservoirs; to determine Q_1, Q_2, and Q_3.

This problem is most conveniently solved by trial. Assume an elevation of the water surface in the piezometer tube at the junction P. This assumed elevation gives at once the losses of head

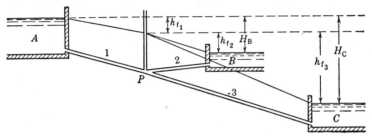

FIG. 112.   Branching pipe connecting three reservoirs.

h_{f_1}, h_{f_2}, and h_{f_3} in pipes 1, 2, and 3. From these losses and the diameters and lengths of the pipes the trial flows Q_1, Q_2, and Q_3 can be obtained by pipe diagram or formula.

The summation of flows is then made to determine the accuracy of this first trial. If the assumed water surface in the piezometer is higher than the middle reservoir B, Q_1 should equal Q_2 plus Q_3; if it is lower, Q_1 plus Q_2 should equal Q_3. The error in the trial Q's indicates the direction in which the assumed piezometric water surface should be moved for the second trial.

A quick indication can be obtained as to the direction of flow in pipe 2 by first assuming the water surface in the piezometer at the level of the water surface in B. If then Q_1 is found to be greater than Q_3, part of Q_1 must flow into B. If, however, Q_1 is found to be less than Q_3, there must be flow out of B. Two or three successive trials should establish Q_1, Q_2, and Q_3 with reasonable accuracy.

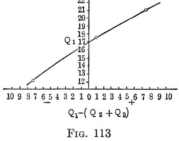

FIG. 113

It may be found helpful in making assumptions to plot computed values of Q_1, Fig. 113, against the error made in each assumption, that is, against Q_1 − $(Q_2 + Q_3)$. The resulting difference may be either plus or minus. If the assumed values of Q_1 are well selected they will define a curve the intersection of which with the Q_1 axis will give the discharge as accurately as is usually required. The points should be on both sides of the Q_1 axis, and preferably one of the points should be quite close to it. Usually not more than three trial solutions will be necessary.

This problem can also be solved analytically. Assuming any formula for pipe friction, as, for example, the Darcy-Weisbach formula, from Fig. 112,

$$H_B = f_1 \frac{L_1}{D_1} \frac{V_1^2}{2g} + f_2 \frac{L_2}{D_2} \frac{V_2^2}{2g} \tag{60}$$

and

$$H_C = f_1 \frac{L_1}{D_1} \frac{V_1^2}{2g} + f_3 \frac{L_3}{D_3} \frac{V_3^2}{2g} \tag{61}$$

Also, since $Q_1 = Q_2 + Q_3$,

$$D_1^2 V_1 = D_2^2 V_2 + D_3^2 V_3 \tag{62}$$

With H_B, H_C, the lengths and diameters of all pipes known, and with assumed values of f, these equations can be solved simultaneously for V_1, V_2, and V_3. Values of f can then be corrected and a second solution made for the velocities.

2. Having given the lengths and diameters of all pipes, Q_1, and the elevations of water surfaces in reservoir A and one of the other reservoirs as B; to determine the elevation of water surface in reservoir C.

Using Q_1, determine the lost head, h_{f_1}, in pipe 1. Then $h_{f_2} = H_B - h_{f_1}$ is the lost head in pipe 2, using which, Q_2 can be computed. Q_2 will be plus or minus depending upon whether the direction of flow in pipe 2 is towards B or P. Then $Q_3 = Q_1 - Q_2$. With Q_3 determined, the head lost in pipe 3 can be computed, and the elevation of water surface in reservoir C obtained.

3. Having given the lengths of all pipes, the elevations of water surfaces in all reservoirs, Q_1, and the diameters of two pipes as D_1 and D_2; to determine D_3.

Determine h_{f_1}, Q_2, and Q_3 as for Case 2. Then with Q_3 and $h_{f_3} = H_C - h_{f_1}$ known, compute D_3.

PROBLEMS

1. Determine the flow into or out of each reservoir in the pipe system shown in Fig. A, using the Manning diagram with $n = 0.011$.

2. Determine the flow into or out of each reservoir in the pipe system shown in Fig. B, using the Hazen-Williams diagram with $C_1 = 120$.

3. If in Fig. B pipe 3 is closed off by a valve at X, determine the amount and direction of flow in pipes 1, 2, and 4.

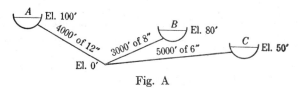

Fig. A

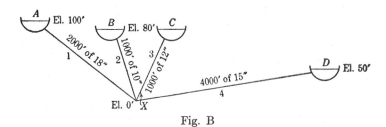

Fig. B

4. A 36-in. concrete pipe 5000 ft long draws water from reservoir A. At its lower end it is joined with a 30-in. concrete pipe 4000 ft long drawing water from reservoir B, both pipes discharging into a 48-in. concrete pipe 8000 ft long. If the 48-in. pipe discharges into a reservoir with water surface 15 ft lower than the water surface in A and 20 ft lower than the water surface in B, determine the discharge.

5. A 48-in. concrete pipe 6000 ft long carries 50 cfs from reservoir A, discharging into two concrete pipes each 4500 ft long and 30 in. in diameter. One of the 30-in. pipes discharges into reservoir B, in which the water surface is 21 ft lower than that in A. Determine the elevation of the water surface in reservoir C, into which the other 30-in. pipe discharges.

6. Pipes 1, 2, and 3, having diameters of 6 in., 8 in., and 10 in., and leading from reservoirs A, B, and C, respectively, join at a common point. All pipes are new cast iron and each is 1000 ft long. If the water level in reservoir C is 6 ft higher than that in A and 16 ft higher than that in B, determine the discharge and direction of flow in pipe 1.

7. The water surface in reservoir A is 20 ft higher than that in B. A 6-in. pipe 900 ft long leads from A to a point where it connects with a 4-in. pipe 500 ft long, leading from B, both discharging into an 8-in.

pipe 2400 ft long. If the 8-in. pipe discharges 1.5 cfs, find the difference in elevation between the end where it discharges into the air and the water surface in reservoir A. ($n = 0.011$ for all pipes.)

8. A new 18-in. cast-iron pipe 2000 ft long carries 9.2 cfs from reservoir A, discharging into two new cast-iron pipes each 1500 ft long. One pipe is 12 in. in diameter and leads to reservoir B, in which the water level is 15 ft lower than that in A. If the water surface in reservoir C is 20 ft lower than that in A, determine the diameter of the pipe leading to C.

118. Pipe Line with Pump. The hydraulics of a pipe line through which a liquid is being drawn from a reservoir by means

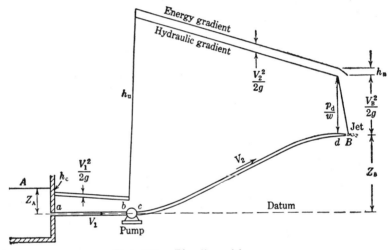

Fig. 114. Pipe line with pump.

of a pump and discharged through a nozzle is illustrated in **Fig. 114.** Selecting the datum at the pump and writing the energy theorem for the entire system from A to B (equation 16, page 97),

$$Z_A + h_U = Z_B + \frac{V_B{}^2}{2g} + (h_c + h_{f_1} + h_{f_2} + h_n) \qquad (63)$$

At the intake from the reservoir the energy gradient drops a distance equal to the contraction loss while the hydraulic gradient drops $V_1{}^2/2g$ farther. The drop in the gradients from a to b represents the frictional loss in pipe 1. At the pump the energy gradient rises a distance equal to the pumping head, or energy per pound of liquid, put into the line by the pump. The hydraulic gradient at c is $V_2{}^2/2g$ below the energy gradient, and the drop in the gradients to d at the base of the nozzle represents the friction loss in pipe 2. The pressure head at the base of the nozzle is represented by p_d/w. In flow through the nozzle the energy gradient drops a distance h_n equal to the nozzle loss, which brings it to a distance $V_B{}^2/2g$ above the nozzle. The hydraulic gradient drops to the elevation of the jet at B where the pressure head is zero.

PROBLEMS

1. The pump in Fig. A draws water from a reservoir. At point A in the suction pipe an open manometer shows a vacuum of 15 in. of mercury, while a pressure gage on the discharge pipe at B reads 60 lb per sq in. Assume that the pipe is new and the elbows are smooth, with $r = D$. If the pump efficiency is 80 per cent, find the required horse-power input to the pump for a discharge of 2.0 cfs.

2. The pump in Fig. A draws water from a reservoir into which the suction pipe projects. Assume that the pipe is new and the elbows are smooth, with $r = D$. If the output power of the pump is 15.2 hp when the discharge is 1.1 cfs, what pressure can be expected in the discharge pipe at B?

3. The pump in Fig. A draws water from a reservoir. At point A in the suction pipe an open manometer shows a vacuum of 11.5 in. of mercury, while a pressure gage at B reads 53.8 ft of water. Assume that the pipe is

Fig. A

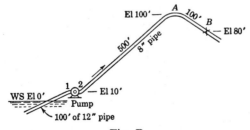

Fig. B

new and the elbows are smooth, with $r = D$. When the measured flow is 1.63 cfs, the input power to the pump is found to be 20.5 hp. Compute the efficiency of the pump.

4. The discharge through the 8-in. pipe shown in Fig. B is throttled to a velocity of 8 ft per sec by means of the valve at B. The pressure head at the summit A is -10 ft of water. All pipes are new steel, and the suction pipe projects into the reservoir. Compute the pressure heads at points 1 and 2; the pumping head; and the horsepower output of the pump. Draw the hydraulic and the energy grade lines showing their elevations (to the nearest tenth of a foot) at strategic points.

5. A pump is discharging 200 gpm of water through 4-in. smooth steel pipe. When the pressure on the discharge side of the pump is 100 lb per sq in., what pressure can be expected in the discharge pipe at a point 1800 ft distant from the pump and 150 ft higher in elevation? Draw the hydraulic and the energy gradients.

6. In Fig. 114, assume that the suction line ab is smooth pipe 8 in. in diameter and 40 ft long with a square-cornered entrance; the discharge line cd is smooth pipe 6 in. in diameter and 500 ft long; and the nozzle is 6 in. in diameter at the base and $1\frac{1}{2}$ in. at the tip, with $C = C_v = 0.95$. Let $Z_A = 5$ ft, $Z_B = 50$ ft. Assuming a pump efficiency of 75 per cent, what horsepower input to the pump is required for discharges of water of: (a) 100 gpm; (b) 500 gpm; and (c) 1000 gpm? Draw the hydraulic and the energy gradients.

7. A pump delivers water through a line of smooth pipe 6 in. in diameter and 800 ft long to a hydrant to which is connected a line of cotton rubber-lined fire hose $2\frac{1}{2}$ in. in diameter. The hose is 300 ft long and terminates in a nozzle with a tip diameter of $1\frac{1}{8}$ in., and $C = C_v = 0.97$. The nozzle is at an elevation 25 ft higher than the pump. Assume a pressure loss in the hydrant of 2 lb per sq in. When the flow is 250 gpm, what pressure in pounds per square inch must be maintained on the discharge side of the pump? If the pressure on the suction side of the pump in the 6-in. suction pipe is 8 lb per 'sq in., what is the horsepower output of the pump?

8. A pump draws water through 50 ft of 12-in. cast-iron pipe from a reservoir in which the water surface is 15 ft higher than the pump, and discharges through 250 ft of 10-in. cast-iron pipe to an elevated tank in which the water surface is 200 ft higher than the pump. The ends of the pipe are square-cornered and there is one 90-degree standard elbow $(r = D)$ in the discharge line. If the discharge is 10 cfs, compute: (a) the pressure heads on the suction and discharge sides of the pump, and (b) the pumping head and the horsepower output of the pump.

GENERAL PROBLEMS

1. Points A and B are 15 miles apart along a new 24-in. steel pipe. A is 35 ft higher than B. With water flowing and with the same pressure at A and B, determine the discharge.

2. Points A and B are 3000 ft apart along a new 6-in. steel pipe. B is 100 ft higher than A. With 1.2 cfs of water flowing from A toward B, compute the necessary pressure at A if the pressure at B is to be 40 lb per sq in.

3. Compute the pressure drop in pounds per square inch per mile of horizontal 24-in. steel pipe carrying 16 cfs of a heavy fuel oil at 60° F.

4. A 24-in. new steel pipe is designed to transport 300,000 barrels of oil per day (1 bbl = 42 gal). Compute the pumping horsepower per mile of pipe required to overcome friction if the oil is (a) a heavy crude (sp gr 0.925, $\mu = 0.002$ lb sec per ft²); (b) a light crude (sp gr 0.855, $\mu = 0.0002$ lb sec per ft²).

5. A new 1-in. pipe brings water from a tank on a hill to another tank at a farmhouse in the valley. The pipe is 500 ft long and both ends are submerged. The difference in level of the water surfaces in the two tanks is 200 ft. The flow in the pipe is controlled by a gate valve in the pipe just outside of the lower tank. With the valve wide open, what flow in gallons per minute can be expected?

6. Determine the discharge through a 10-in. concrete pipe, 250 ft long, if the difference in elevation between the water surface in the supply reservoir and the end of the pipe where it discharges into the air is 20 ft, the entrance to the pipe being sharp-cornered.

7. A 2-in. pipe 50 ft long extends vertically downward from the bottom of an elevated tank and discharges into air. The entrance from tank to pipe is square-cornered. When the water in the tank is 10 ft deep over the entrance to the pipe, what is the discharge? Determine the pressure head in the pipe at a point 5 ft below the tank and at the vena contracta, assuming a coefficient of contraction of 0.65 and neglecting lost head between tank and vena contracta.

8. A cleaning solvent at a temperature of 60° F flows by gravity from one tank to another through 2-in. wrought-iron pipe, both ends of the pipe being square-cornered and submerged. The difference in elevation of the liquid surfaces in the tanks is 7.5 ft. The pipe line is 27 ft long and contains three 90° standard elbow bends. The radius of the center line of the bend is 2 in. Flow is controlled by a 2-in. gate valve near the lower tank. With the valve one-fourth open, determine the discharge in gallons per minute, assuming the same minor loss coefficients as for water.

9. A new 6-in. cast-iron pipe 100 ft long is connected in series with a 4-in. new cast-iron pipe 50 ft long. If the 4-in pipe discharges into the air at a point 16 ft lower than the water level in the supply tank, determine the rate of discharge, all connections being sharp-cornered.

10. A new 4-in. cast-iron pipe 100 ft long siphons water from a tank, discharging into the air at a point 12 ft lower than the water level in the tank. Determine the gage pressure at the highest point in the siphon, which is 20 ft above and 50 ft from the discharge end of the pipe.

11. A pump draws water from a river through a new 12-in. cast-iron pipe, the entrance being rounded so that $K_c = 0.08$. Determine the gage pressure at a point in the suction line 8 ft above the water level in the stream and 90 ft from the entrance to the pipe when $Q = 7.1$ cfs.

12. A pump draws water through a new 6-in. horizontal cast-iron pipe 100 ft long. To the end of this pipe is connected a standard 90-degree elbow and a 6-in. vertical riser pipe 10 ft long, the lower end of which projects 4 ft below the water surface in the supply reservoir. If it is not permissible for the absolute pressure at the pump to drop below 5 lb per sq in., determine the maximum discharge.

13. A concrete pipe culvert 90 ft long and 3 ft in diameter is built through a road embankment. The culvert is laid on a grade of 1 ft per 100 ft. Water is 5 ft deep above the top of the pipe at the entrance, and at the outlet the top of the pipe is submerged to a depth of 2 ft. Assume sharp-cornered inlet and outlet. What is the discharge?

14. A horizontal concrete pipe culvert 50 ft long is to be built through a road embankment, both ends of the pipe being sharp-cornered and submerged. If the difference in elevation of water surfaces at the ends of the pipe can not exceed 4.0 ft, what diameter of pipe will be required for a discharge of 100 cfs? Compare with the answer obtained by considering the culvert as a short tube and using the data on page 148.

16. Three smooth rubber-lined fire hose, each 200 ft long and $2\frac{1}{2}$ in. in diameter and having 1-in. nozzles, are connected to a 6-in. fire hydrant. If for the nozzles $C_c = 1$ and $C_v = 0.97$, determine the necessary pressure in the hydrant in order to throw streams 100 ft high, the nozzles being 10 ft above the hydrant.

17. Two smooth rubber-lined fire hose, each 300 ft long and $2\frac{1}{2}$ in. in diameter and having 1-in. nozzles, are connected to a 6-in. fire hydrant. $C_c = 1$ and $C_v = 0.97$. It is necessary to throw streams 80 ft high, the nozzles being 20 ft above the hydrant. Determine the horsepower that must be supplied at the hydrant.

18. Two smooth rubber-lined fire hose, each 300 ft long and $2\frac{1}{2}$ in. in diameter and having 1-in. nozzles, are connected to a 6-in. fire hydrant. If, for the nozzles, $C_c = 1$ and $C_v = 0.97$, what height of stream can be thrown when the pressure in the hydrant is 70 lb per sq in., the nozzles being 10 ft above the hydrant?

19. It is desired to pump crude oil at 50° F through a long line of 18-in. welded steel pipe at a mean velocity of 2 ft per sec. The specific gravity of the oil is 0.925, and the absolute viscosity is 0.0028 lb sec per sq ft. Compute the pumping horsepower per mile of pipe required to overcome friction.

20. If the velocity of the oil in problem 19 is tripled to reduce the time of delivery, how many times as much pumping horsepower would be required?

21. A 12-in. welded steel pipe 5200 ft long conducts water from a reservoir to a nozzle, the jet from which drives a turbine. The water surface in the reservoir is at elevation 450 ft. The pipe leaves the reservoir at elevation 430, running thence 1200 ft on level grade, thence 3500 ft on straight grade to elevation 0, thence 500 ft on level grade to the nozzle. The nozzle has a tip diameter of 3 in., with $C = C_v = 0.96$. Determine the discharge and the horsepower in the jet, and draw the hydraulic and the energy gradients.

22. A concrete mixer used in highway work is supplied with water at the rate of 10 gpm through 1-in. pipe (actual $D = 1.049$ in.) $\frac{1}{2}$ mile long. The pond from which the water is taken is 30 ft below the point of delivery. The pipe discharges into a tank at atmospheric pressure. What horsepower must be furnished to the pump which forces the water through the pipe if the pump efficiency is 60 per cent?

23. A centrifugal pump draws water through a 15-in. pipe from a reservoir in which the water surface is 10 ft lower than the pump, and discharges through a 12-in. pipe. At a point in the discharge pipe 8 ft above the pump a pressure gage reads 42 lb per sq in. When the discharge is 8.0 cfs, the head lost in the suction pipe is 0.5 ft, the head lost in the discharge pipe between pump and gage is 3.5 ft, and the power input to the pump is 124 hp. Determine the efficiency of the pump.

24. A straight 6-in. new cast-iron pipe 2000 ft long joins two reservoirs which have a difference of water surface elevation of 15 ft. Both ends of the pipe are submerged 10 ft. A pump is to be placed in the line to increase the flow into the lower reservoir to three times the flow which would be produced by gravity alone. (*a*) Where should the pump be placed if the pressure head in the pipe is not to be less than -15 ft of water? (*b*) Compute the pumping head and the horsepower output of the pump. (*c*) Draw the hydraulic gradient, neglecting minor losses.

25. A 36-in. riveted-steel-pipe penstock ($k = 0.003$) 500 ft long leads from a reservoir to a turbine which discharges through a draft tube into a tail race. The difference in water surface levels in reservoir and tail race is 100 ft. The turbine is 6 ft above the tail race. When the discharge is 30 cfs, compute: (*a*) the pressure head in the penstock just before it reaches the turbine, and (*b*) the output horsepower of the turbine, assuming 80 per cent efficiency and neglecting the head lost in the draft tube.

26. A 6-in. pipe supplies water to a turbine, as shown in Fig. A, at a rate of 4.45 cfs. The pressure gage 2 ft above point A in the supply pipe reads 20 lb per sq in. At point C, 4 ft lower than the turbine, the discharge pipe is 12 in. in diameter and the pressure is shown by an open mercury manometer. The head lost from B to C is 2 ft. Compute the drop in hydraulic energy through the turbine (from A to B) and the horsepower being delivered to the turbine.

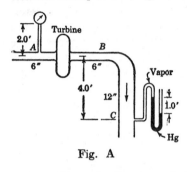

Fig. A

27. A pump delivers water through a line of smooth pipe 6 in. in diameter and 800 ft long to a hydrant to which are connected three lines of cotton rubber-lined fire hose $2\frac{1}{2}$ in. in diameter. Each hose is 300 ft long and terminates in a nozzle with a tip diameter of $1\frac{1}{8}$ in. and $C = C_v = 0.97$. The nozzles are at an elevation 25 ft higher than the pump. Assume a pressure loss in the hydrant of 5 lb per sq in. When the flow in each hose is that of a standard firestream for business districts (250 gpm), what pressure must be maintained on the discharge side of the pump? If the pressure on the suction side of the pump in the 6-in. suction pipe is 8 lb per sq in., what is the horsepower output of the pump?

28. A 10-in. cast-iron pipe line $ABCD$ in Fig. B is supplied with water from a reservoir at A. When the flow from A to D is 3.0 cfs,

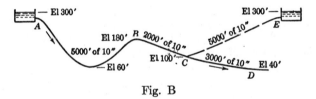

Fig. B

what is the pressure in the pipe at B? (Assume $C_1 = 120$.) In order to maintain larger pressures in the pipe at B, a second reservoir is to be constructed at E and connected by 10-in. cast-iron pipe to the original line at C. With the same flow of 3.0 cfs in CD, but now drawing from both reservoirs, what pressure can be expected at B?

29. A cubical tank 10 ft on each edge, filled with water, discharges through a new 2-in. cast-iron pipe 200 ft long. Determine the time required to empty the tank if the discharge end of the pipe is 10 ft lower than the bottom of the tank. Assume f to be constant and equal to 0.03.

30. A reservoir 60 ft by 100 ft contains water to a depth of 10 ft. Determine the diameter of concrete pipe 3000 ft long that will be required to empty the reservoir in 1 hr, the discharge end of the pipe being 12 ft lower than the bottom of the reservoir. Assume $f = 0.02$.

31. A new 12-in. cast-iron pipe 100 ft long, having a sharp-cornered entrance, draws water from the bottom of a reservoir and discharges into the air. The bottom of the reservoir is 60 ft square, and each of the four sides has a slope of 1 vertical to 4 horizontal. If the water is 10 ft deep at the beginning and the discharge end of the pipe is 2 ft below the bottom of the reservoir, determine the time required to empty.

32. Water flows from a to e through the pipe system shown in Fig. C. It is known that the flow in pipe $bfgc$ is 500 gpm, and that the pressure head at a is 200 ft. Assuming that $C_1 = 120$, and that all the pipes are level, determine the total flow from a to e and the pressure heads (to the nearest foot) at b, c, d, and e.

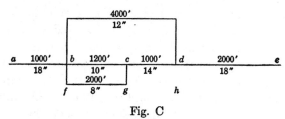

Fig. C

33. The flow from a to e through the pipe system shown in Fig. C is 10 cfs. If $n = 0.011$ for all the pipes, determine the loss of head between a and e. How much would this loss of head be reduced by the installation of an additional line consisting of 1400 ft of 12-in. pipe from g via h to d?

Chapter VIII

OPEN CHANNELS

119. Description. An open channel is a conduit in which a liquid flows with a free surface. As contrasted with liquid in a pipe, in which flow normally occurs under pressure, the liquid conveyed by an open channel exerts no pressure other than that caused by its own weight and the pressure of the atmosphere.

The general theory in this chapter applies to all liquids, but since there are few test data available on open-channel flow of liquids other than water at natural temperatures, the empirical coefficients cited apply only to water.

120. Uses of Open Channels. Open channels may be either natural or artificial. Natural water channels vary in size from tiny side-hill rivulets through brooks, small rivers, and large rivers, to tidal estuaries. Underground streams in caves are open channels as long as they have a free surface. Natural channels are usually irregular in cross section and alignment and in character and roughness of stream bed. Streams in erodible material may frequently or continuously shift their location and cross section. Such irregularities and changes in natural streams introduce engineering problems, for instance, in navigation and flood control, beyond the scope of this book, which treats only of flow in fixed channels of uniform roughness.

Artificial channels are built for various purposes:

1. Water-power development: Water is brought from streams or reservoirs to headworks above power plants.

2. Irrigation: Water is brought from streams or reservoirs to storage ponds or tanks or directly to land to be irrigated.

3. City water supply: Water is brought from streams or storage reservoirs to ponds supplying city distribution systems.

4. Sewerage: City sewers, although usually covered conduits or pipes, ordinarily are designed as open channels because they are not supposed to flow full but to have a free surface under atmospheric pressure.

5. Drainage: Low-lying, swampy, or waterlogged lands are frequently made productive by draining them through open ditches or by laying and covering pipe which may or may not flow full.

6. Flood control: Protection of cities or valuable lands from floods often requires improving a natural channel by straightening, cleaning, or paving to increase its capacity, or by building additional flood channels on new locations.

121. Distribution of Velocities.   The flow of water in open channels is ordinarily turbulent.   The exceptional case of laminar flow occurs so seldom that it will not be considered here.

As in pipes (Art. 92), velocities in open channels are retarded near the conduit surface, and if there were no other influences the

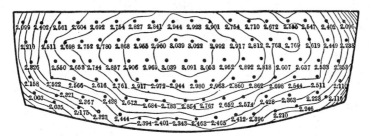

Fig. 115.   Velocities in Sudbury conduit.

maximum velocity in a cross section of an open channel would occur at the water surface.   Surface tension, however, produces a resistance to flow and causes the maximum velocity to occur at some distance below the surface.   Under ideal conditions, where there are no disturbing influences of any kind, the distribution of velocities in a regular channel will be uniform and similar on either side of the center.   There are, however, sufficient irregularities in every channel to prevent a uniform distribution of velocities.   The lines of equal velocity plotted from a large number of measurements for the Sudbury conduit near Boston, Fig. 115, show a more regular distribution of velocities than will be found in most channels.

122. Wetted Perimeter and Hydraulic Radius.   The wetted perimeter P of any conduit is the line of intersection of its wetted surface with a cross-sectional plane (Art. 100).   In Fig. 116 the wetted perimeter is the length of the broken line $abcd$.   In a circular conduit flowing partly full the wetted perimeter is the arc of a circle.

The hydraulic radius R is the area of cross section of the stream divided by the wetted perimeter.

Open-channel formulas express the velocity as a function of the hydraulic radius.

123. Steady, Uniform, and Continuous Flow. The definitions of these three conditions of flow, as

Fig. 116.   Trapezoidal channel.

given in Arts. 45 to 47, apply in open-channel flow and should be reviewed. The equation of continuity

$$Q = A_1 V_1 = A_2 V_2 = A_3 V_3 \cdots \text{etc.} \qquad (1)$$

also applies to any open stream in which flow is steady and continuous past points 1, 2, 3, ... etc.

124. Energy in an Open Channel. The principles stated in the general discussion in Art. 48 of the energy contained in a stream of fluid are applicable to open-channel flow.

Kinetic Energy. In open channels where velocities in different parts of a cross section are not the same, the kinetic energy per pound of liquid flowing past any cross section such as AB in Fig. 117 is $\alpha V^2/2g$, where V is the mean velocity in the cross section and α is a coefficient depending for its value upon the distribution of velocities but always being greater than unity. The range of variation of α is not well determined but ordinarily lies between 1.1 and 1.2. However, velocity head in open channels is com-

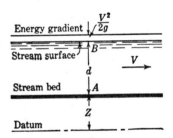

Fig. 117.   Energy in an open channel.

monly taken as the head due to the mean velocity, that is, α is assumed to equal unity. Precise experimental studies may require a measurement of the distribution of velocities in the cross section and a determination of α.

Pressure Energy. Pressure energy in open-channel flow is ordinarily computed with reference to the bed of the channel. If the stream filaments past section AB (Fig. 117) are approximately straight, the pressure head at A equals d, the depth of flow. This is the usual condition in open-channel problems. If, however, the stream filaments have vertical curvature, as, for instance, in flow over dams or weirs, centrifugal force causes the pressure on the stream bed to be appreciably different from that corresponding to the depth.

Elevation Energy. Elevation energy can be referred to any selected datum, usually the low point of the stream bed in any particular problem. With the datum as shown in Fig. 117, the elevation head at point A is Z foot-pounds per pound of liquid passing section AB.

Total Energy. The total energy at section AB is therefore

$$H = \frac{V^2}{2g} + d + Z \tag{2}$$

and is measured as before in foot-pounds per pound of liquid passing the cross section. The location of the *energy gradient*, assuming straight stream filaments so that the pressure head equals the depth, is $V^2/2g$ above the surface of the stream, whereas the *hydraulic gradient* coincides with the stream surface.

PROBLEMS

1. Compute A, P, and R for each of the channel cross sections shown in Fig. A. Depth $d = 5$ ft. Bottom width B in (b) 10 ft; in (c) 3 ft.

2. If $Q = 200$ cfs in problem 1, compute V for each cross section.

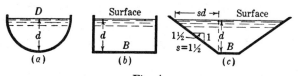

Fig. A

3. If a flow of 50 cfs is to be carried in a flume at a velocity of 6 ft per sec, compute the dimensions of the cross section if it is: (a) semicircular; (b) rectangular with the width equal to twice the depth; (c) trapezoidal with $B = d$ and side slope $s = 3$ horizontal to 4 vertical.

4. Compare the wetted perimeters and hydraulic radii of the three different forms of cross section of problem 3. Which of the three forms carries the given flow with the least area of channel lining?

5. A conduit is to have a cross-sectional area of 10 sq ft. Compute the wetted perimeter and hydraulic radius if the section is: (a) circular flowing full (pipe); (b) semicircular open channel.

Uniform Flow

125. Lost Head. Figure 118 represents an open channel of constant width with steady, uniform flow in the reach of horizontal length L from point 1 to point 2. By definition of uniform flow, the depth d and the mean velocity V are constant throughout the reach. As a result, the bed of the channel, the water surface, and the energy gradient are all parallel.

Since the energy gradient is a graphical representation of total energy or total head, the drop in the gradient is a measure of lost head. In Fig. 118 the loss of head due to friction in length L is indicated by h_f. The ratio h_f/L, or the loss of head in feet per foot,

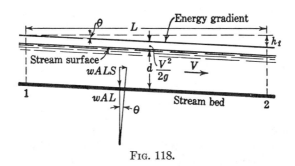

Fig. 118.

is the slope of the energy gradient, and is represented by the symbol S. With uniform flow, the slope of the energy gradient S, the slope of the stream surface S_w, and the slope of the stream bed S_0 are all equal.

Since frictional losses in open channels and pipes are of the same nature they are governed by the same laws. To make the general laws as stated for pipes (Art. 96) apply to open channels it is necessary only to substitute, respectively, the words *channel* and *hydraulic radius* for pipe and diameter. The base formulas for pipes therefore apply equally to open channels. Equation 28, page 197, is in the form generally used for open channels. For convenience of reference it is repeated here:

$$V = C'' R^y S^z \tag{3}$$

The further consideration of frictional losses in open channels is largely empirical. Numerical values of coefficients and exponents must be derived from experimental data. A few of the

more commonly used open-channel formulas are given in thefollow-
ing pages.

126. The Chezy Formula. This fundamental law of turbulent
flow applies either to open channels or to pipes (Art. 103), though
it was originally used for open channels. It is derived mathematically by assuming: (1) that the prism of liquid shown in Fig.
118 moves downhill without change in shape, and (2) that each
unit of area of the stream bed offers resistance to flow proportional
to the square of the velocity.

Assuming uniform flow, the depths, areas, velocities, pressure
heads, and velocity heads are the same at the two ends of the prism.
The lost head h_f is therefore equal to the drop in elevation of the
channel in the horizontal distance L. The slopes of open channels
are usually so small that it may be assumed that the slope length
equals the horizontal length and that $\sin \theta = \tan \theta = h_f/L = S$.

Since the velocity is constant, the component of the force of
gravity parallel to the channel bed must equal the total force
resisting the flow. The gravity component is

$$wAL \sin \theta = wALS \tag{4}$$

A being the area of cross section. The unit resisting force τ
(tau) is assumed to vary as some power of the velocity so $\tau = KV^x$.
The total resisting force is therefore $(KV^x)PL$, where P is the
wetted perimeter. Thence

$$wALS = (KV^x)PL \tag{5}$$

and, since $A/P = R$,

$$wRS = KV^x \tag{6}$$

Chezy as early as 1775 is reported to have recommended an
equation of this form with $x = 2$. If $\sqrt{w/K}$ is replaced by a
general coefficient C, equation 6 reduces to

$$V = C \sqrt{RS} \tag{7}$$

which is known as the Chezy formula.

The value of the coefficient C varies with the characteristics of
the channel. The formula is not dimensionally correct since, for
a given channel, C and S having numerical values, V varies as
$\sqrt{R}$, and the dimensional equation is ft/sec $= \sqrt{\text{ft}}$. To make
the equation dimensionally correct, therefore, C must be a function
of length and time as well as of the roughness of the channel bed.

Many years of effort have been devoted to attempts to obtain a simple, comprehensive expression for C in the Chezy formula. Out of the variety of results only three equations are used to any great extent in modern hydraulics. These three methods of determining the Chezy C will now be discussed.

127. Formulas for Determining the Chezy C. *Kutter Formula.* An elaborate investigation of all available records of measurements of flow in open channels including the Mississippi River was made by Ganguillet and Kutter,[1] Swiss engineers, in 1869. As a result of their study they deduced the following empirical formula, commonly called the Kutter formula, for determining the value of C in the Chezy formula:

$$C = \frac{41.65 + \dfrac{0.00281}{S} + \dfrac{1.811}{n}}{1 + \dfrac{n}{\sqrt{R}}\left(41.65 + \dfrac{0.00281}{S}\right)} \tag{8}$$

In this formula, C is expressed as a function of the hydraulic radius R, and the slope S, as well as the coefficient of roughness n, the value of which increases with the degree of roughness of the channel. Values of n were published by the authors of the formula but not all construction materials were covered, and later experimental data have shown the need of revising and supplementing them.

The solution of the Kutter formula is given by tables which usually accompany the formula. The use of the Chezy formula with the Kutter coefficient thus becomes much simplified. A short table of values of C corresponding to different values of R, S, and n is given on page 250. Interpolations may be necessary in using this table, although with the uncertainty which exists in any hydraulics problem as to the exact value of n, interpolation closer than the nearest whole number is unwarranted.

Manning Formula. In a treatise published in 1890, Manning[2] stated that a study of the experimental data then available led him to the conclusion that the values of the exponents y and z in equation 3 which best represented the law of flow in open channels

---

[1] E. Ganguillet and W. R. Kutter, *Flow of Water in Rivers and Other Channels*, translation by R. Hering and J. C. Trautwine, Jr., John Wiley & Sons.

[2] Robert Manning, "Flow of Water in Open Channels and Pipes," *Trans. Inst. Civil Engrs. (Ireland)*, vol. 20, 1890.

were, respectively, $\frac{2}{3}$ and $\frac{1}{2}$; and that there appeared to be a close correspondence between C'' and the reciprocal of Kutter's n. Stated in metric units, this formula is

$$V = \frac{1}{n} R^{\frac{2}{3}} S^{\frac{1}{2}} \tag{9}$$

which, in foot-pound-second units, is

$$V = \frac{1.486}{n} R^{\frac{2}{3}} S^{\frac{1}{2}} \tag{10}$$

This can be considered as the Chezy formula with

$$C = \frac{1.486}{n} R^{\frac{1}{6}} \tag{11}$$

The coefficient of roughness n is to be given the same value as n in the Kutter formula.

Bazin Formula. This formula, first published[1] in 1897, considers C to be a function of R but not of S. Expressed in foot-pound-second units the formula is

$$C = \frac{157.6}{1 + \dfrac{m}{\sqrt{R}}} \tag{12}$$

in which m is a coefficient of roughness. Values of m proposed by Bazin are given below.

Description of Channel	Value of m
Smooth cement or planed wood	0.109
Planks, ashlar and brick	0.290
Rubble masonry	0.833
Earth channels of very regular surface	1.54
Ordinary earth channels	2.36
Exceptionally rough channels	3.17

128. Determination of Roughness Coefficient. The roughness coefficients, n in the Kutter and Manning formulas, and m in the Bazin formula, are supposedly dependent only on the nature of the stream bed. For a given kind of channel lining, for instance, concrete of uniform smoothness, n and m are assumed to be con-

[1] *Annales des ponts et chaussées*, 1897.

stant for all depths of flow and widths of channel. The extent to which this assumption is true needs further research, but probably the variation of the roughness coefficients with form of cross section is relatively small.

Values of n based on recommendations by Scobey,[1] Horton,[2] and Ramser[3] are given in the table below, average values being given for the various descriptions of channel bed. Slight variations in methods of construction of artificial channel linings materially affect the values. Growth of vegetation and deposits of algae cause values of n to increase. A winding channel has higher values of n than a straight channel with the same lining, while a channel containing debris and obstructions has its capacity reduced not only because of increase in roughness coefficient but also because of decrease in cross-sectional area.

VALUES OF ROUGHNESS COEFFICIENT n FOR USE IN
MANNING AND KUTTER FORMULAS

DESCRIPTION OF CHANNEL	AVERAGE VALUE OF n
Small straight flumes of best planed timber, or laboratory flumes lined with glass or brass	0.009
Lowest value to be used for straight planed-timber flumes under ideal conditions in field service (for R up to 2 ft)	.010
New planed timber or wood-stave flumes under excellent conditions; channels lined with smoothly finished concrete; smooth straight metal flumes without projecting bands or joints (for R up to 5 ft)	.012
Vitrified sewer pipe, larger sizes, well laid; glazed brickwork; good concrete pipe	.013
For conservative design of painted metal, wood-stave, or concrete flumes under normal conditions; plank flumes with longitudinal battens; gunite lining troweled with cement mortar; common clay drain tile	.014
Conduits lined with brick of average workmanship; vitrified sewer pipe and concrete pipe in fair condition, ashlar masonry	.015
Metal flumes with heavy compression bars; general value for other flumes in poor condition	.017

[1] Fred C. Scobey, " The Flow of Water in Flumes," *U. S. Department of Agriculture, Technical Bulletin* 393, 1933.

[2] Robert E. Horton, " Some Better Kutter's Formula Coefficients," *Eng. News.*, Feb. 24, May 4, 1916.

[3] C. E. Ramser, " Flow of Water in Drainage Channels," *U. S. Department of Agriculture, Technical Bulletin* 129, 1929.

VALUES OF ROUGHNESS COEFFICIENT n FOR USE IN
MANNING AND KUTTER FORMULAS (*Continued*)

DESCRIPTION OF CHANNEL	AVERAGE VALUE OF n
Straight unlined canals in earth, in best condition	.020
Conservative design value for unlined canals in earth and gravel, with some curves, in good condition	.0225
Corrugated metal flumes and culverts; unlined canals in earth and gravel, winding, in fair condition	.025
Canals with rough stony beds or with weeds on earth banks; dry rubble; dredged ditches, clean and straight; natural streams with clean, straight banks, full stage without sudden changes in cross section or alignment	.030
Winding natural streams, clean but with some pools and shoals; dredged ditches in fair condition	.035
Rivers of irregular cross section at low stages; streams and canals obstructed with vegetation and debris	.040 to .100
Overbank flow on flood plains	.100 to .175

129. Comparison of Open-channel Formulas. The Kutter, Manning, and Bazin formulas for C in the Chezy formula are the best known and most widely used. A large number of other flow formulas have been published, many of which doubtless possess merit. It is not ordinarily advisable, however, to use any but a commonly accepted formula unless there is good reason for so doing. The successful use of any open-channel formula requires an accurate knowledge of conditions and judgment in the selection of coefficients. Even the most experienced engineers may make considerable error in selecting coefficients, with corresponding error in their results.

The Kutter formula has for many years been the most widely used of all open-channel formulas. The use of the Manning formula, which was first used on irrigation works in Egypt, India, and Australia, has spread to practically all parts of the world. Engineers generally are recognizing its advantages over the more cumbersome Kutter formula.

The Kutter formula shows C to be a function of the slope S, while the Manning formula does not. If 3.28 ft is substituted for R, equations 8 and 11 both reduce to $C = 1.811/n$. Since S does not appear in this equation, it follows that, with the same value of n, when R equals 1 meter, the Kutter and the Manning formulas give identical results and, moreover, the Kutter formula gives the same value of C for all slopes.

VALUES OF C FROM THE KUTTER FORMULA

Slope S	n	Hydraulic Radius R in Feet														
		0.2	0.3	0.4	0.6	0.8	1.0	1.5	2.0	2.5	3.0	4.0	6.0	8.0	10.0	15.0
.00005	0.010	87	98	109	123	133	140	154	164	172	177	187	199	207	213	220
	.012	68	78	88	98	107	113	126	135	142	148	157	168	176	182	189
	.015	52	58	66	76	83	89	99	107	113	118	126	138	145	150	159
	.017	43	50	57	65	72	77	86	93	98	103	112	122	129	134	142
	.020	35	41	45	53	59	64	72	80	84	88	95	105	111	116	125
	.025	26	30	35	41	45	49	57	62	66	70	78	85	92	96	104
	.030	22	25	28	33	37	40	47	51	55	58	65	74	78	83	90
.0001	0.010	98	108	118	131	140	147	158	167	173	178	186	196	202	206	212
	.012	76	86	95	105	113	119	130	138	144	148	155	165	170	174	180
	.015	57	64	72	81	88	92	103	109	114	118	125	134	140	143	150
	.017	48	55	62	70	75	80	88	95	99	104	111	118	125	128	135
	.020	38	45	50	57	63	67	75	81	85	88	95	102	107	111	118
	.025	28	34	38	43	48	51	59	64	67	70	77	84	89	93	98
	.030	23	27	30	35	39	42	48	52	55	59	64	72	75	80	85
.0002	0.010	105	115	125	137	145	150	162	169	174	178	185	193	198	202	206
	.012	83	92	100	110	117	123	133	139	144	148	154	162	167	170	175
	.015	61	69	76	84	91	96	105	110	114	118	124	132	137	140	145
	.017	52	59	65	73	78	83	90	97	100	104	110	117	122	125	130
	.020	42	48	53	60	65	68	76	82	85	88	94	100	105	108	113
	.025	30	35	40	45	50	54	60	65	68	70	76	83	86	90	95
	.030	25	28	32	37	40	43	49	53	56	59	63	69	74	77	82
.0004	0.010	110	121	128	140	148	153	164	171	174	178	184	192	197	198	203
	.012	87	95	103	113	120	125	134	141	145	149	153	161	165	168	172
	.015	64	73	78	87	93	98	106	112	115	118	123	130	134	137	142
	.017	54	62	68	75	80	84	92	98	101	104	110	116	120	123	128
	.020	43	50	55	61	67	70	77	83	86	88	94	99	104	106	110
	.025	32	37	42	47	51	55	60	65	68	70	75	82	85	88	92
	.030	26	30	33	38	41	44	50	54	57	59	63	68	73	75	80
.001	0.010	113	124	132	143	150	155	165	172	175	178	184	190	195	197	201
	.012	88	97	105	115	121	127	135	142	145	149	154	160	164	167	171
	.015	66	75	80	88	94	98	107	112	116	119	123	130	133	135	141
	.017	55	63	68	76	81	85	92	98	102	105	110	115	119	122	127
	.020	45	51	56	62	68	71	78	84	87	89	93	98	103	105	109
	.025	33	38	43	48	52	55	61	65	68	70	75	81	84	87	91
	.030	27	30	34	38	42	45	50	54	57	59	63	68	72	74	78
.01	0.010	114	125	133	143	151	156	165	172	175	178	184	190	194	196	200
	.012	89	99	106	116	122	128	136	142	145	149	154	159	163	166	170
	.015	67	76	81	89	95	99	107	113	116	119	123	129	133	135	140
	.017	56	64	69	77	82	86	93	99	103	105	109	115	118	121	126
	.020	46	52	57	63	68	72	78	84	87	89	93	98	102	105	108
	.025	34	39	44	49	52	56	62	65	68	70	75	80	83	86	90
	.030	27	31	35	39	43	45	51	55	58	59	63	67	71	73	77

The terms involving S in the Kutter formula were introduced to make the formula agree with the measurements of flow of the Mississippi River by Humphreys and Abbott. These measurements have since been shown to have a possible error of as much as 10 or 15 per cent. Later experiments do not verify the conclusions of Ganguillet and Kutter regarding the influence of slope on the value of the coefficient, and it appears that the Kutter formula would be more satisfactory with the S terms omitted.

The Manning formula is simpler to use than the Kutter formula and, with the same value of n, gives about the same results except for flat slopes. Although, as noted above, the variation of n with form of cross section is relatively small, it appears that for a channel of a given roughness and slope there is slightly less variation of n with hydraulic radius in the Manning formula than in the Kutter formula. For the smoother channels the value of m in the Bazin formula appears to vary considerably with the hydraulic radius.

PROBLEMS

1. Determine the discharge of a new planed-timber flume of rectangular cross section 4 ft wide, flowing 2 ft deep, with a fall of 2.5 ft per 1000 ft.

2. Determine the discharge of a concrete-lined canal of a trapezoidal cross section, bottom width 5 ft, depth of flow 5 ft, side slopes $1\frac{1}{2}$ horizontal to 1 vertical, slope 3 ft per mile.

3. A circular concrete sewer 5 ft in diameter and flowing half full has a slope of 4 ft per mile. Determine the discharge.

4. A dredged river channel with sandy bottom has a bottom width of 140 ft, depth of 10 ft, and side slopes of 3 horizontal to 1 vertical. If the fall is 6 in. per mile, determine the discharge.

5. An earth canal carries a depth of water of 6 ft. The canal is 20 ft wide on the bottom and has side slopes of 1.5 horizontal to 1 vertical. $S = 0.0002$. Using a value of n of 0.025, compute the discharge by the Manning formula, and with this discharge determine: (a) the value of n in the Kutter formula; (b) the value of m in the Bazin formula.

6. What should be the slope of a planed-timber flume, rectangular in cross section, to carry 120 cfs if the width is 6 ft and the depth of flow is to be 4 ft?

7. What should be the slope of a semicircular smooth steel flume with smooth joints to carry 100 cfs, the diameter of the flume being 8 ft?

8. An earth canal is to carry 400 cfs at a mean velocity of 2.2 ft per sec. The side slopes are to be 2 horizontal to 1 vertical. The depth of water is to be one-fourth of the bottom width. Assuming that the canal is to be kept in good condition, find the necessary slope.

9. A V-shaped channel, built of unplaned lumber, carries 30 cfs at a velocity of 8 ft per sec. Each of the sides makes an angle of 45 degrees with the horizontal. Determine the slope.

10. A rectangular concrete-lined canal is to carry 50 cfs. The bottom width is to be twice the depth. What should be the dimensions of the stream cross section if the slope is to be 0.001?

11. What would be the probable depth of flow of a river at low stage 300 ft wide carrying 10,000 cfs with a fall of 1 ft per mile. Assume $R = d$.

12. Determine the proper size of a semicircular wood-stave flume to carry 500 cfs across a valley 3000 ft wide with a drop of 2 ft.

13. An earth canal in good condition, having a bottom width of 12 ft and side slopes of 2 horizontal to 1 vertical, is designed to carry 200 cfs. If the slope of the canal is 2.1 ft per mile, determine the depth of water.

14. A smooth-metal flume of semicircular cross section has a diameter of 6 ft and a slope of 0.005. What diameter of corrugated metal flume will be required to have the same capacity?

130. Alternate Stages of Flow. The channel illustrated in Fig. 119 carries water at a depth d and mean velocity V. The total

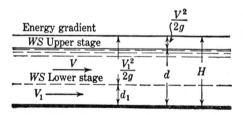

Fig. 119. Two stages of equal energy.

head H, or the energy per pound of water, measured above the bed of the channel is

$$H = d + \frac{V^2}{2g} \tag{13}$$

whence

$$V = \sqrt{2g(H - d)} \tag{14}$$

If A is the area of cross section, the discharge of the channel is

$$Q = A \sqrt{2g(H - d)} \tag{15}$$

For a rectangular cross section, the discharge per foot width of channel is

$$q = d \sqrt{2g(H - d)} \qquad (16)$$

The variation of q with d, from equation 16, with H given a constant value of 10 ft, and for values of d from zero to H, is shown in Fig. 120. It appears from the curve that, within the limits indicated, there are two depths (illustrated also by d and d_1, Fig. 119) at which any given discharge will flow with the same energy content, that is, with the same total head H. For example, with $H = 10$ ft, 50 cfs per foot width of channel will flow at depths of 2.2 and 9.6 ft.

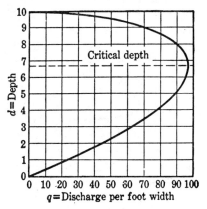

FIG. 120. Variation of discharge with depth for constant energy ($H = 10$ ft.).

The two depths of flow are called alternate stages, and are spoken of as the upper stage and the lower stage. Because of the smaller cross sectional area, the velocity is greater at the lower stage than at the upper stage.

Also, from Fig. 120, for $H = 10$ ft, q reaches a maximum at a value of d between 6 and 7 ft. The value of d for maximum q is called critical depth and can be determined by differentiating equation 16. Thus

$$\frac{dq}{dd} = \sqrt{H - d} - \frac{1}{2}\frac{d}{\sqrt{H - d}} = 0 \qquad (17)$$

from which

$$H = \tfrac{3}{2}d_c \quad \text{or} \quad d_c = \tfrac{2}{3}H \qquad (18)$$

where d_c denotes *critical depth.*

Critical depth is the depth at which, for a given total head, the discharge is a maximum, or, conversely, the depth at which a given flow occurs with a minimum content of energy.

Substituting $\tfrac{3}{2}d_c$ for H in equation 16, when $d = d_c$,

$$q = \sqrt{g}\, d_c^{3/2} \quad \text{or} \quad d_c = \sqrt[3]{\frac{q^2}{g}} \qquad (19)$$

With $H = 10$ ft, $d_c = 6.67$ ft, and $q = 97.6$ cfs. Critical depth for any q in a rectangular channel can be computed from the second form of equation 19.

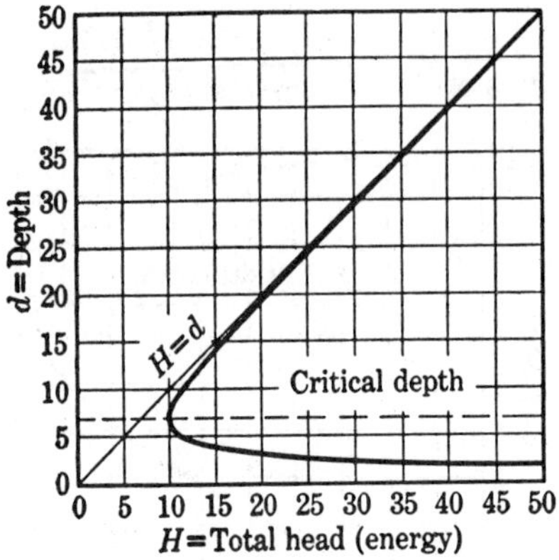

FIG. 121. Variation of energy with depth for constant discharge ($q = 97.6$ cfs).

If q in equation 16 is given the value 97.6 and H is then plotted against d, the curve of Fig. 121 is obtained. The upper leg of the curve is asymptotic to the line $H = d$, since as the depth increases the velocity head approaches zero. The lower leg is asymptotic to the H axis, since as the depth decreases the velocity head rapidly increases, reaching ∞ at zero depth. The depth of minimum energy for the given q is critical depth and as shown in equation 18 has the value $\frac{2}{3}H$. It is important to note that the velocity head then equals $\frac{1}{3}H$. Hence

$$\frac{V_c{}^2}{2g} = \frac{d_c}{2} \quad \text{or} \quad V_c = \sqrt{gd_c} \tag{20}$$

Equation 20 provides a simple criterion for determining whether a given stream is flowing at upper or lower stage. If the velocity head is less than half the depth, flow is at upper stage, while if the velocity head is greater than half the depth, flow is at lower stage. If the velocity head equals half the depth, flow is occurring at critical depth.

It will be observed from Fig. 121 that, at or near critical depth, a relatively large change in depth corresponds to a relatively small change in energy. Flow in this region is therefore quite unstable and is usually indicated by excessive turbulence and characteristic water-surface undulations.

131. Critical Slope. For any given discharge and cross section of channel there is always a slope just sufficient to maintain flow at critical depth. This is termed the critical slope. The numerical value of this critical slope can be computed by the Manning formula, or other open-channel formula, after the critical depth corresponding to the given discharge has been determined. Because of the wavy stream surface which is characteristic of uniform flow at or near critical depth, the design of channels with slopes near the critical should be avoided as far as possible.

PROBLEMS

1. Compute critical depth for a rectangular flume 10 ft wide carrying 400 cfs. Draw the depth-energy curve for values of d from 2 to 8 ft.

2. Critical depth in a rectangular flume 5 ft wide is 2.61 ft. Compute the discharge.

3. What should be the slope of a planed-timber flume, rectangular in cross section, to carry 30 cfs if the width is 4 ft and the depth of flow is to be 1.5 ft? Is the flow at upper or lower stage? Determine critical depth and critical slope for the given discharge and width.

4. Water is flowing in a rectangular flume with a velocity of 12.5 ft per sec and a depth of 3.20 ft. Is the flow at upper or lower stage? What is the alternate depth of flow with equal energy?

5. Determine the critical slope of a rectangular smooth concrete flume 20 ft wide which is to carry 50 cfs per foot of width.

132. Cross Section of Greatest Efficiency. The most efficient cross section of an open channel, from a hydraulic standpoint, is the one which, with a given slope, area, and roughness factor, will have the maximum capacity. This cross section is the one having the smallest wetted perimeter, as can be seen from an examination of one of the open-channel formulas. There are usually practical objections to using cross sections of greatest hydraulic efficiency, but the dimensions of such cross sections should be known and adhered to as closely as conditions appear to justify.

Of all open-channel cross sections having a given area, the semicircle has the smallest wetted perimeter, and it is therefore the cross section of highest hydraulic efficiency. Only a few engineering materials of construction, however, are adapted to a semicircular cross section, sheet metal or steel plates which take tensile stress readily and can be supported along the sides of the flume being best adapted to it. Wood staves, supported by metal tie rods, are adapted to the semicircular cross section. Semicircular canals have frequently been lined with brick or stone masonry and have also been built using precast concrete sections.

The semicircular cross section is not well adapted to flumes built of planks and structural timber or of concrete poured in place, nor to unlined canals in earth or gravel because these soils will not stand on a steep slope. For such construction materials, therefore, a rectangular or trapezoidal cross section is ordinarily used.

Timber flumes are ordinarily built with vertical sides. Concrete lining can be placed on any slope up to 1 to 1 without forms but for steeper slopes, forms are required. Asphalt and tar mixes have also been used in lining trapezoidal channels. Unlined earth banks will ordinarily not stand on slopes greater than $1\frac{1}{2}$ horizontal to 1 vertical, whereas canals in sandy material and river levees may require side slopes as flat as 3 to 1.

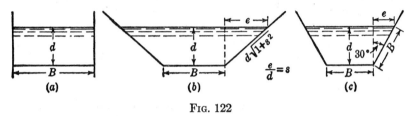

Fig. 122

Properties of trapezoidal sections and methods of determining sections of greatest efficiency are shown in the following analysis:

In Fig. 122, let the side-slope ratio $e/d = s$. Then

$$A = Bd + sd^2 \tag{21}$$

from which

$$B = \frac{A}{d} - sd \tag{22}$$

Also

$$P = B + 2d\sqrt{1 + s^2} = \frac{A}{d} - sd + 2d\sqrt{1 + s^2} \tag{23}$$

Assuming A and s constant and equating to zero the first derivative of P with respect to d,

$$\frac{dP}{dd} = -\frac{A}{d^2} - s + 2\sqrt{1 + s^2} = 0 \tag{24}$$

Substituting for A from equation 21,

$$\frac{Bd + sd^2}{d^2} = 2\sqrt{1 + s^2} - s \tag{25}$$

or

$$B = 2d(\sqrt{1 + s^2} - s) \tag{26}$$

from which the relation between depth of water and bottom width of canal of the most efficient trapezoidal cross section can be obtained for any value of s.

From equations 21 and 23,

$$R = \frac{A}{P} = \frac{Bd + sd^2}{B + 2d \sqrt{1 + s^2}} \tag{27}$$

Substituting B from equation 26 and reducing,

$$R = \frac{d}{2} \tag{28}$$

or, the trapezoidal cross section of greatest efficiency has a hydraulic radius equal to one-half the depth of water.

Expressing d in terms of A and s from equations 22 and 26,

$$d = \sqrt{\frac{A}{2\sqrt{1 + s^2} - s}} \tag{29}$$

Substituting this value in equation 23

$$P = 2 \sqrt{A} \sqrt{2 \sqrt{1 + s^2} - s} \tag{30}$$

Equating to zero the first derivative with respect to s and reducing,

$$s = 1/\sqrt{3} = \tan 30° \tag{31}$$

and the section becomes a half hexagon (Fig. 122c). Thus, of all the trapezoidal sections (including the rectangle), for a given area, the half hexagon has the smallest perimeter and is therefore the most efficient cross section.

A semicircle having its center in the middle of the water surface can always be inscribed within a cross section of maximum efficiency. This is illustrated for a trapezoidal cross section in Fig. 123. OA, OB, and OC are drawn from a point O on the center line of the water surface perpendicular to the sides of the channel EF, FG, and GH, respectively. Let $EF = GH = x$; $FG = B$; $OA = OC = r$; and $OB = d$. As before, A = area of section and P = wetted perimeter. Then from the figure

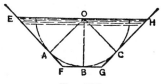

FIG. 123

$$A = xr + \tfrac{1}{2}Bd \quad \text{and} \quad P = 2x + B$$

Since $R = \frac{1}{2}d$,

$$R = \frac{A}{P} = \frac{xr + \frac{1}{2}Bd}{2x + B} = \frac{d}{2}$$

from which

$$r = d$$

That is, OA, OB, and OC are all equal, and a semicircle with center at O is tangent to the three sides.

133. Circular Sections. The maximum discharge from a channel of circular cross section occurs at a little less than full depth, as can be seen from an examination of open-channel formulas. In the investigation of a particular channel n and S are constant. From Fig. 124, r being the radius of the circle,

$$P = \frac{360 - \theta}{360} \times 2\pi r \tag{32}$$

and

$$A = \frac{360 - \theta}{360} \times \pi r^2 + \frac{1}{2}r^2 \sin \theta \tag{33}$$

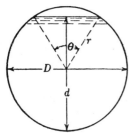

FIG. 124. Circular channel.

With these equations an expression for $AR^{\frac{2}{3}}$ can be written, differentiating which and equating to zero, the value of θ which makes Q a maximum is found to be 57° 40'. The corresponding depth of water is $d = 0.938D$. Other open-channel formulas give substantially the same result. This means that a pipe carrying water not under pressure, when free from obstructions and laid on a true grade, will not flow full. Since there is always likelihood of slight backwater, it is usual to assume that the maximum capacity occurs with the conduit flowing full.

134. Irregular Channels. Open-channel formulas do not apply accurately to natural streams since the channel sections and slope of water surface vary and the flow is non-uniform. At the lower stages, streams usually contain alternating reaches of riffles and slack water. During high stages this condition largely disappears, and the water surface becomes approximately parallel to the average slope of the bottom of the channel. The degree of roughness of natural streams varies greatly within short reaches and even within different parts of the same cross section. This can be seen from Fig. 125, which illustrates a stream in flood stage. The channel of normal flow, *abc*, will probably have a different coefficient of roughness from that of the floodplain *cde*. Also the portion of the left-hand bank lying above ordinary high water may be covered with trees or other vegetation and have a higher coefficient than the lower portion. Rocks and other channel irregularities cause varying conditions of turbulence, and their effect on the coefficient of roughness is difficult to determine.

There are times when the engineer must estimate as well as he can the carrying capacity of a natural channel. Under such circumstances it is usual to make a survey of a selected reach of channel to determine a slope of water surface and an average cross section to use with an open-channel formula. The degree of accuracy to be secured from results obtained in this manner will depend largely upon the ability of the engineer to judge of the effect of the various factors

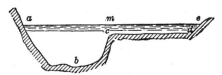

Fig. 125. Channel with irregular cross section.

upon the coefficient of roughness. Better results are obtained where streams have fairly straight and uniform channels and are relatively free from conditions causing turbulence. As the stage increases, and the water surface assumes a more regular slope, open-channel formulas apply more accurately.

Where an open-channel formula must be applied to an irregular section such as that indicated in Fig. 125, *the cross section is divided into two portions by a vertical line mc,* and the discharges are computed for each portion separately. As the two portions of the channel differ in roughness, different coefficients should be selected for each.

135. Obstructions and Bends. The most common types of obstructions in open channels are submerged weirs, gates, and bridge piers. Losses of head resulting from weirs and gates are treated in an earlier chapter.

Bridge piers, Fig. 126, constrict a channel and obstruct the flow. The loss of head, h_g, or the amount that the water will be backed up, will not usually be of any importance except where velocities are comparatively high. The most important case arises in determining the backing-up effect of bridge piers during flood stages of streams. The total loss of head is made up of three parts: a loss of head due to contraction of the channel at the upstream end of the

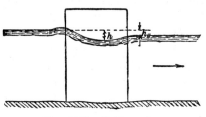

FIG. 126.   Bridge pier.

piers, a loss of head due to enlargement of the channel at the downstream end, and an increase in loss of head due to friction resulting from the increase in velocity in the contracted portion of the channel. On account of the higher velocity, the surface of the water between the piers is depressed, the vertical distance, h, measuring the increase in velocity head plus the loss of head. The distance $h - h_g$ is a measure of the velocity head reconverted into static head.

Piers should be so designed as to permit changes in velocity to occur with a minimum amount of turbulence. Figure 127 represents two horizontal sections of piers. Section A will cause less turbulence and consequently less loss of head than section B.

Curves or *bends* in the alignment of a channel cause a loss of head in addition to the loss that would occur if the channel were straight but otherwise unchanged. This loss is small for low velocities such as occur in earth canals, and ordinarily no allowance is made for it unless the curves are frequent and sharp. Where velocities are relatively high, a greater slope should be provided for curved reaches than for tangents. Common practice is to allow for loss

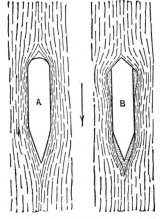

FIG. 127.   Effect of shape of piers on turbulence.

of head at curves in selecting the coefficient of roughness for the open-channel formula.

136. Velocity Distribution in Natural Streams. The distribution of velocities in a river of irregular cross section, as determined from measurements with a current meter, is shown in the upper portion of Fig. 128. The numerals show velocities at the points where measurements were made, and the irregular lines are interpolated equal velocity lines.

The curves in the lower portion of Fig. 128 show the distribution of velocities in vertical lines. These curves are called vertical velocity curves, and the velocities from which they are plotted are

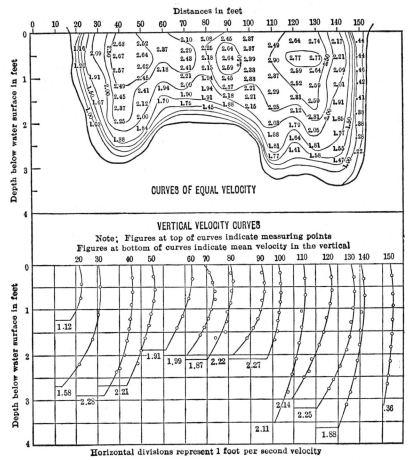

FIG. 128. Velocities in natural stream.

called velocities in the vertical. The following properties of vertical velocity curves have been determined from measurements of velocities of a large number of streams and a study of curves plotted from them.

1. In general, the maximum velocity occurs somewhere between the water surface and one-third of the depth, the distance from

the surface to the point of maximum velocity being proportionally greater for greater depths of water. For shallow streams the maximum velocity is very near to the surface; for very deep streams it may lie at about one-third of the depth. A strong wind blowing either upstream or downstream will affect the distribution of velocities in the vertical.

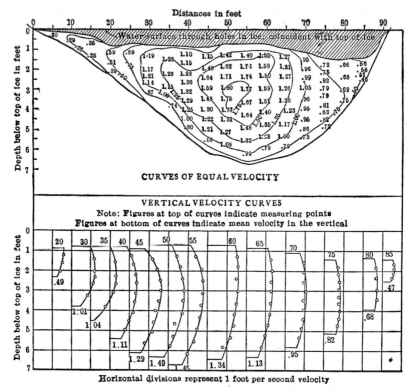

FIG. 129. Velocities in ice-covered stream.

2. The mean velocity in the vertical is ordinarily at 0.55 to 0.65 of the depth. The velocity at 0.6 depth is usually within 5 per cent of the mean velocity.

3. The mean of velocities at 0.2 depth and 0.8 depth usually gives the mean velocity in the vertical within 2 per cent.

4. The mean velocity in the vertical is ordinarily 0.80 to 0.95 of the surface velocity. The smaller percentage applies to the shallower streams.

These four properties of vertical velocity curves are made use of in measuring the discharge of streams.   Mean velocities in successive verticals are first obtained by measuring the velocity at 0.6 of the depth in each vertical or, where greater accuracy is required, by taking the mean of the velocities at 0.2 and 0.8 of the depth. The mean of velocities in any two adjacent verticals is considered to be the mean velocity between the verticals.   The area between the verticals having been determined, the discharge through this portion of the cross section of the stream is the product of this area and the mean velocity.   The sum of all discharges between successive verticals is the total discharge.

The distribution of velocities in an ice-covered stream, Fig. 129, is modified by the retarding influence of the ice.   The amount of this retardation exceeds the skin friction of a free water surface, and the maximum velocity therefore occurs nearer mid-depth. The mean velocity in the vertical for an ice-covered stream is not at 0.6 depth, but the mean of velocities at 0.2 and 0.8 depth gives approximately the mean velocity the same as for a stream with a free surface.

PROBLEMS

1. If the most efficient of all cross sections can be used, what shape and size of open channel would you recommend to carry 300 cfs with a velocity of 7 ft per sec?

2. An open sewer is to be of most efficient cross section and lined with brick, well laid, on a grade of 3 ft per mile.   The flow capacity is to be 1000 cfs.   Determine the proper cross section.

3. If the most efficient of all trapezoidal cross sections can be used, what shape and size of open channel would you recommend to carry 300 cfs with a velocity of 7 ft per sec.?   Compare resulting wetted perimeter with that of problem 1.

4. What should be the width and the depth of flow of a rectangular planed-timber flume of most efficient cross section to carry 90 cfs, with a velocity of 5 ft per sec?   What grade should the flume have?   What percentage less flow would this flume carry if $d = 2$ ft and $B = 9$ ft?

5. What should be the bottom width and the depth of flow in a concrete-lined canal of most efficient trapezoidal cross section with side slopes $\frac{1}{2}$ horizontal to 1 vertical to carry 400 cfs on a grade of 3 ft per mile?

6. Find the most efficient cross section and the required grade of a trapezoidal canal in clean earth with good alignment to carry 470 cfs at a velocity of 3 ft per sec, assuming side slopes of 2 horizontal to 1 vertical.

7. What should be the bottom width and the depth of flow for a concrete-lined canal of most efficient cross section with side slopes $\frac{3}{4}$ horizontal to 1 vertical to carry 1200 cfs on a grade of 5 ft per mile? What is the velocity of flow?

8. Determine the slope in feet per mile that a circular concrete sewer, 5 ft in diameter, must have when flowing at its maximum capacity if the mean velocity is 8 ft per sec.

9. A trapezoidal concrete-lined canal is to have side slopes of $\frac{1}{2}$ horizontal to 1 vertical and a bottom width of 8 ft. What will be the depth of flow for best hydraulic efficiency and what will be the capacity of the canal if the grade is 2 ft per mile?

10. A planed-timber flume is to have a rectangular cross section of best efficiency. The velocity is not to exceed 8 ft per sec on a grade of 5 ft per mile. What should be the dimensions of the cross section and the capacity of the canal? Is flow at upper or lower stage?

11. An earth canal in good condition is 60 ft wide on the bottom and has side slopes of 2 horizontal to 1 vertical. One side slope extends to an elevation of 20 ft above the bottom of the canal. The other bank, which is a practically level meadow at an elevation of 6 ft above the bottom of the canal, extends back 500 ft from the canal and then rises abruptly The meadow is covered with short grass and weeds. If the slope of the canal is 2.2 ft per mile, determine the discharge when the water is 8 ft deep in the canal.

12. The river at flood stage shown in Fig. 125 has an average slope of 0.001. For the main channel $n = 0.030$ and for the flood plain $n = 0.040$. Area $abcm = 3000$ sq ft, area $mcde = 900$ sq ft. Wetted perimeter $abc = 200$ ft and $cde = 300$ ft. Compute the discharge when the depth of flow on the flood plain is 3 ft. Compute the false discharge obtained by applying the same formula directly to the entire cross section, assuming an average value of n of 0.035.

Non-uniform Flow

137. General. Uniform flow is approached closely in long flumes and conduits of uniform cross section and straight grade but is difficult and sometimes impossible to secure in short flumes. Moreover, non-uniform flow occurs where the stream enters and leaves the channel; at obstructions such as dams, weirs, or bridge piers; and at changes in the form of cross section which may be necessitated by natural conditions of soil and topography.

Changes in cross section in open-channel flow may be either gradual or abrupt. Gradual changes in which the flow is either accelerated or retarded may be analyzed by means of the energy theorem in conjunction with a formula for open channel flow. Abrupt changes, with the accompanying secondary effects of vertical curvature of the stream lines when the flow is accelerated, and excessive turbulence, waves, and surface rollers when the flow is retarded are sometimes more difficult to analyze.

The variety of conditions encountered in open-channel flow is greater than in pipe flow both because of the existence of the free surface and also because of the two alternate stages of flow with equal energy. Only a few of the most frequently occurring cases can be considered here.

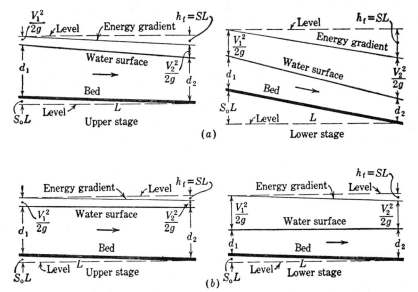

FIG. 130. Gradually accelerated and retarded flow.

138. Gradually Accelerated and Retarded Flow. The mathematical analysis of the problem of gradually accelerated and retarded flow is given in this article; the conditions which produce non-uniform flow of this type are described in Arts. 139 and 140. Gradual acceleration and retardation of flow can occur at either upper or lower stage, but flow does not ordinarily pass gradually from one stage to the other. Such a change is usually abrupt.

The same principles apply to both accelerated and retarded flow. They can be best investigated by considering the channel divided into reaches. In Fig. 130, (a) shows accelerated flow at both upper and lower stages and (b) shows retarded flow. The length of reach in each case is L, and the slope of the bottom of the channel is S_o. The loss of head in the reach is $h_f = SL$, where S is the slope of the energy gradient. Mean velocities at sections 1 and 2, respectively, the upstream and the downstream ends of the reach, are V_1 and V_2, and the corresponding depths are d_1 and d_2.

The mean velocity, V_m, area, A_m, wetted perimeter, P_m, and hydraulic radius, R_m, in the reach are considered to be the means of the respective values at the two ends of the reach. Slightly different results are obtained if d_m is assumed to be $\frac{1}{2}(d_1 + d_2)$ and the corresponding V_m and R_m are used. Both methods are approximations, but the error introduced by their use can be kept within any desired range by properly limiting the velocity change in the reach.

Assuming the datum to be the bed of the channel at the downstream section for any of the four cases illustrated in Fig. 130, the energy equation is

$$\frac{V_1{}^2}{2g} + d_1 + S_oL = \frac{V_2{}^2}{2g} + d_2 + SL \tag{34}$$

whence

$$L = \frac{\left(\dfrac{V_2{}^2}{2g} + d_2\right) - \left(\dfrac{V_1{}^2}{2g} + d_1\right)}{S_o - S} \tag{35}$$

But by the Manning formula

$$S = \left(\frac{nV_m}{1.486R_m{}^{\frac{2}{3}}}\right)^2 \tag{36}$$

With S_o and n known and the velocity and cross section at either end of the reach given, the distance L to the cross section corresponding to any other depth can be computed directly. If the length of reach and one cross section are given, the depth at the other end of the reach can be obtained by trial solutions of equations 34 and 36. The latter method is practically always used in the solution of problems such as backwater in natural streams which are irregular in cross section, slope, and alignment.

In measurements of the value of roughness factor in channels where, as frequently happens, it is not possible to obtain uniform flow, use must be made of the foregoing non-uniform flow theory. The discharge must be measured as well as the cross-sectional dimensions at each end of the test reach, the length of the reach, and the slope of the stream bed.   It is usually advisable to determine the cross-sectional dimensions at several intermediate points in order to obtain more precise values of V_m and R_m.   From these data S can be computed from equation 36 and inserted along with computed V_m and R_m in the Manning or the Kutter-Chezy formula for computation of n.

139. Conditions Producing Accelerated and Retarded Flow. Various conditions producing accelerated and retarded flow are shown in Figs. 131 to 136.   Figure 131 shows a canal with a uniform slope which is flatter than the critical slope for a given dis-

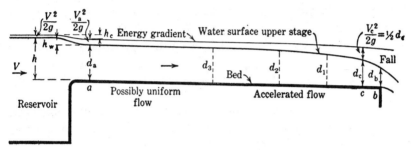

FIG. 131.   Channel with accelerated flow at upper stage.

charge (Art. 131).   Water enters the canal from a reservoir and discharges over a fall.   Critical depth d_c in this case occurs a short distance upstream from the brink of the fall.   O'Brien found[1] that, for a channel with level bed, the location of critical depth was approximately $12d_c$ upstream from the brink, the distance increasing as the slope of the channel increased.

For rectangular channels the discharge q per foot width of stream and the critical depth are related by the equation

$$d_c = \sqrt[3]{\frac{q^2}{g}} \qquad (19)$$

If q is known, the depth of flow just above the fall can be computed. Rouse[2] has shown that for slopes less than the critical the depth

---

[1] M. P. O'Brien, " Analyzing Hydraulic Models for Effects of Distortion," *Eng. News-Record*, Sept. 15, 1932.

[2] H. Rouse, " Discharge Characteristics of the Free Overfall," *Civil Engineering*, April 1936, p. 257.

d_b at the brink is $0.715d_c$, or $d_c = 1.40d_b$. If d_b can be measured, q can therefore be computed from the relation

$$q = \sqrt{g \times d_c^3} = 9.4d_b^{3/2} \tag{37}$$

Proceeding upstream from the fall the depth gradually increases but at a decreasing rate. If the flume is relatively long compared to the depth of flow, the depth in the upper reach of the channel may become practically constant, that is, the flow may be uniform. The equations of uniform flow apply to such a reach while the principles of Art. 138 apply to the gradually accelerated flow approaching the fall.

The conditions at entrance to the channel can be analyzed by writing the energy equation from the reservoir to point a. With the channel bed at entrance as the datum, the energy equation is

$$h + \frac{V^2}{2g} = d_a + \frac{V_a^2}{2g} + h_c \tag{38}$$

where h_c is the loss of head due to contraction of the stream. This loss is similar in nature to that occurring at a contraction in a pipe and can be similarly expressed as

$$h_c = K_c \frac{V_a^2}{2g} \tag{39}$$

Values of K_c for open-channel contractions have not been as well determined as for pipes, but it appears that they are quite similar. The maximum value of K_c for square-cornered contraction may thus be taken as 0.5 with smaller values for rounded or tapered contractions. With care in design the value of K_c may be reduced nearly to zero.

It is important to note the drop h_w in the water surface at the entrance. This drop would occur even if there were no loss of head since a portion of the elevation head in the reservoir is changed to velocity head at a. The amount of this drop can be determined by writing the energy equation with the water surface at a as datum. Thus

$$h_w + \frac{V^2}{2g} = \frac{V_a^2}{2g} + h_c \tag{40}$$

or

$$h_w = \frac{V_a^2}{2g} - \frac{V^2}{2g} + h_c \tag{41}$$

The determination of the capacity of a given flume of this type involves simultaneous agreement between: (1) entrance conditions, (2) uniform flow in the flume, (3) gradually accelerated flow in the flume, and (4) critical depth relations at the fall. Such a problem is usually best solved by trial, as shown in the following example.

EXAMPLE. Determine the capacity of the wooden flume of rectangular cross section illustrated in Fig. 131 if the length is: (a) 5000 ft; (b) 300 ft. The entrance is rounded with $K_c = 0.1$. The width of flume is 10 ft and the slope of the bed is 0.001. The flume takes water from a large reservoir with the water surface 8 ft above the flume bed at entrance. There is a free fall at the discharge end. Assume $n = 0.012$.

Solution. (a) With so long a flume the flow for some distance below the entrance should be approximately uniform. The problem can be solved by assuming Q, determining by trial the depth of uniform flow for that Q, then testing the solution by the energy equation at entrance, neglecting velocity head in the reservoir,

$$8 = d_a + 1.1 \frac{V_a^2}{2g}$$

The table form is convenient for the solution:

Q	d_a	A	P	R	Manning V_a	Q	$V_a^2/2g$	$d_a + 1.1 V_a^2/2g$
cfs	ft	sq ft	ft	ft	ft per sec	cfs	ft	ft
Try 500	Try 6.0	60	22	2.73	7.64	460		
	6.5	65	23	2.82	7.83	510		
	6.4	64	22.8	2.81	7.80	500	0.95	7.44 Too low
Try 550	Try 7.0	70	24	2.91	8.00	560		
	6.9	69	23.8	2.90	7.98	550	0.99	7.99 OK

With $Q = 550$ cfs, $q = 55$ cfs, and by equation 19, $d_c = 4.54$ ft. With as small a slope as 0.001 the location of critical depth at $12d_c$ (page 267) or about 54 ft upstream from the fall is sufficiently accurate. Above this point the flow is at upper stage and the surface curve can be traced by successive solutions of equation 35 for the distance L from a point of known depth to an upstream point of assumed depth.

With $d_c = 4.54$ ft, $V_c^2/2g = \frac{1}{2}d_c = 2.27$ ft and $H_c = d_c + V_c^2/2g = 6.81$ ft. Find distance L to point at which depth is, say, 5.00 ft. With $d_1 = 5.00$ ft, $V_1 = 11.0$ ft per sec, $V_1^2/2g = 1.88$ ft, and $H_1 = 6.88$ ft. Mean values in the reach are used to determine the slope of the energy gradient. Thus $d_m = 4.77$ ft, $A_m = 47.7$ sq ft, $V_m = 11.53$ ft per sec,

$P_m = 19.54$ ft, $R_m = 2.44$ ft, and by Manning's formula $S = 0.00264$.
By equation 35,

$$L = \frac{6.81 - 6.88}{0.001 - 0.00264} = \frac{-0.07}{-0.00164} = 43 \text{ ft}$$

and ΣL from brink $= 54 + 43 = 97$ ft.

Similar solutions for successive reaches give these lengths:

From depth of	To depth of	L ft	ΣL ft
d_b	$d_c = 4.54$ ft	54	54
$d_c = 4.54$ ft	$d_1 = 5.00$	43	97
$d_1 = 5.00$	$d_2 = 5.50$	165	262
$d_2 = 5.50$	$d_3 = 6.00$	426	688
$d_3 = 6.00$	$d_4 = 6.50$	1052	1740
$d_4 = 6.50$	$d_5 = 6.70$	1060	2800
$d_5 = 6.70$	$d_6 = 6.80$	1170	3970

As depth of uniform flow, 6.9 ft, is approached the difference between S_0 and S approaches zero and length L approaches ∞ for even a small change in depth. The water surface curve actually is asymptotic to the line $d = 6.9$, but practically uniform flow exists for several hundred feet downstream from the entrance.

(b) With the short length of 300 ft it is doubtful that uniform flow exists in any reach of appreciable length. One method of solution is to assume a value of critical depth, trace the water surface curve upstream to the entrance, and check the energy equation at that point. Since the capacity of the short flume should be greater than that of the long flume in (a) the critical depth should be somewhat larger.

Try $d_c = 4.70$ ft. Then $A_c = 47.0$ sq ft, $V_c^2/2g = 2.35$ ft, $V_c = 12.3$ ft per sec, and $Q = 578$ cfs. Tracing the water surface upstream as in (a) gives the following values

From depth of	To depth of	L ft	ΣL ft
d_b	$d_c = 4.70$ ft	56	56
$d_c = 4.70$ ft	$d_1 = 5.20$	43	99
$d_1 = 5.20$	$d_2 = 5.70$	169	268

With only 32 ft of length left, the depth at entrance would evidently be about 5.8 ft, with $V_a = 9.97$ ft per sec, $V_a^2/2g = 1.55$ ft, and $d_a + 1.1 V_a^2/2g = 5.8 + 1.7 = 7.5$ ft, which is too low.

Try $d_c = 5.00$ ft. Then $A_c = 50.0$ sq ft, $V_c^2/2g = 2.50$ ft, $V_c = 12.7$ ft per sec, and $Q = 635$ cfs.

From depth of	To depth of	L ft	ΣL ft
d_b	$d_c = 5.00$ ft	60	60
$d_c = 5.00$ ft	$d_1 = 5.50$	43	103
$d_1 = 5.50$	$d_2 = 6.00$	144	247
$d_2 = 6.00$	$d_3 = 6.10$	55	302

With $d_a = 6.10$ ft, $V_a{}^2/2g = 1.69$ ft and $d_a + 1.1V_a{}^2/2g = 6.10 + 1.86 = 7.96$ ft, which is close enough agreement. The discharge is therefore about 640 cfs.

Figure 132 shows a canal with a slope which is steeper than the critical. Critical depth occurs close to the end of the drop-down curve at the head of the flume. From that point, since the slope

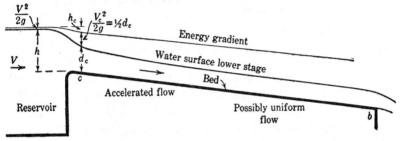

FIG. 132.   Channel with accelerated flow at lower stage.

is steeper than the critical, the flow accelerates gradually but at a decreasing rate. If the flume is long a condition of practically uniform flow may be established in the lower part.

Entrance conditions are again represented by writing the **energy** equation with respect to the stream bed at c as datum. Thus

$$h + \frac{V^2}{2g} = d_c + \frac{V_c{}^2}{2g} + h_c \qquad (42)$$

Since from the theory of critical depth $d_c = 2\dfrac{V_c{}^2}{2g}$, and since $h_c = K_c\dfrac{V_c{}^2}{2g}$, equation 42 can be written

$$h + \frac{V^2}{2g} = (3 + K_c)\frac{V_c{}^2}{2g} \qquad (43)$$

Gradually retarded flow at lower stage is illustrated in Fig. 133. Water enters the canal through a sluice gate (see page 136). The loss of head at the gate is shown by the drop h_o in the energy gradient. For the conditions shown, flow from the gate is at lower stage, that is, $d_a < 2V_a{}^2/2g$. If the slope of the canal is less than the critical, the flow is gradually retarded downstream from point a. If the same slope is maintained for a sufficient distance, a hydraulic jump will occur. (See Art. 141.)

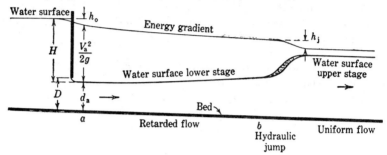

FIG. 133.   Channel with retarded flow at lower stage.

If the water surface downstream from the jump is raised sufficiently to move the jump upstream against the gate, the flow through the gate becomes submerged, as shown in Fig. 66b.

Gradually retarded flow at upper stage, known as backwater, is illustrated in Fig. 134. The normal stage of flow, for the given discharge, slope, and roughness of channel, is indicated by the

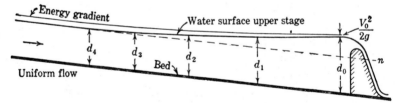

FIG. 134.   Channel with retarded flow at upper stage (backwater).

dashed line n. The overflow dam raises the water surface to an elevation depending on the height of the dam and the head required to produce discharge, as determined by a suitable weir formula. With a known elevation of the water surface just upstream from the dam, the principles of Art. 138 can be applied to successive reaches to determine the water-surface elevation at any desired distance upstream.

140. Effect of High Stage Downstream. If the water-surface elevation in the channel or reservoir downstream from the fall in

Fig. 131 is raised above the brink b the fall is submerged and a wave of increased depth (Art. 144) moves upstream into the flume. The height of this wave as it moves upstream gradually diminishes, but the depth of flow may be increased throughout the entire flume and even in the approach channel or head reservoir. After flow

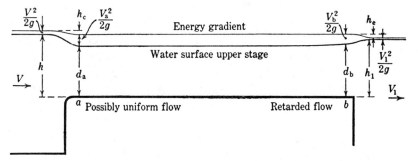

FIG. 135.   Channel with small slope — effect of high stage downstream.

conditions are stabilized the profile of flow will be similar to that shown in Fig. 135, with flow at upper stage at all points.

The energy relations at entrance are again

$$h + \frac{V^2}{2g} = d_a + \frac{V_a{}^2}{2g} + h_c \qquad (44)$$

Downstream from the drop-down curve, if the flume is long, there may be a reach of approximately uniform flow, following which the depth gradually increases and the flow is gradually retarded.

At the discharge end of the flume an enlargement of cross-sectional area occurs, with a loss of head due to enlargement which can be expressed as

$$h_e = K_e \frac{V_b{}^2}{2g} \qquad (45)$$

where V_b is the mean velocity in the smaller channel. As with pipes the value of K_e is larger for a sudden increase of cross-sectional area than the value of K_c for a sudden contraction. Experiments indicate that for carefully designed transitions in which the velocity change is made to take place gradually and at an approximately constant rate K_e may be made as small as 0.2 or, under particularly favorable conditions, even less, but for abrupt changes practically all the velocity head in the smaller channel will be lost, that is, K_e is from 0.9 to 1.0.

The flow at the enlargement can be analyzed by writing the energy theorem from one side of the transition to the other with respect to the stream bed at b as datum:

$$d_b + \frac{V_b{}^2}{2g} = h_1 + \frac{V_1{}^2}{2g} + h_e \qquad (46)$$

If the water surface downstream from a flume with steeper than critical slope (Fig. 132) is raised above the brink of the fall, a wave again tends to move upstream. However, the velocity of flow in the flume is greater than the velocity with which the wave tends to advance (Art. 144). The result is a hydraulic jump (Fig. 136) which forms in the flume at a position (Art. 142) depending on the

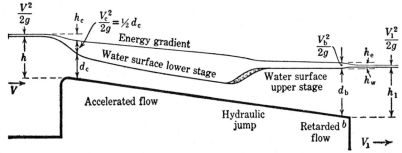

FIG. 136. Channel with large slope — effect of high stage downstream.

relative velocities and depths above and below the jump. The effect of the higher stage downstream is not transmitted upstream farther than the jump, and as long as the discharge remains unchanged, there is no change in the stage of flow upstream from the jump.

PROBLEMS

1. A smooth concrete channel of rectangular cross section 10 ft wide having a level grade terminates in a fall. The discharge is 400 cfs. (*a*) What is the depth of water 1000 ft upstream from the fall? (*b*) How far upstream to a depth of 5 ft?

2. A concrete spillway channel 1500 ft long with rectangular cross section 12 ft wide having a level grade receives water through a rounded entrance from a reservoir and terminates in a fall. When the water surface in the reservoir is 6 ft higher than the bottom of the channel, what is the discharge?

3. A smooth wooden flume with vertical sides is 8 ft wide and has a grade of 0.002. The discharge is 190 cfs, the depth of water being 3 ft.

It is proposed to construct a weir in the flume which will raise the water surface 1 ft at a section 1000 ft upstream. Assuming the water to have a plane surface, what will be the depth of water 500 ft upstream from the weir?

4. A discharge of 60 cfs per ft width of channel leaves the spillway of a dam with a velocity of 50 ft per sec and passes over a concrete apron 100 ft wide having a grade of 0.005. Determine the depth of water 25 ft downstream from the spillway.

5. A concrete-lined channel with side slopes of 1 to 1 is 10 ft wide on the bottom and has a grade of 0.0025. The discharge is 275 cfs. It is proposed to construct a diversion gate in the canal which will back up the water sufficiently to make the depth 4 ft just upstream from the gate. What will be the depth of water 500 ft upstream?

6. A smooth concrete-lined chute of rectangular cross section 6 ft wide has a slope of 1 in 50 and carries 300 cfs. At section A the depth of water is 3.8 ft. What will be the depth 40 ft downstream from A?

7. A smooth wooden flume with vertical sides is 8 ft wide and has a grade of 2 ft per 1000 ft. There is a sharp-crested weir 3.5 ft high extending across the flume. When the head on the weir is 3.1 ft, how much deeper is the water 200 ft upstream from the weir than it would be with the same quantity of water flowing but with the weir removed?

8. The downstream face of a spillway has a slope of 45 degrees and joins a level apron at its lower end. The crest of the spillway is 35 ft higher than the apron. There is a discharge per foot length of 50 cfs over the spillway, the head on the crest being 6.0 ft. Determine the velocity of the water at the bottom of the spillway where it enters upon the apron.

9. A semicircular concrete-lined channel, flowing full and having a radius of 3 ft and a grade of 8 ft per mile, changes abruptly to an earth canal in good condition having a bottom width equal to twice the depth, a grade of 2.25 ft per mile, and side slopes of 2 horizontal to 1 vertical. If the concrete-lined channel is 2000 ft long and the earth canal is 8000 ft long, determine the drop in the water surface in the total length of 10,000 ft.

10. An earth canal in good condition carrying 300 cfs and having side slopes of 2 horizontal to 1 vertical has a bottom width of 16 ft and a depth of 5 ft. This section continues from Sta. 0 to Sta. 45 and again from Sta. 48 to Sta. 80. From Sta. 45 to Sta. 48 an open, rectangular wooden flume, 300 ft long and having a width of 8 ft and a depth of 4 ft, carries the water across a ravine. Assuming abrupt changes in section, determine the drop in water surface from Sta. 0 to Sta. 80.

11. An earth canal in good condition with side slopes of 2 horizontal to 1 vertical carries 200 cfs at a velocity of 2 ft per sec. The depth of water is one-third of bottom width of canal. This canal discharges into a flume with a tapered entrance, the conditions being such that the loss of head at entrance is one-half of what it would be for an abrupt change in section. The flume is 7 ft wide and has vertical sides. The slope of the flume is such that it carries 200 cfs at a uniform depth of 3.5 ft. Determine how much the bottom of the flume should be above or below the bottom of the canal.

12. A planed-timber flume of rectangular cross section 16 ft wide having square-cornered ends connects two reservoirs 300 ft apart. The bottom of the flume is level and is 5 ft below the water surface in one reservoir and 2 ft below the surface in the other reservoir. Determine the discharge.

141. Hydraulic Jump. The hydraulic jump is an abrupt rise in water surface which results from retarding water flowing at the

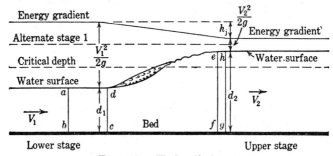

FIG. 137.   Hydraulic jump.

lower stage. The change in stage is from a depth less than critical depth to one greater than critical depth, but because of the loss of head in the jump, the depth after the jump is less than the alternate stage of equal energy before the jump.

The hydraulic jump could be considered along with the abrupt translatory wave (Art. 144) since each is a manifestation of the same phenomenon. It is convenient, however, inasmuch as the hydraulic jump has many useful applications requiring special consideration, to investigate it independently. The pressure-momentum theory of the hydraulic jump in channels of rectangular cross section is given in this article; the flow conditions producing the jump and affecting its location are described in Arts. 140 and 142.

The hydraulic jump is illustrated in Fig. 137. A channel width of 1 ft is assumed. Mean velocities before and after the jump are respectively V_1 and V_2. The depth, in changing from d_1 to d_2, passes through the stage of critical depth. In the region of the jump there is a surface roller with characteristic turbulence and boiling of the water and accompanying loss of head. This loss of head is indicated by h_j, the drop in the energy gradient.

Consider that in a short interval of time the mass of water *abcd*, Fig. 137, moves to *efgh*. In changing positions the water loses momentum. The unbalanced force acting to retard the mass must equal the rate of change of momentum. (See Chapter IX.) If the unbalanced force F acts upon the mass M for the time t,

$$F = \frac{MV_1 - MV_2}{t} = \frac{qw}{g}(V_1 - V_2) \tag{47}$$

q being the discharge per unit width of channel and w the unit weight of water. The unbalanced force is assumed to be the difference between hydrostatic pressures corresponding to the depths d_2 and d_1, or

$$F = \frac{d_2{}^2 w}{2} - \frac{d_1{}^2 w}{2} \tag{48}$$

Equating 47 and 48,

$$\frac{q}{g}(V_1 - V_2) = \frac{d_2{}^2 - d_1{}^2}{2} = \frac{(d_2 - d_1)(d_2 + d_1)}{2} \tag{49}$$

Substituting q/d_1 for V_1 and q/d_2 for V_2, and reducing

$$\frac{q^2}{g} = d_1 d_2 \frac{d_1 + d_2}{2} \tag{50}$$

The usual form of solution is with q and one of the depths given to solve the resulting quadratic for the other depth. The loss of head in the jump is then the difference in total heads before and after the jump, neglecting the effects of loss of head due to friction or those of channel slope, but in the short distance required for the transition the influence of these factors is comparatively unimportant. Experiments from a number of sources indicate that equation 50 holds within the limits of error in making measurements.

It should be noted that as d_1 increases and d_2 decreases to reach a common value d_c, equation 50 becomes

$$q = \sqrt{g}\, d_c{}^{3/2} \tag{19}$$

which is the equation for discharge at critical depth.

142. Position of Hydraulic Jump. Frequently in engineering work it is important to know where a hydraulic jump will occur, as, for instance, in the design of a spillway. As water discharges over an overflow dam, most of its original potential energy is converted into kinetic energy. Unless means are provided for the dissipation of part of this kinetic energy in frictional loss, together with the reconversion of a certain amount into potential energy, these high velocities are likely to cause erosion at the toe of the dam and result in failure of the structure. One of the most efficient methods of converting kinetic energy into frictional loss is by means of the hydraulic jump. This conversion, however, must occur on the apron of the dam, and therefore the determination of the location of the jump becomes a matter of prime importance.

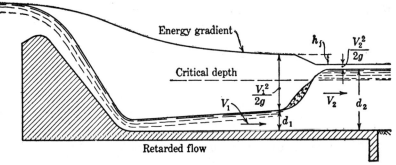

FIG. 138.   Hydraulic jump on apron of dam.

An overflow dam with a hydraulic jump on the apron is illustrated in Fig. 138. It is assumed that the discharge q per foot width is given and that the corresponding downstream depth d_1 and velocity V_1 are known. The energy gradient indicates loss of head at a gradually changing rate between the crest of the dam and the jump. At the jump there is a loss of head h_j, indicated by an abrupt drop in the energy gradient.

The velocity at the toe of the dam and also the profile of water surface over the apron can be determined by applying the principles of accelerated and retarded flow (Art. 138). As shown in the figure the slope of the apron is insufficient to overcome friction at the existing velocity, and the flow is therefore retarded. Under conditions that commonly exist, d_2 is considered constant and the apron has a uniform slope. Computations can then start with a known velocity at the toe of the dam, and the distance to the depth d_1 as determined from equation 50 can be determined from equations 35 and 36. If the velocity change is considered great enough to require it, the distance can be computed in two or more reaches. In making these computations a coefficient of roughness for the apron must be estimated, and since there is always considerable uncertainty regarding its proper value, the position where the jump will occur cannot be determined exactly. Other things being equal, the rougher the surface of the apron, the nearer the dam the jump will occur.

A method of determining the position of the jump where depths upstream and downstream from the jump are both variable is illustrated in Fig. 139. Water is shown coming down a channel with slope steeper than the critical, with accelerated flow. At some point downstream a dam or other obstruction backs up the water to a stage greater than critical depth. It is assumed that all water-surface

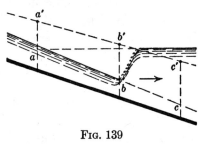

FIG. 139

profiles both with and without backwater have been determined. At any selected sections such as a, b, c, etc., the value of d_2 corresponding to d_1 at the section is computed from equation 50, and the points a', b', c', etc., indicating these depths, are plotted in their proper positions. The place where the line joining these points intersects the backwater surface gives the position of the jump.

PROBLEMS

1. If a discharge of 10 cfs per ft width of channel has a velocity of 12 ft per sec, to what depth can it jump?

2. A stream having a discharge per foot width of channel of 10 cfs has a depth after jump of 3.0 ft. Determine the velocity of the water before the jump.

3. In a flume of rectangular cross section 5 ft wide, water flowing at a depth of 1 ft jumps to a depth of 3 ft. Determine the discharge.

4. Water upon leaving the spillway of a dam passes over a level concrete apron 200 ft wide. Conditions are such that a hydraulic jump will form on the apron. When the discharge is 50 cfs per ft width of channel the velocity where the water leaves the spillway is 45 ft per sec and the depth after the jump is 10 ft. Determine the distance downstream from the dam to the place where the jump occurs.

5. Water upon leaving the spillway of a dam passes over a concrete apron 200 ft wide having a slope of 1 in 50. Conditions are such that a hydraulic jump will form on the apron. When the discharge is 50 cfs per ft width of channel the velocity where the water leaves the spillway is 48 ft per sec, and the water surface after the jump is 10 ft higher than the upstream end of the apron. Determine the distance downstream from the dam to the place where the jump occurs.

143. Translatory Waves. A sudden change in the quantity of water entering or leaving a channel causes the consequent readjustments in velocity and depth to occur in a wave or series of waves. If discharge at the intake is suddenly increased a sufficient amount an abrupt accelerating wave forms immediately and travels downstream; an abrupt decelerating wave travels upstream when discharge from the outlet is suddenly reduced a sufficient amount. In each case there is an increase in depth, and each wave in appearance and in fact is a moving hydraulic jump. In order for an abrupt wave to form it is necessary that the depths and *relative* velocities before and after the wave satisfy the conditions of the hydraulic jump (equation 50). The corresponding waves which accompany a reduction in depth produced either by decreased discharge at the intake or increased discharge at the outlet have sloping faces and are apparent only by a gradual lowering of the water surface.

144. The Abrupt Wave. It is assumed that the wave illustrated in Fig. 140 has been produced by instantaneously increasing the discharge through gate G from q_1 per unit width of channel to q_2. In a short reach, the modifying effects of frictional loss and channel slope are comparatively small, and they will be neglected. Between the wave and the gate, each section has the same depth d_2 and the same velocity V_2. Downstream from the wave. the depth d_1 and the velocity V_1 remain the same as before the additional water was admitted. The figure illustrates conditions after one second of increased flow.

The wave which travels with a velocity v_w is, at the end of 1 sec, a distance v_w below the gate. The volume of water per foot of width entering the channel in 1 sec is $q_2 = d_2 V_2$, shown in the figure by the area *abcd*. The increase in volume or $q_2 - q_1$ is represented by the area *aefg*, or, expressed algebraically,

$$q_2 - q_1 = v_w(d_2 - d_1) \tag{51}$$

Substituting V_2d_2 and V_1d_1 respectively for q_2 and q_1 and transposing,

$$V_2 = v_w - \frac{d_1}{d_2}(v_w - V_1) \tag{52}$$

The mass of water $dchg$ has had its velocity increased from V_1 to V_2, and its momentum has thereby been increased. Calling the

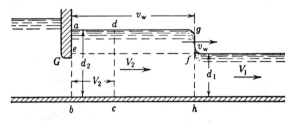

FIG. 140. Abrupt wave traveling downstream.

mass M, the unbalanced force required to change the momentum in one second is

$$F = M(V_2 - V_1) = \frac{(v_w - V_2)d_2w(V_2 - V_1)}{g} \tag{53}$$

The unbalanced force is equal to the difference in hydrostatic pressures corresponding to the depths d_2 and d_1, or

$$F = \frac{d_2{}^2w}{2} - \frac{d_1{}^2w}{2} \tag{54}$$

Equating the values of F in 53 and 54 and reducing,

$$\frac{g}{2d_2}(d_2{}^2 - d_1{}^2) = (V_2 - V_1)(v_w - V_2) \tag{55}$$

Substituting V_2 from equation 52 and making algebraic transformations,

$$(v_w - V_1)^2 = \frac{gd_2}{2d_1}(d_2 + d_1) \tag{56}$$

and

$$(v_w - V_1) = \pm\sqrt{\frac{gd_2}{2d_1}(d_2 + d_1)} \tag{57}$$

Equation 57 is general and applies to all cases of the abrupt wave. $(v_w - V_1)$ is the velocity of the wave with respect to the water in the shallower portion of the stream.

The solution of equation 57 for v_w gives

$$v_w = \pm \sqrt{\frac{gd_2}{2d_1}(d_2 + d_1)} + V_1 \tag{58}$$

The plus sign applies to waves traveling downstream (Fig. 140) and the minus sign to waves traveling upstream (Fig. 141), such as would be produced by instantaneously reducing the opening of

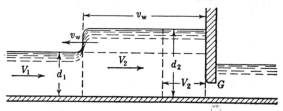

Fig. 141.   Abrupt wave traveling upstream.

gate G.   Considering motion to the right positive and to the left negative, equation 58 follows directly from Fig. 141 as readily as from Fig. 140.

When the wave has zero velocity it becomes a hydraulic jump. This is seen by placing $v_w = 0$ and $V_1 = q/d_1$ in equation 58. The equation then reduces to equation 50.

As d_1 increases and d_2 decreases to reach a common value d_c, equation 58 becomes

$$v_w = \pm \sqrt{gd_c} + V_1 \tag{59}$$

The term $\sqrt{gd_c}$ represents critical velocity (equation 20).

When the flow of a channel is suddenly increased or decreased by a given amount, neither the new depth d_2 which the water will assume nor the velocity v_w of the wave is known.   Neglecting the effects of frictional losses and channel slope, these are given by the simultaneous solution of equations 51 and 58.   If the depth of water d_1 is suddenly increased to a new depth d_2, the velocity of the wave is given by equation 58, and this value can be substituted in 51 for determining the new discharge q_2.

A good example of the effects of sudden increase in depth is afforded by the *bore*.   This type of abrupt wave may occur in a tidal stream when a rapidly rising tide enters its mouth.   Bores several feet in height have been observed many miles above the mouth of some of the larger tidal streams.

As a bore moves upstream, v_w is negative, and the direction of the new velocity (Fig. 141) as given by equation 52 may be either upstream or downstream, depending upon the discharge of the stream and the amount that the depth is increased. V_2 will equal zero when

$$V_1 = (d_1 - d_2) \sqrt{\frac{g}{2} \frac{d_1 + d_2}{d_1 d_2}} \tag{60}$$

This equation is obtained by placing $V_2 = 0$ in equation 52 and eliminating v_w between equations 52 and 58.

145. The Sloping Wave. It has been shown in the preceding article that a sudden change in discharge acting to increase the depth may cause the formation of an abrupt wave. An example of the sloping wave which occurs whenever there is a sudden reduc-

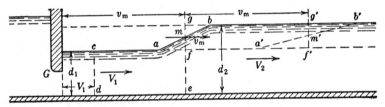

Fig. 142. Sloping wave.

tion in depth is illustrated in Fig. 142. The opening of gate G is assumed to be instantaneously decreased, the depth of water being reduced thereby from d_2 to d_1. For all cases of sloping wave the same as for the abrupt wave, d_2 is the greater depth, q_2 and V_2 being respectively discharge per unit width of channel and velocity at this depth. q_1 and V_1 are the corresponding discharge and velocity at the lesser depth d_1. The figure indicates conditions at the end of 1 sec.

The face of the wave is shown by the sloping line ab. The line fg, representing the mean position of the wave, is so drawn that area $amf = bmg$. The mean velocity of the wave is v_m. During the first second of reduced discharge the volume of water which entered the channel was $q_1 = V_1 d_1$. The decrease in volume was

$$q_2 - q_1 = v_m(d_2 - d_1) \tag{61}$$

If it is assumed that the mass of water represented in the figure by the rectangle $cdef$ has had its velocity reduced from V_2 to V_1, following the procedure given in the preceding section for abrupt waves,

$$v_m = \pm \sqrt{\frac{gd_1}{2d_2}} (d_2 + d_1) + V_2 \tag{62}$$

Contrasted with the assumptions just mentioned, the actual conditions are as follows: upstream from a the water has been decelerated from V_2 to V_1, downstream from b no change in velocity has occurred, while between a and b deceleration is still in progress. Because of this difference between actual and assumed conditions, equation 62 is not generally applicable. For waves of very small height, however, discrepancies introduced by erroneous assumptions disappear, and the velocity of the wave becomes

$$v_w = \pm \sqrt{gd} + V \tag{63}$$

d and V being respectively depth and velocity of water at the wave.

The sloping wave shown in Fig. 142 can be assumed to be made up of a large number of very small waves. The bottom wave will then have a velocity $v_a = \pm \sqrt{gd_1} + V_1$, and the top wave will have a velocity $v_b = \pm \sqrt{gd_2} + V_2$. Assuming a straight-line variation between v_a and v_b, the mean velocity of the wave is

$$v_m = \tfrac{1}{2}(v_a + v_b) = \tfrac{1}{2}(\pm \sqrt{gd_1} + V_1 \pm \sqrt{gd_2} + V_2) \tag{64}$$

Since d_2 is greater than d_1, v_b is greater than v_a, and the slope of the face of the wave becomes progressively flatter. At the end of 2 sec the wave has the position $a'b'$, and after a short time it becomes discernible only by a gradual lowering of the water surface. This discussion assumes instantaneous change of stage, which will not actually occur. The time consumed in reducing the depth decreases still more the slope of the wave.

If, in Fig. 142, motion to the right is considered positive and in the opposite direction negative, the signs before the radicals in equation 64 are positive. For a sloping wave traveling upstream, such as would be produced by suddenly increasing instead of decreasing the gate opening in Fig. 141, the signs before the radicals are negative.

In deriving equation 64 a straight-line variation between v_a and v_b was assumed. In reality the wave has a curved face, and the formula is therefore not exact. It is easy to apply, however, and will give results close enough in ordinary problems. The simultaneous solution of equations 61 and 64 will give v_m and d_1 when q_2, d_2, and q_1 are given. The velocity of the top wave is given directly from equation 63. If the water is lowered a known amount, that is, if q_2, d_2, and d_1 are given, the velocities of the top and the bottom of the wave can be obtained from equation 63, the mean of these velocities being v_m. A direct solution of (61) for q_1 can now be made.

146. Effect of Frictional Loss on Waves. The theory of the two preceding articles does not take into consideration loss of head due to friction or channel slope. In a short and fairly deep reach of channel the modifying effect of these factors may be comparatively slight, but as the channel extends in length or as the depth of water decreases they become increasingly important. Back of a wave the flow may be non-uniform and unsteady. A complete investigation of the water-surface profile therefore requires that computations be made in reaches at the end of successive time intervals.

A starting point in calculations can be obtained by determining conditions at the end of 1 sec with frictional losses neglected. Results thus obtained can be modified to include friction if this is thought advisable. Wherever frictional losses are to be considered, it is necessary to satisfy the requirements of the Manning formula (or some other open-channel formula) in addition to the formulas applying specifically to the wave. The principles involved in computing the water-surface profile back of a wave are comparatively simple, but the calculations may become extremely long and tedious.

The effect of frictional loss on the height and velocity of the wave increases greatly as d_1 decreases. Equation 59 shows v_w to increase as d_1 decreases and to become infinite for $d_1 = 0$. Equation 54 shows that, as d_1 decreases, V_2 approaches v_w and becomes equal to it under limiting conditions. For very small values of d_1, the frictional loss due to the large value of V_2 given by equation 54 completely transforms the actual conditions of flow. To take an extreme example, assume that $d_2 = 10$ and $d_1 = 0.1$, with $V_1 = 0$. Substituting these values in equation 58, $v_w = 127.4$, and then from equation 52, $V_2 = 126.1$. Since frictional losses at such high velocities are enormous, it will be impossible to maintain more than a small fraction of the V_2 given by equation 52.

An abrupt wave may form in a river channel after a heavy rainfall or if a dam fails and suddenly releases a large volume of impounded water. Such waves moving down channels which contain little or no water are retarded by friction to velocities which have been found to be no greater than about 15 miles per hour (22 ft per sec).

PROBLEMS

1. A channel of rectangular cross section 15 ft wide is carrying 200 cfs of water at a velocity of 3 ft per sec. If the discharge is suddenly increased by the addition of 100 cfs, what new depth will the water in the channel assume?

2. A channel of rectangular cross section 15 ft wide, carrying 200 cfs of water at a velocity of 3 ft per sec, has its discharge suddenly increased sufficiently to raise the water surface 1 ft. Determine the new rate of discharge.

3. A tidal stream having an average depth of 10 ft is discharging into the ocean with a velocity of 2 ft per sec. Assuming an instantaneous rise in tide of 3 ft, determine approximately the velocity of the wave and the new velocity (upstream or downstream) of water.

4. Water is flowing in a channel of rectangular cross section 20 ft wide with a velocity of 2 ft per sec and depth of 4 ft. What instantaneous increase in flow will be required to produce a wave of acceleration that will travel down the channel with a velocity of 15 ft per sec?

GENERAL PROBLEMS

1. The river channel cross section shown in Fig. A is symmetrical about the center line. The bed and banks are of firm gravel and well-sodded earth, respectively, with fairly straight alignment. The slope

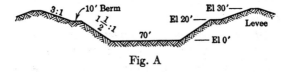

Fig. A

is 2.5 ft per mile. Assuming $n = 0.035$ for the entire cross section, compute the capacity of the channel for depths of 16, 20, 24, and 28 ft.

2. Determine the discharge of a semicircular planed wood-stave flume 8 ft in diameter laid on a grade of 6 ft per mile.

3. Determine the probable flow of a river at full stage 1000 ft wide, average depth 30 ft, with a fall of 0.7 ft per mile. Assume $R = d$. The river bed is of uniform cross section with good alignment.

4. An earth canal, in good condition, has a bottom width of 12 ft and side slopes of 2 horizontal to 1 vertical. If the grade of the canal is 1.5 ft per mile, determine its capacity for depths of 4, 6, and 8 ft.

5. What is the capacity of a 36-in. vitrified sewer pipe, well laid, on a straight grade of 3 ft per mile, when it is flowing 2 ft deep?

6. What slope in feet per mile should a smoothly finished concrete-lined canal have to carry 1300 cfs, if the bottom width is 10 ft, the depth 8 ft, and the side-slope ratio $\frac{3}{4}$ horizontal to 1 vertical?

7. What fall per mile should a river have to discharge 500,000 cfs with a width of 2400 ft and an average depth of 40 ft? Assume $R = d$. The river is at bank-full stage with fairly stable bed.

8. A semicircular concrete channel having a 5-ft radius carries 120 cfs when flowing full. Determine the grade in feet per mile.

9. Determine the critical slope for a smooth wooden flume of rectangular cross section, 12 ft wide, carrying 100 cfs.

10. An earth canal in good condition having a bottom width of 12 ft and side slopes of 2 horizontal to 1 vertical is designed to carry 180 cfs at a mean velocity of 2.25 ft per sec. What is the necessary grade of the canal?

11. An earth canal in good condition with side slopes of $1\frac{1}{2}$ horizontal to 1 vertical carries 600 cfs at a velocity of 2.5 ft per sec. If the bottom width is twice the depth, determine the grade of the canal.

12. An earth canal in good condition with some curves is to carry 200 cfs with a trapezoidal cross section, side slopes 2 to 1, and the bottom width half the depth of flow. Compute the dimensions of the canal if the grade is 3 ft per mile.

13. A canal is to be constructed in earth and maintained in good condition. If it is given a grade of 1.5 ft per mile, side slopes of 2 horizontal to 1 vertical, and a bottom width that is ten times the depth when the discharge is 2000 cfs, determine the bottom width.

14. A dredged ditch with side slopes of 2 horizontal to 1 vertical is designed to have a bottom width equal to the depth. If it is to carry 600 cfs and have a grade of 2 ft per mile, determine the bottom width.

15. An earth canal in good condition is to be constructed with side slopes of $1\frac{1}{2}$ horizontal to 1 vertical and a fall of 2 ft per mile. Determine the depth and the bottom width of the most efficient section if the discharge is 600 cfs.

16. What should be the bottom width and the depth of flow for a canal of most efficient trapezoidal cross section, side slopes 2 horizontal to 1 vertical, to carry 200 cfs with a velocity of 2.5 ft per sec? What should be the slope of the canal if it is to be in earth and kept in good condition?

17. Determine the bottom width of an earth canal, in good condition, having side slopes of 2 horizontal to 1 vertical, if it is to have the most efficient section and carry 200 cfs with a grade of 2 ft per mile.

18. An earth ditch having side slopes of 2 horizontal to 1 vertical is to have a grade of 1.8 ft per mile and carry 80 cfs when maintained in good condition. What per cent greater will be the cross-sectional area and wetted perimeter if the bottom width is made twice the depth than if the most efficient section is used?

19. A sharp-crested weir, having a crest 20 ft long and 3.5 ft above the bottom of the canal, is installed in the middle of an earth canal, in good condition, having a bottom width of 12 ft and side slopes of 2 horizontal to 1 vertical.   If the measured head over the weir is 1.5 ft and the depth of water downstream from the weir is 3.0 ft, determine the grade of the canal in feet per mile.

20. A V-shaped flume of planed lumber carries 50 cfs from a reservoir with uniform flow at a velocity of 8 per sec.   Each side of the flume makes an angle of 45 degrees with the horizontal.   The frictional loss at entrance is 0.3 of the velocity head.   Determine the difference in elevation between the water surface in the reservoir and that in the flume at a point 1000 ft from the entrance.

21. An earth canal in good condition, 30 ft wide on the bottom, having side slopes of 2 horizontal to 1 vertical, and carrying water 4 ft deep, has a grade of 2 ft per mile.   At Sta. 0 the bottom of the canal is at elevation 592.5.   At Sta. 80 it changes abruptly to a concrete-lined section having a bottom width equal to the depth of water, side slopes of 1 horizontal to 2 vertical, and a grade of 1 ft in 1000 ft.   Determine the elevation of the bottom of the canal at Sta. 150, the stations being at 100-ft intervals.

22. An earth canal containing weeds and grass has a bottom width of 15 ft and side slopes of 2 horizontal to 1 vertical.   The depth of water is 4 ft and the slope is 2.75 ft per mile.   It is desired to change the section to a semicircular concrete-lined channel having a slope of 1.5 ft in 1000 ft.   Determine the radius of the semicircular channel if it flows full.   If the change in section is abrupt and sharp-cornered, what will be the drop in water surface where the change in section occurs?

23. A V-shaped channel, built of unplaned lumber and having a slope of 2 ft per 1000 ft, carries water from a reservoir.   Each side of the channel makes an angle of 45 degrees with the horizontal, and the lowest point in the notch is 5 ft below the water surface in the reservoir.   If the frictional loss at entrance is equal to half the velocity head, determine the discharge.

24. A long rectangular concrete-lined channel 20 ft wide has a slope of 2 ft per 1000 ft.   If the bottom of the channel at the entrance is 4 ft below the water surface in the reservoir and the frictional loss at entrance is equal to one-half the velocity head, determine the discharge.

25. An earth canal in good condition, having a bottom width of 30 ft and side slopes of 2 horizontal to 1 vertical, carries water at a depth of 5 ft with a velocity of 2.5 ft per sec. This canal changes abruptly to a rectangular concrete-lined section having a width of 10 ft and a grade of 1 ft per 1000 ft. Determine the difference in elevation between the bottom of the earth canal 1000 ft upstream from the change in section and the bottom of the concrete-lined canal 1000 ft. downstream from the change, assuming uniform flow in each section.

26. A trapezoidal earth canal in good condition has a bottom width of 100 ft, side slopes of 2 horizontal to 1 vertical, a slope of 0.00025, and flows with a depth of 5 ft. This canal discharges into two rectangular concrete-lined channels, A and B, having widths of 20 ft and 40 ft, respectively, and each having a slope of 0.001. If the bottom of channel A is flush with the bottom of the earth canal at the junction, determine the relative elevation of channel B so as to maintain uniform flow in each of the three canals. Assume that the frictional loss at the entrance to each of the concrete channels is 0.2 times the respective velocity heads in those channels.

27. A canal, 58,000 ft long (580 stations), is to be constructed with a capacity of 300 cfs. The canal diverts from a river and terminates at a reservoir into which it discharges. The water surface in the river at the point of diversion is to be maintained at an elevation of 770 ft.

(a) Water is to be diverted through six head gates, having rectangular openings each 2 ft by 5 ft. Determine the head required to force the water through these openings, assuming a coefficient of discharge of 0.80.

(b) From Sta. 0 to Sta. 425 the canal is in earth section, having side slopes of 2 horizontal to 1 vertical, and a depth of water of 0.3 of the bottom width of the canal. Velocity of water is to be 2.1 ft per sec. Assuming $n = 0.0225$, determine the slope of the canal.

(c) Between Sta. 425 and Sta. 500 the canal is in rock and is to have a semicircular section lined with concrete. The grade of the canal is to be 2 ft per 1000 ft with $n = 0.014$. Determine the head lost at entrance, assuming $K_c = 0.18$. Also determine diameter of canal section.

(d) From Sta. 500 to Sta. 580 the section of canal is the same as from Sta. 0 to Sta. 425. At the reservoir end of the canal (Sta. 580) a broad-crested weir is to be constructed in order that a uniform depth of water may be maintained throughout the entire length of earth section. Length of this weir is to be equal to the bottom width of the earth canal. The weir has a rectangular section, with horizontal crest 10 ft broad and rounded entrance. Determine height of crest above bottom of the canal. Assume $K_e = 1.00$ at Sta. 500.

(e) Tabulate the elevations of the water surface, to nearest 0.1 ft, at the following stations: Sta. 0 + 10, Sta. 424 + 90, Sta. 425 + 10, Sta. 500, and Sta. 579 + 90.

28. Water discharges from a large reservoir through a rounded entrance into a smooth timber flume 10 ft wide, having a rectangular cross section and a grade of 2 in 100. If the bottom of the flume at the intake is 4 ft below the water surface in the reservoir, determine the distance downstream from the intake to the section at which the depth of water will be 2.3 ft.

29. A concrete-lined canal of trapezoidal cross section 10 ft wide has side slopes of 1.5 horizontal to 1 vertical and a grade of 0.002. The canal carries 400 cfs of water at a depth of 3 ft. A dam is to be built in the canal that will increase the depth to 9 ft. If the elevation of water surface at the dam is 100, what will be the elevation of water surface 3000 ft upstream from the dam?

30. A canal carries 300 cfs of water at a depth of 5.5 ft and velocity of 2 ft per sec. This water is to be conveyed to a lower elevation through a smooth concrete channel with rounded entrance having a rectangular cross section and grade of 1 in 10. The depth of water is to be kept at 3 ft throughout, the width to vary as required to maintain this depth. Determine the width of channel at entrance, also the distance from the entrance to widths of 8 ft, 6 ft, and 4 ft, respectively. What is the minimum width possible for this depth of water?

31. A planed timber flume of rectangular cross section 16 ft wide having square-cornered ends connects two reservoirs 1000 ft apart. The bottom of the flume is level and is 5 ft below the water surface of one reservoir and 4 ft below the surface of the other. Determine the discharge.

32. A smooth wooden flume of rectangular cross section 8 ft wide having a grade of 0.02 carries 240 cfs of water under conditions of uniform flow. A weir is to be constructed which will back up the water in the flume to a depth of 6.5 ft. How far upstream from the weir will a jump occur?

33. A channel of rectangular cross section has its outlet controlled by a gate which discharges freely into the air. The gate has the same width as the channel, and the sill of the gate is flush with the bottom of the channel. The depth of water in the channel is 8 ft, and the velocity is 4 ft per sec. Assuming the height of gate opening to be instantaneously reduced from 2 ft to 1 ft, the coefficient of discharge of the gate remaining constant, determine: (a) the new rate of discharge in the canal, (b) the new depth of water, and (c) the velocity of the decelerating wave.

Chapter IX

HYDRODYNAMICS

147. Fundamental Principles. Newton's laws of motion form the basic principles of the subject of hydrodynamics. These laws may be briefly stated as follows:

I. Any body at rest or in motion with a uniform velocity along a straight line will continue in that same condition of rest or motion until acted upon by some external force.

II. The rate of change in the momentum of a moving body is proportional to the force producing that change, which occurs along the same straight line in which the force acts.

III. To every action there is always an equal and opposite reaction.

These three laws of Newton's are frequently referred to as the laws of inertia, force, and stress, respectively. On account of the fundamental importance of these laws it is essential that a clear conception be had of their full significance. As an aid in acquiring this conception the following discussion is presented.

148. Interpretation of Newton's Laws. Newton's first law of motion is merely a statement that matter possesses no ability, *per se*, to change its condition of rest, or motion, and that any such change must be brought about through the action of some external force.

Newton's second law states the fundamental principle of mechanics that when an unbalanced force is applied to a body, the resulting acceleration is proportional to the force. With proper consideration for units as indicated in Art. 3, this principle is stated as

$$F = Ma \tag{1}$$

Since the average rate of acceleration $a = (v_2 - v_1)/t$,

$$F = M \frac{(v_2 - v_1)}{t} = \frac{M}{t} (v_2 - v_1) \tag{2}$$

If a continuous stream of fluid is having its velocity changed from v_1 to v_2 by a vane or other object, by letting $t = 1$, the

291

quantity M/t becomes the mass per unit of time, M_1. The continuous force exerted *by the vane on the fluid* is then

$$F = M_1(v_2 - v_1) \tag{3}$$

From Newton's third law, the force exerted *by the fluid against the vane* is equal but opposite in direction, and can therefore be written

$$F = M_1(v_1 - v_2) \tag{4}$$

The same equation can be obtained by starting from the Bernoulli energy theorem. Equation 14, page 96, is the mathematical statement of the principle that, neglecting friction, the total head, or the total amount of energy per unit of weight, is the same at every point in the path of flow.

If, however, some of the energy is extracted from the fluid, or, in other words, if the fluid is made to do work upon some machine, as in a turbine, a term h_I must be added to the energy equation, which then becomes

$$\frac{v_1^2}{2g} + \frac{p_1}{w} + z_1 = \frac{v_2^2}{2g} + \frac{p_2}{w} + z_2 + h_I \tag{5}$$

In considering the dynamic action of jets that are free and unconfined, throughout which the pressure is atmospheric and the elevation is practically unchanged, this equation becomes

$$\frac{v_1^2}{2g} = \frac{v_2^2}{2g} + h_I \tag{6}$$

Each term in equations 5 and 6 represents either energy or work expressed in foot-pounds per pound of fluid. Multiplying by the total weight of fluid W, substituting M for W/g, and transposing, the total work is

$$G = \tfrac{1}{2}Mv_1^2 - \tfrac{1}{2}Mv_2^2 \tag{7}$$

This equation states the law of mechanics that the change in kinetic energy is equal to the work done. Substituting Fl for G, since work equals force times distance,

$$Fl = \frac{M}{2}(v_1^2 - v_2^2) = \frac{M}{2}(v_1 - v_2)(v_1 + v_2) \tag{8}$$

In this equation l is the distance through which a constant force F is exerted by a mass M while its velocity is being changed from v_1 to v_2. If the acceleration is constant,

$$l = \frac{v_1 + v_2}{2} t$$

and equation 8 becomes

$$Ft = M(v_1 - v_2) \qquad (9)$$

which states another law of mechanics that impulse equals change in momentum. Letting $t = 1$ sec,

$$F = M_1(v_1 - v_2) \qquad (4)$$

where M_1 represents the mass per second having its velocity changed from v_1 to v_2.

149. Vectors. A vector is a quantity that may be considered as possessing direction as well as magnitude, whereas a scalar possesses magnitude but no direction. Examples of scalars are mass, time, volume, and energy. Scalar quantities may be added algebraically.

Examples of vector quantities are velocity, acceleration, and force. Such quantities cannot be added algebraically except when their direction is the same. For instance, a mass having a velocity of 10 ft per sec in a certain direction, if given an additional velocity of 10 ft per sec at right angles to the original direction, will have a resultant velocity of 14.14 ft per sec instead of 20 ft per sec.

Although force is given as an example of a vector quantity, in reality it belongs to a special group of vector quantities which possess an additional attribute other than magnitude and direction. In order to determine fully the effect that a given force will have upon an object, the location of the line of action must also be known. Vectors representing such quantities are known as limited or constrained vectors to distinguish them from free vectors.

The sum or difference of two vectors may be found by drawing both from a common origin and then completing the parallelogram. That diagonal which passes through the common origin is the sum of the two vectors; the other diagonal is their difference.

150. Relative and Absolute Velocities. Strictly speaking, all motion is relative. No object in the universe is known to be fixed in space. An airplane is said to be flying 300 miles per hour, but this is its velocity only with respect to the surface of the earth beneath it.   The earth's surface itself is moving at a tremendous speed both with respect to its axis and to the sun, each of which is whirling through space at a still greater rate.

It is nevertheless convenient in connection with this subject to consider all motion with respect to the earth's surface as *absolute* motion.   The airplane above referred to, which will be called *A*, has therefore an absolute velocity of 300 miles per hour.   Another plane in pursuit, which will be referred to as *B*, may have an absolute velocity of 350 miles per hour, but its relative velocity with respect to *A* is only 50 miles per hour.   If the two planes were to fly in opposite directions, each retaining its same absolute

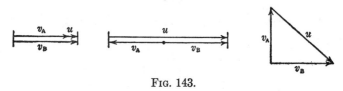

FIG. 143.

velocity, the relative velocity between them would be 650 miles per hour.   If they were to fly at right angles to each other their relative velocities would be $\sqrt{300^2 + 350^2} = 461$ miles per hour.

Since velocities are vector quantities, these results may be obtained graphically as in Fig. 143 in which

v_A = the absolute velocity of *A*.
v_B = the absolute velocity of *B*.
u = the relative velocity of *B* with respect to *A*.

Expressed as a vector sum

$$v_A + u = v_B \tag{10}$$

or as a vector difference

$$v_B - v_A = u \tag{11}$$

A verbal statement of equation 10 is frequently of assistance in solving problems of absolute and relative velocities:

The absolute velocity of A plus the velocity of B relative to A equals the absolute velocity of B.

From this explanation it is apparent that the relative velocity of a moving object *B* with respect to another moving object *A* may be found by drawing from a common origin two vectors representing the absolute velocities of A and B and then drawing a third vector from the terminus of A to the terminus of B.

Furthermore, if one object impinges upon another the resulting force of impact depends upon the relative velocities of the two objects rather than upon their absolute velocities. In other words, the force exerted is the same as it would be if the absolute velocities of both were changed by the same amount, so long as their relative velocities remain unchanged.

151. Force Exerted by a Jet. Since force and velocity are vector quantities, it follows from equation 4 that, if a jet of fluid impinges against a vane which is either moving or at rest and thereby has its velocity in any direction changed, a force F is exerted upon the vane the magnitude of which in any direction is equal to the change in momentum per second that the jet undergoes in the same direction. In other words, the force F is equal to the mass impinging per second times the change in velocity in the direction of the force. X and Y components of the force exerted by a jet the path of which lies in the XY plane will therefore be

F_x = Mass impinging per second $\times$ Change in velocity along the X axis.

F_y = Mass impinging per second $\times$ Change in velocity along the Y axis.

The tangent of the angle α which the resultant force makes with the X axis is F_y/F_x, and the resultant force $F = F_y/\sin \alpha$.

The change in velocity may be either positive or negative. In the case of a decrease in velocity the dynamic force exerted by the fluid on the vane is in the same direction as flow, whereas in the case of an increase in the velocity the dynamic force exerted on the vane is opposed to the direction of flow. *In general, the direction of the force exerted by the jet on the vane is opposite to the direction of acceleration of the jet.*

Consider the vane shown in Fig. 144 to be moving with a uniform velocity v' in the original direction of the jet. The absolute velocity of the jet as it impinges at A is V, and its relative velocity with respect to the vane is $V - v' = u$. As the vane moves through the successive positions 1, 2, 3, and 4, a particle entering at A moves across the vane and leaves it at B. Neglecting friction, the relative velocity of the jet with respect to the vane remains unchanged while the jet flows from A to B so that the jet leaves the vane at B with a relative velocity u in a tangential direction.

The actual path of the jet as it moves across the vane is shown by the dotted line. The vector diagrams at the four positions of the vane show the relation between the constant absolute velocity v' of the vane, the constant relative velocity u of fluid to vane, and the absolute velocity V, V_2, V_3, V_B of the fluid. At 1, with the jet striking the vane tangentially, the three velocity vectors are parallel. At other points the vector diagram is a parallelogram in which

$$\text{Vector } V = \text{vector } v' + \text{vector } u$$

If the vane is moving with a velocity component parallel to the jet, the volume Q' striking the vane each second is different from

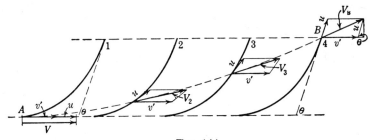

FIG. 144

Q, the discharge of the jet, and depends on the cross-sectional area A of the jet and the relative velocity of jet to vane. Thus $Q' = Au$ and the mass impinging per second is

$$M_1 = \frac{Q'w}{g} = \frac{Auw}{g} \tag{12}$$

The change in velocity of the jet is best determined by graphical analysis of each problem. For example, in Fig. 144 the initial velocity of the jet along the X axis is V while the final velocity is the X component of V_B. By the geometry of the vector diagram at B, this component is $v' + u \cos \theta$, where θ is the deflection angle of the vane. The change in X component of velocity is thus

$$\Delta V_x = V - (v' + u \cos \theta) = u(1 - \cos \theta) \tag{13}$$

Since in this case the X component of the velocity decreases as the jet passes across the vane, or in other words, the X component of acceleration is to the left, the force F_x exerted by the jet against the vane is to the right. Moreover, the Y component of the force, F_y, is directed downward since the Y acceleration is upward, the change in velocity in the Y direction being from 0 at A to $u \sin \theta$ at B.

If a jet is directed against a plate as shown in Fig. 145a or a double-cusped vane as shown in Fig. 145b so that the jet is deflected symmetrically with respect to the X axis, F_y will be zero since the Y components balance, being equal and opposite in direction.

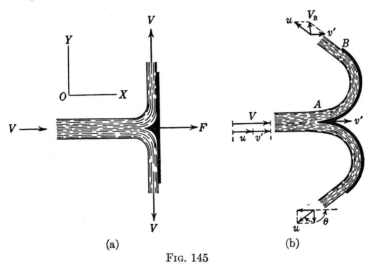

(a)   (b)

Fig. 145

If a series of vanes are so arranged on the periphery of a wheel that the entire jet, directed tangentially to the circumference, is striking either one vane or another successively, the mass impinging per second becomes

$$M_1 = \frac{Qw}{g} = \frac{AVw}{g} \tag{14}$$

It should be noted that, when F_y is radial, F_x is the only component of the force tending to produce rotation.

EXAMPLE. A jet of water 2 in. in diameter (Fig. 146) moving with a velocity of 100 ft per sec strikes a vane which is moving in the same direction as the jet with a velocity of 60 ft per sec. The deflection angle of the vane is 135°. Find: (a) the X and Y components of the force exerted by the jet on the vane; and (b) the direction and absolute velocity of the water leaving the vane.

Solution. (a) The vector diagram at A shows that $u = 40$ ft per sec.

The mass impinging per second is thus

$$M_1 = \frac{0.0218 \times 40 \times 62.4}{g} = 1.69 \text{ slugs}$$

The vector diagram at B shows components of V_B as follows:

X component $= 60 - u \cos 45° = 60 - 28.3 = 31.7$ ft per sec

Y component $= u \sin 45°$ $\qquad\qquad = 28.3$ ft per sec

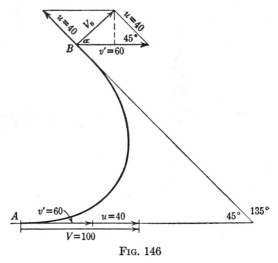

FIG. 146

Therefore $F_x = 1.69 (100 - 31.7) = 115$ lb to the right

$\qquad F_y = 1.69 \times 28.3 \qquad = 48$ lb downward

(b) $\tan \alpha = 28.3/31.7 = 0.892$, $\alpha = 41.8°$, and $V_B = 28.3/\sin \alpha = 42.5$ ft per sec.

PROBLEMS

1. A jet of water 1 in. in diameter and having a velocity of 25 ft per sec strikes against a plate as in Fig. 145a. Determine the force on the plate: (a) if the plate is fixed; (b) if the plate is moving in the same direction as the jet with a uniform velocity of 10 ft per sec.

2. A jet having a diameter of 2 in. and a velocity of 40 ft per sec is deflected through an angle of 60 degrees by a fixed, curved vane. Determine X and Y components of the force exerted.

3. A nozzle discharges 2 cfs with a velocity of 50 ft per sec. The nozzle is inclined downward so that as the jet strikes a fixed curved vane it is directed 30 degrees down from the horizontal. The jet is deflected upward 90 degrees, making an angle of 60 degrees with the horizontal as it leaves the vane. Determine X and Y components of the force exerted.

4. A jet having a diameter of 2 in. and a velocity of 50 ft per sec is deflected by a vane which is curved through an angle of 60 degrees and which is moving with a velocity of 20 ft per sec in the same direction as the jet. Determine X and Y components of the force exerted and direction and velocity of the water leaving the vane.

5. A horizontal nozzle discharges 4.0 cfs with a velocity of 80 ft per sec. The jet strikes an unsymmetrical two-cusped vane that is moving in the same direction as the jet with a velocity of 20 ft per sec. Half of the jet is deflected upward by a cusp that has a deflection angle of 90 degrees, and the other half is deflected downward by a cusp with a deflection angle of 45 degrees. Determine the horizontal and vertical components of the force exerted on the vane.

6. A 2-in. nozzle having a coefficient of contraction of 0.92 discharges 1.0 cfs. Determine the force required to move a single flat plate toward the nozzle with a velocity of 20 ft per sec, the jet impinging normally on the plate.

7. A horizontal jet of 2.0 cfs strikes tangentially on one or another of a series of moving curved vanes, each having a deflection angle of 60 degrees. The water as it leaves the vane has an absolute velocity of 40 ft per sec and makes an angle of 30 degrees with the horizontal. Determine: (*a*) the horizontal force exerted on the vanes, and (*b*) the horsepower developed.

8. A nozzle discharges 2.0 cfs horizontally, the jet striking tangentially on one or another of a series of curved vanes each having a deflection angle of 150 degrees. The velocity of the jet is 60 ft per sec, and the vanes have a velocity of 30 ft per sec in the same direction as the jet. Determine: (*a*) the absolute velocity of the water as it leaves the vanes, and (*b*) the horsepower developed.

152. Work Done on Moving Vanes. Since work is equal to force times distance it is apparent that, for a jet to do any work upon a vane, the vane must be moving with a velocity between zero and the velocity of the jet, since at these limiting velocities either the distance or the force is equal to zero. The question then arises as to what velocity the vane should have, for any given velocity of jet, to perform the maximum amount of work.

The amount of work done per second is the product of the force acting in the direction of motion and the distance through which it acts. Assuming that the direction of motion of the vane is parallel with the direction of the jet, the force acting is (Art. 151)

$$F_x = \frac{wA(V - v')^2}{g}(1 - \cos\theta) \tag{15}$$

and the distance through which it acts per second is equal to the velocity of the vane, v'. The work done in foot-pounds per second is therefore

$$G = \frac{wA(V - v')^2}{g}(1 - \cos\theta)v' \tag{16}$$

Considering v' as the variable in this expression and equating the first derivative to zero, the relation between V and v' may be determined for which G is a maximum.

$$\frac{dG}{dv'} = \frac{wA(1 - \cos\theta)}{g}(V^2 - 4Vv' + 3v'^2) = 0$$

from which

$$v' = V \quad \text{and} \quad v' = \frac{V}{3} \tag{17}$$

When $v' = V$, no work is done since the force exerted is then zero and this value represents a condition of minimum power. For maximum power with a single vane, therefore, $v' = V/3$.

In a series of vanes so arranged that the entire jet strikes either one vane or another successively, the force exerted in the direction of motion, which is assumed parallel with the direction of the jet, is (Art. 151)

$$F_x = \frac{wAV}{g}(V - v')(1 - \cos\theta) \tag{18}$$

and

$$G = \frac{wAV}{g}(V - v')(1 - \cos\theta)v' \tag{19}$$

Differentiating, and equating to zero,

$$\frac{dG}{dv'} = \frac{wAV(1 - \cos\theta)}{g}(V - 2v') = 0$$

and for maximum power with a series of vanes

$$v' = \frac{V}{2} \tag{20}$$

Substituting this value of v' in 19, and noting that $wAV/g = M_1$,

$$G = \tfrac{1}{2}M_1V^2 \frac{1 - \cos \theta}{2} \tag{21}$$

which is $(1 - \cos \theta)/2$ times the total kinetic energy available in the jet. For $\theta = 180°$ this expression equals unity, and

$$G = \tfrac{1}{2}M_1V^2 \tag{22}$$

the total kinetic energy of the jet being converted into work. This also appears from considering that the relative velocity of the jet as it leaves the vane is $V/2$, which is also the velocity of the vane. These two velocities being equal and opposite in direction have a resultant of zero. The fluid thus leaves the vane with zero absolute velocity, signifying that all its original energy has been utilized in performing work.

These principles are utilized in the design of impulse turbines, which consist of a series of vanes attached to the periphery of a wheel. The angle θ must be somewhat less than 180 degrees so that the jet in leaving a vane will not interfere with the succeeding vane. Making the angle θ equal to 170 degrees in place of 180 degrees reduces the force applied to the wheel by less than 1 per cent.

153. Forces Exerted upon Closed Channels. In the preceding articles of this chapter the discussion has been restricted to forces exerted by jets impinging against flat and curved surfaces. As it was always considered that the flow was free and unconfined the only forces acting were dynamic. Consideration will now be given to the longitudinal thrust exerted upon a closed channel by fluid flowing through it under pressure.

Under conditions of steady flow through a curved channel of either constant or varying diameter, there is a thrust exerted upon the channel that is the resultant of a dynamic force and the total pressures exerted upon the end sections of the fluid contained in the channel.

In Fig. 147 is shown a curved channel having a diameter decreasing from AB to CD and a deflection angle θ. Let p_1, A_1, and V_1 represent respectively the pressure, area, and mean velocity at AB, and p_2, A_2, and V_2 the corresponding values at CD. In flowing from AB to CD the fluid is accelerated from V_1 to V_2, and the force R producing this acceleration is the resultant of all the component forces acting on the mass $ABCD$. These forces consist of the pressures on the sections AB and CD, the pressure exerted by the channel walls $ACBD$, and the force of gravity. By assuming

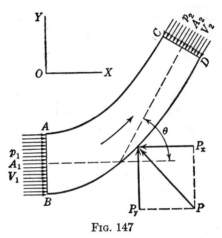

FIG. 147

that the center line of the channel lies in a horizontal plane so that the action of gravity is normal to the direction of flow, this latter force may be ignored.

Letting P_x and P_y represent the X and Y components of the forces exerted by the channel upon the fluid, the resultant X and Y components of the force producing acceleration are

$$R_x = A_1p_1 - A_2p_2 \cos\theta - P_x = \frac{Qw}{g}\,(V_2 \cos\theta - V_1) \quad (23)$$

$$R_y = -A_2p_2 \sin\theta + P_y = \frac{Qw}{g}\,V_2 \sin\theta \quad (24)$$

the right-hand members in these equations representing the increase in momentum along the X and Y axes resulting from the accelerating forces.

From equations 23 and 24,

$$P_x = A_1p_1 - A_2p_2 \cos\theta + \frac{Qw}{g}\,(V_1 - V_2 \cos\theta) \quad (25)$$

$$P_\flat = A_2p_2 \sin\theta + \frac{Qw}{g}\,V_2 \sin\theta \quad (26)$$

If the channel is one of constant diameter throughout, $A_1 = A_2$, $V_1 = V_2$, and $p_1 = p_2$ (approximately), and the equations reduce to

$$P_x = \left(Ap + \frac{AV^2w}{g}\right)(1 - \cos \theta) \qquad (27)$$

$$P_y = \left(Ap + \frac{AV^2w}{g}\right)\sin \theta \qquad (28)$$

If the angle θ equals 90° these equations become

$$P_x = P_y = Ap + \frac{AV^2w}{g} \qquad (29)$$

If the angle θ equals zero, or, in other words, if the channel is straight but of varying diameter, equation 25 reduces to

$$P_x = A_1 p_1 - A_2 p_2 - \frac{Qw}{g}(V_2 - V_1) \qquad (30)$$

Considering a straight channel of constant diameter throughout, equation 30 reduces to

$$P_x = A(p_1 - p_2) \qquad (31)$$

PROBLEMS

1. A jet 1 in. in diameter and having a velocity of 40 ft per sec strikes normally against a flat plate moving in the same direction as the jet. Determine: (a) the velocity of the plate if the jet is to perform the maximum amount of work; (b) the corresponding amount of work in foot-pounds per second.

2. A nozzle discharges 2 cfs with a velocity of 40 ft per sec. The entire jet strikes a series of vanes that are moving in the same direction as the jet with a velocity of 10 ft per sec. Each vane has a deflection angle of 150 degrees. Determine: (a) the work done per second; (b) the kinetic energy per second in the water leaving the vanes; and (c) the kinetic energy per second in the jet before it strikes the vanes.

3. A $1\frac{1}{2}$-in. nozzle has a coefficient of velocity of 0.97 and a coefficient of contraction of unity. The base of the nozzle has a diameter of 4 in., at which point the gage pressure is 80 lb per sq in. The jet strikes a double-cusped vane which has a deflection angle of 150 degrees and a velocity in the direction of the jet of 30 ft per sec. Determine: (a) the pressure exerted on the vane, and (b) the amount of work done, expressed in foot-pounds.

4. A jet having a diameter of $1\frac{1}{2}$ in. and a velocity of 60 ft per sec strikes a series of vanes so arranged on the periphery of a wheel that the entire jet strikes the vanes. The deflection angle of the vanes is 170 degrees. Determine the maximum amount of work that can be done and the direction and absolute velocity of water leaving the vanes.

5. A horizontal straight pipe gradually reduces in diameter from 12 in. to 6 in. If, at the larger end, the gage pressure is 40 lb per sq in. and the velocity is 10 ft per sec, what is the total longitudinal thrust exerted on the pipe? Neglect friction.

6. Water flows from a 36-in. pipe through a reducer into a 24-in. pipe. If the gage pressure at the entrance to the reducer is 60 lb per sq in. and the velocity is 7 ft per sec, determine the resultant thrust on the reducer. Assume the frictional loss in the reducer to be 5 ft.

7. A bend in a pipe line gradually reduces from 24 in. to 12 in. The deflection angle is 60 degrees. If at the larger end the gage pressure is 25 lb per sq in. and the velocity is 8 ft per sec, determine X and Y components of the dynamic thrust exerted on the bend. Also determine X and Y components of the total thrust exerted on the bend, neglecting friction.

154. Resistance to Object Moving through Fluid. The previous articles have dealt with forces exerted on vanes, plates, or closed channels by streams of fluid of finite cross-sectional area. A brief discussion will now be given of the resistance force which is encountered when an object moves through a fluid, the boundaries of which are so far removed that the cross section can be considered of infinite extent. A similar resistance force is set up when a fluid of relatively large cross-sectional area moves past a stationary object.

The principle that force is equal to the rate of change of momentum is used to derive the basic equation for the resistance. It is considered that, as a body moves through a fluid, it imparts to a certain mass of the fluid each second a certain velocity.

The mass of fluid affected per second by the motion of the body is considered to be

$$M_1 = k_1 \frac{wAV}{g}$$

where k_1 is a proportionality factor, w is the unit weight of the fluid, A is the area of the body projected onto a plane normal to the direction of motion, and V is the velocity of the body. It is also considered that the velocity imparted to this mass of fluid is proportional to the velocity of the body, or

$$v = k_2 V$$

k_2 being a proportionality factor.

The resistance is then the product of mass affected per second and change in velocity, or

$$D_f = k_1 k_2 \frac{w}{g} A V^2 \qquad (32)$$

This resistance is commonly called the drag force, or drag. Dividing and multiplying by 2, and substituting a coefficient of drag C_D for $2k_1 k_2$, the general equation for drag becomes

$$D_f = C_D w \frac{V^2}{2g} A \qquad (33)$$

The similarity should be noted between this equation and the equation for hydrostatic pressure on a plane area (Art. 28),

$$P = w\bar{h}A$$

The latter equation defines total pressure as equal to the intensity of pressure at the center of gravity times the area. By equation 33, the resistance to a body moving through a fluid, or the force exerted on a body held stationary in a moving stream of fluid, is equal to the stagnation point pressure (Art. 55) corresponding to the velocity of the fluid, multiplied by the projected area of the body, and modified by a coefficient.

Since $w/g = \rho$, the mass density, equation 33 is often written

$$D_f = C_D \frac{\rho V^2}{2} A \qquad (34)$$

Here $\rho V^2/2$ represents the stagnation point pressure.

The application of equation 33 or 34 requires a knowledge of the value of the drag coefficient C_D. Theoretical and experimental studies have shown that C_D varies with the Reynolds number as well as with the form of the body. Values for a few of the more common geometrical forms are shown in Fig. 148 and are discussed here.

Sphere. The Reynolds number is $N_R = DV/\nu$, where D is the diameter of the sphere, V is its velocity relative to the fluid, and ν

is the kinematic viscosity of the fluid. At small values of the Reynolds number, as with flow in pipes, viscous forces predominate and the drag coefficient is defined by the equation

$$C_D = \frac{24}{N_R} \qquad (35)$$

Since $N_R = DV\rho/\mu$, from equation 34,

$$D_f = \frac{24\mu}{DV\rho} \times \frac{\rho V^2}{2} \times \frac{\pi}{4} D^2$$

whence,

$$D_f = 3\pi DV\mu \qquad (36)$$

which is the usual form of what is known as Stokes' law for resistance of objects in " creeping motion."

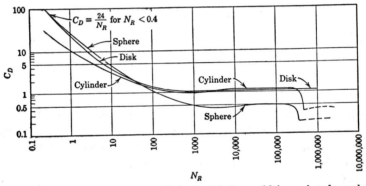

FIG. 148. Variation of drag coefficient with Reynolds' number for sphere, disk, and cylinder. *Source:* F. Eisner, "Das Widerstandsproblem," *Proc. 3d Intern. Congr. Applied Mech.*, Stockholm, 1930, Vol I, pp. 23–42.

As the Reynolds number for the sphere increases beyond about 0.4, the drag coefficient decreases less rapidly and is approximately constant at a value of about 0.5 for Reynolds' numbers between 1000 and 300,000. The sudden drop in value of the coefficient at N_R equal to about 300,000 is thought to result from a change from laminar to turbulent flow in the boundary layer on the surface of the sphere, causing a narrower wake behind the sphere and thus reducing the drag.

Circular Disk. The disk is considered to be very thin and to be moving relative to the fluid with the plane of the disk normal to the direction of motion. The diameter of the disk is used in computing the Reynolds number.

The disk follows Stokes' law up to a value of Reynolds' number of about 2, after which the coefficient decreases less rapidly and reaches a constant value of about 1.1 for values of Reynolds' numbers larger than about 1000.

Circular Cylinder. The cylinder is considered to be moving relative to the fluid with the axis of the cylinder normal to the direction of motion. The cylinder is considered to have infinite length. The diameter is used in computing the Reynolds number.

The cylinder does not follow Stokes' law as closely at low values of the Reynolds number as do the sphere and the disk, although the coefficient varies inversely with N_R. At larger values of N_R the coefficient for the cylinder agrees closely with that for the circular disk, until a value of N_R of about 300,000 is reached. Here the change in the boundary layer flow is thought to occur, and the coefficient shows a drop similar to that of the sphere.

Flat Rectangular Plate. The plate is considered to be moving relative to the fluid with the plane of the plate normal to the direction of motion. The drag coefficient has been found to vary slightly with the ratio of length to breadth. For a square plate, Dodge and Thompson[1] give a value of C_D of about 1.10, and show this value increasing to 1.25 for a rectangle having a ratio of 10 to 1, and to approximately 2.0 for a ratio of infinity.

Values of C_D for flat plates have been used to determine wind pressures on buildings with flat sides. Since the wind moves past only the sides and top of the building, the ratio of height to width of building should be doubled in order to determine the value of C_D. There is some indication that an increase in the thickness of the building parallel to the direction of the wind reduces the coefficient somewhat, but results are on the side of safety if this reduction is neglected.

155. Terminal Velocity. A body falling freely through a fluid is being acted upon by three vertical forces: (1) its weight, acting downward; (2) the drag force, acting upward; and (3) the buoyant force of the fluid acting upward. If the body falls freely for a sufficient length of time, the sum of the two upward forces may become equal to the weight of the body. In this case the summation of the vertical forces has become zero and the body has acquired a constant velocity of fall, called terminal velocity.

[1] R. A. Dodge and M. J. Thompson, *Fluid Mechanics*, McGraw-Hill Book Co., p. 333.

For a solid body falling through a gas, such as air, the buoyant force is usually negligible. The terminal velocity is reached in that case when the drag force equals the weight.

PROBLEMS

1. The projected area of an automobile is 30 sq ft and its drag coefficient is 0.45. If the automobile is moving through still air at 60° F and standard barometric pressure, compute the resistance force and the horsepower required to overcome this force at speeds of 20, 40, and 80 miles per hour?

2. A smooth steel ball 2 in. in diameter is moving horizontally through air at 30° F and 14.4 lb per sq in. barometric pressure with a velocity of 600 ft per sec. The specific gravity of the ball is 7.85. The absolute viscosity of the air is 0.00000036 lb-sec per sq ft. Compute the drag force and the horizontal deceleration of the ball.

3. A man is descending in a parachute at an altitude of 5000 ft where the air temperature is 40° F. The man and his equipment weigh 250 lb. The parachute is 18 ft in diameter. Assuming that the drag coefficient for the parachute is the same as for a circular disk, compute the man's terminal velocity.

4. Determine the average pressure in pounds per square foot on the face of a building 200 ft wide and 400 ft high when the wind velocity normal to the face of the building is 60 miles per hour. Assume standard barometric pressure and an air temperature of 60° F.

5. A submarine when running submerged in sea water has a drag coefficient of 0.15 and a projected area of 65 sq ft. Determine the drag force for a velocity of 15 miles per hour and the horsepower required to overcome this force.

156. Water Hammer in Pipe Lines. In Fig. 149 is shown a pipe line leading from a reservoir, A, and discharging into the air at B near which is located a gate valve. If the valve is suddenly closed, a dynamic pressure called water hammer is exerted in the pipe in excess of the normal static pressure. The magnitude of this pressure is frequently much greater than that of any static pressure to which the pipe may ever be subjected, and the possibility of the occurrence of such pressure must therefore be investigated in connection with the design of pipe lines.

This dynamic pressure is the result of a sudden transformation of the kinetic energy of the moving mass of water within the pipe into pressure energy. Since force equals mass times acceleration, or

$$F = M \frac{dv}{dt} \qquad (37)$$

it follows that, if the velocity of the mass M could be reduced from v to zero instantaneously, this equation would become

$$F = M \frac{v}{0} \qquad (38)$$

or, in other words, the pressure resulting from the change would be infinite. Such an instantaneous change is impossible, however.

FIG. 149

The following nomenclature will be used, all units being expressed in feet and seconds except as noted:

b = thickness of pipe walls.
D = inside diameter of pipe.
A = cross-sectional area of pipe in square feet.
e = modulus of elasticity of pipe walls in pounds per square foot.
e' = modulus of elasticity of water in pounds per square foot.
E = modulus of elasticity of pipe walls in pounds per square inch.
E' = modulus of elasticity of water in pounds per square inch
g = acceleration of gravity.
h = head due to water hammer (in excess of static head).
H = total head producing discharge through valve.
L = length of pipe line.
T = time of closing valve.
V_v = mean velocity of water through valve.

V = mean velocity of water in pipe before closure of valve.

v_w = velocity of pressure wave along pipe.

Consider the conditions within the pipe immediately following the closure of the valve. Let l_1, l_2, l_3, $\cdots l_n$ represent infinitesimally short sections of pipe, as shown in Fig. 149. The instant the valve is closed, the water in section l_1 is brought to rest, its kinetic energy is transformed into pressure energy, the water is somewhat compressed, and the pipe expands slightly as a result of the increased stress to which it is subjected. Because of the enlarged cross-sectional area of l_1 and the compressed condition of the water within it, a greater mass of water is now contained within this section than before the closure. It is evident then that a small volume of water flowed into section l_1 after the valve was closed. An instant later a similar procedure takes place in l_2 and then in l_3, so that evidently a wave of increased pressure travels up the pipe to the reservoir. When this wave reaches the reservoir the entire pipe is expanded and the water within it is compressed by a pressure greater than that due to the normal static head.

There is now no longer any moving mass of water within the pipe, the conversion of whose kinetic energy into pressure energy serves to maintain this high pressure, and therefore the pipe begins to contract and the water to expand with a consequent return to normal static pressure. This process starts at the reservoir and travels as a wave to the lower end. During this second period some of the water stored within the pipe flows back into the reservoir, but on account of the inertia of this moving mass an amount flows back greater than the excess amount stored at the end of the first period so that the instant this second wave reaches the valve the pressure at that point drops not only to the normal static pressure but below it.

A third period now follows during which a wave of pressure less than static sweeps up the pipe to the reservoir. When it reaches the reservoir the entire pipe is under pressure less than static, but since all the water is again at rest the pressure in l_n immediately returns to the normal static pressure due to the head of water in the reservoir. This starts a fourth period marked by a wave of normal static pressure moving down the pipe. When the valve is reached, the pressure there is normal and for an instant the conditions throughout the pipe are similar to what they were when the valve was first closed. The velocity of the water (and the resultant water hammer) is now, however, somewhat less than it was at the time of closure because of friction and the imperfect elasticity of the pipe and the water.

Instantly another cycle begins similar to the one just described, and then another, and so on, each set of waves successively diminishing, until finally the waves die out from the influences mentioned.

Equation 38 shows that for instantaneous closure of valve the pressure created would be infinite if the water were incompressible and the pipe inelastic. Since it is impossible to close a valve instantaneously, it is apparent that a series of pressure waves is created, similar to the one just described, causing an increasing pressure at the valve. If the valve is completely closed before the first pressure wave has time to return to the valve as a wave of low pressure, or, in other words, if T is less than $2L/v_w$, it is evident that the pressure has been continually increasing up to the time of complete closure and that the resulting pressure is the same as if the valve had been instantaneously closed. But if T is greater than $2L/v_w$, then before the valve is completely closed the earlier pressure waves have returned as waves of low pressure and tend to reduce the rise of pressure resulting from the final stages of valve closure.

Hence if T is equal to or less than $2L/v_w$, h will be the same as for instantaneous closure, but if T is greater than $2L/v_w$ then h will be diminished as T increases.

157. Rise in Pressure when $T \lessgtr 2L/v_w$. The theory of water hammer is based upon the law of conservation of energy. The amount of kinetic energy contained in the moving column of water within the pipe is $MV^2/2 = wALV^2/2g$. This energy is used up in doing work in compressing the water and in stretching the pipe walls.

If the resulting pressure head is h feet above normal the compression of the water column absorbs $\dfrac{(wh)^2}{2}\dfrac{AL}{e'}$ foot-pounds of energy, since the final intensity of pressure is wh, the average total pressure is $whA/2$, the unit compression is wh/e', and the total compression or distance through which the average total pressure acts is whL/e'.

In a similar manner the total work done in stretching the pipe walls is $\dfrac{(wh)^2 LAD}{2eb}$ foot-pounds, since the unit stress in the pipe walls is $whD/2b$, the average total stress in the pipe walls is $whDL/4$, the unit elongation is $whD/2be$, and the total elongation or distance through which the force acts is $wh\pi D^2/2be$.

The energy equation therefore becomes

$$\frac{wALV^2}{2g} = \frac{(wh)^2 AL}{2e'} + \frac{(wh)^2 LAD}{2eb} \tag{39}$$

from which

$$h = \frac{V}{g\sqrt{\dfrac{w}{g}\left(\dfrac{1}{e'} + \dfrac{D}{eb}\right)}} \tag{40}$$

Reducing, so that the moduli are expressed in pounds per square inch, by dividing each by 144, this equation becomes

$$h = \frac{12V}{g\sqrt{\dfrac{w}{g}\left(\dfrac{1}{E'} + \dfrac{D}{Eb}\right)}} = \frac{4660V}{g\sqrt{1 + \dfrac{E'D}{Eb}}} \tag{41}$$

It will now be shown that $\dfrac{4660}{\sqrt{1 + \dfrac{E'D}{Eb}}}$ represents the velocity at which the pressure wave travels up the pipe line. In the time t, a column of water of length $v_w t$ is brought to rest. The rate of change of momentum is therefore

$$\frac{MV}{t} = \frac{wAv_w tV}{gt} = \frac{wAv_w V}{g}$$

and this must equal the force exerted by the increased pressure which is whA. Therefore,

$$whA = \frac{wAv_w V}{g}$$

from which

$$h = \frac{v_w V}{g} \tag{42}$$

Comparing equations 41 and 42, it is apparent that

$$v_w = \frac{4660}{\sqrt{1 + \dfrac{E'D}{Eb}}} \qquad (43)$$

Since sound is transmitted by means of pressure waves, v_w is the velocity of sound through water in that particular pipe.

The foregoing theory was first derived and experimentally verified by Professor N. Joukovsky,[1] of Moscow, Russia, in 1898.

158. Rise in Pressure when $T > 2L/v_w$. Numerous formulas have been derived for the determination of the rise in pressure in a pipe line resulting from the slow closure of a valve or of turbine gates, but most of them are unreliable, or else are true only under special conditions. The method that will be followed here is that of arithmetic integration, as proposed by Norman R. Gibson and published in vol. 83 of the *Transactions of the American Society of Civil Engineers* for 1919. It is based on the foregoing theory of pressure waves and consists of a method of tracing the action of these waves instant by instant. It is claimed that the accuracy of this method has been verified by many careful experiments.

Assume that the valve, instead of being closed in a continuous motion, is closed by a series of small instantaneous movements. Each small movement of the valve will destroy a small portion, ΔV, of the velocity V; and since this destruction is instantaneous the resulting increase in pressure head will be, from equation 42,

$$\Delta h = \frac{v_w \Delta V}{g} \qquad (44)$$

There will thus be transmitted up the pipe line a series of small pressure waves which, when added together, will give the total excess pressure produced.

If it is assumed, as usual, that each small movement of valve produces the same reduction in area of valve opening, it will be necessary to determine the resulting reduction in velocity in the pipe, for, obviously, the instant the valve starts closing, the pressure behind it starts rising, and this rise in pressure increases the velocity through the opening and diminishes the rate of retardation

---

[1] Miss O. Simin, " Water Hammer, with special reference to the researches of Professor N. Joukovsky," *Proc., Am. Water Works Assoc.,* 1904.

of the velocity in the pipe. This reduced retardation has an important bearing on the problem and must be taken into account in determining the resulting rise in pressure.

The velocity through the valve opening may be expressed by the formula for discharge through orifices

$$V_v = C_v \sqrt{2gH} \tag{45}$$

This formula may be written in the form

$$V_v = K' \sqrt{H} \tag{46}$$

in which K' may be determined in any given problem in which the velocity in the pipe V, the total head H, and the ratio β of the area of the valve opening to the area of the pipe are all known, since

$$V_v = \frac{V}{\beta} = K' \sqrt{H} \tag{47}$$

Equation 47 may be written

$$V = \beta K' \sqrt{H} \tag{48}$$

or

$$V = K \sqrt{H} \tag{49}$$

in which

$$K = \beta C_v \sqrt{2g} \tag{50}$$

As the valve is closed the value of β, or of K, will become smaller and smaller, with corresponding increments, Δh, in the head producing discharge through the valve, and with simultaneous decrements, ΔV, in the velocity in the pipe, so that equation 49 may be written

$$V - \Delta V = K \sqrt{H + \Delta h} \tag{51}$$

The solution of the problem consists in finding those values of ΔV and Δh that will satisfy equations 44 and 51 for the different values of K.

The method may best be explained by the solution of a problem.

EXAMPLE. Determine the rise in pressure that will occur in a penstock leading to a power plant if the turbine gates are closed in 2.1 sec. $H = 165$ ft, $L = 820$ ft, $V = 11.75$ ft per sec, and $v_w = 4680$ ft per sec.

For convenience, T has been taken as an even multiple of $2L/v_w =$

$2 \times 820/4680 = 0.35$ sec. To simplify the computations as much as possible, friction will be neglected Also, for convenience, it will be assumed that the gates are closed in $2.1/0.35 = 6$ successive instantaneous movements. Each intervening interval will therefore be just long enough to allow the pressure wave resulting from one movement to travel up the penstock and return as a low-pressure wave at the instant that the next movement takes place.

From equation 49,

$$K = \frac{V}{\sqrt{H}} = \frac{11.75}{\sqrt{165}} = 0.915$$

For each successive movement K will be reduced by one-sixth of its original value, or by 0.1525, since the movements are assumed to be of uniform magnitude.

The first three columns of the table following may now be filled in, and also the initial values of columns 4 and 5.

The remaining values are obtained in the following manner.

Assume a value for ΔV caused by the first movement of the gate. From equation 44,

$$\Delta h = \frac{4680}{32.2} \Delta V = 145\Delta v \tag{52}$$

Substituting the assumed value for ΔV in this equation, a trial value for Δh is found. It is now necessary to determine whether or not

1	2	3	4	5	6	7	8
Interval	Time	K	Head	Velocity V	ΔV	Δh	$\Sigma(\Delta h)$
0	0.00	0.915	165	11.75	0.39	57	
1	0.35	0.762	222	11.36	1.18	171	57
2	0.70	0.610	279	10.18	1.91	277	114
3	1.05	0.457	328	8.27	2.48	360	163
4	1.40	0.305	362	5.79	2.82	410	197
5	1.75	0.152	378	2.97	2.97	430	213
6	2.10	0.000	382	0.00			217

these trial values satisfy equation 51. This may best be done by substituting this trial value for Δh and solving for ΔV. If this value for ΔV does not check the assumed value, a new value for ΔV must be assumed and the computations repeated until the two equations are satisfied. The correct value for ΔV will always be found to lie between the assumed value and the computed value and will usually be found to be much nearer the former than the latter.

For the initial gate movement, ΔV is found to be 0.39 ft per sec and $\Delta h = 57$ ft. Hence, during the first interval of 0.35 sec the total head acting is 222 ft and the velocity in the penstock is 11.36 ft per sec.

The computation will now be carried through for the second interval. First assume that $\Delta V = 1.00$. Then $\Delta h = 145 \times 1.00 = 145$. Substituting in equation 51,

$$11.36 - \Delta V = 0.610\sqrt{165 + 145 - 57}$$

from which $\Delta V = 1.65$.

In substituting for $(H + \Delta h)$ in equation 51, it must be remembered that this quantity represents the total effective head producing discharge through the gates during that interval. The effect of the preceding pressure waves must therefore be taken into account. At the beginning of the second interval the first pressure wave will have returned to the gates as a wave of low pressure and will reduce the effective head during the second interval by the same amount that it increased it during the first interval.

The correct value of ΔV is now known to lie between 1.00 and 1.65. Next assume $\Delta V = 1.20$. Then $\Delta h = 145 \times 1.20 = 174$, and, substituting in equation 51 and solving, $\Delta V = 1.12$. Finally, assuming $\Delta V = 1.18$, $\Delta h = 171$. These values are found to satisfy equation 51.

During the third interval the first pressure wave will have returned as a wave of high pressure, while the second wave will be one of low pressure. Therefore

$$H + \Delta h = 165 + 57 - 171 + \Delta h$$

and, during the fourth interval,

$$H + \Delta h = 165 - 57 + 171 - 277 + \Delta h$$

and so on.

Graphical representation of the behavior of each individual pressure wave throughout the period of closure, as well as a curve showing the algebraic sum of the effects of these waves, may aid in visualizing the solution of this problem.

It is possible to derive a series of algebraic equations, one for each time interval, that would constitute a direct solution of the problem of water hammer and would lead to practically the same results as are obtained by the preceding trial-and-error method of arithmetic integration. The equations are long and cumbersome, however, and must be solved by means of logarithms, whereas all computations in the solution outlined here may easily be made by slide rule.

For the analytical method, as well as for a more complete discussion of the method given here, the original paper should be consulted.

GENERAL PROBLEMS

1. A 1-in. nozzle has a coefficient of velocity of 0.97 and a coefficient of contraction of 1.00. The base of the nozzle has a diameter of 3 in., at which point the gage pressure is 80 lb per sq in. The jet strikes a vane which has a deflection angle of 150° and a velocity in the direction of the jet of 30 ft per sec. Determine: (*a*) the pressure exerted on the vane; (*b*) the amount of work done per second; and (*c*) the velocity of the water leaving the vane.

2. A tank 4 ft wide and 6 ft long, supported on frictionless rollers, contains water having a depth of 6 ft. In one end of this tank is an orifice having an area of 24 sq in. and a coefficient of discharge of 0.60. If the center of the orifice is 4.0 ft below the water surface, what force will be required to hold this tank stationary?

3. A large tank contains water having a depth of 4 ft. In the bottom of the tank is a 6-in.-diameter orifice having a coefficient of discharge of 0.60 and a coefficient of contraction of 0.62. A vane is placed 5 ft below the plane of the orifice in such manner that the jet from the orifice strikes the vane and is deflected through an angle of 150°. What force is required to hold the vane stationary?

4. A 3-in. fire hose discharges through a 1-in. nozzle which has a coefficient of contraction of 1.0 and a coefficient of velocity of 0.95. If the gage pressure at the base of the nozzle is 40 lb per sq in., determine the longitudinal thrust on the nozzle. If the axis of the nozzle coincides with that of the hose will the connection between the hose and nozzle be in tension or compression?

5. A man weighing 200 lb is descending in a parachute through air at 80° F and normal sea-level pressure. Determine the proper diameter of parachute in order that the man's terminal velocity shall be no greater than the velocity he would acquire in jumping freely through a height of 10 ft. Assume that the drag coefficient of the parachute is the same as that of a circular disk.

6. Compute the terminal settling velocity of a spherical particle of sand 1 millimeter in diameter and having a specific gravity of 2.7 when it is dropping through (*a*) water at 60° F; (*b*) a heavy fuel oil at 60° F.

7. A 24-in. cast-iron pipe $\frac{3}{4}$ in. thick and 6000 ft long discharges water from a reservoir under a head of 80 ft. What is the pressure due to water hammer resulting from the instantaneous closure of a valve at the discharge end?

8. A 24-in. cast-iron pipe, $\frac{3}{4}$ in. thick and 2000 ft long, discharges water from a reservoir under a head of 80 ft. If a valve at the discharge end is closed in 6 sec determine the magnitude of the resulting water hammer.

Chapter X

HYDRAULIC SIMILITUDE AND DIMENSIONAL ANALYSIS

159. Introduction. The principles of hydrokinetics developed in the previous chapters are based on mathematical theory, but the accuracy of the results obtained by their application to practical engineering problems frequently depends on experimental data obtained in field and laboratory. There is evidence that Leonardo da Vinci studied principles of hydraulic design by means of small models of structures and machines, but until rather recently hydraulic experimentation was usually carried on to a full scale — that is, on weirs, channels, pipe lines, and dams, as constructed in the field.

Within the last half century, however, methods have been developed whereby, as a result of experiments conducted on a scale model, it is possible to predict the behavior of a full-scale structure, or prototype. The principles on which this procedure is based comprise the theory of *hydraulic similitude*. The analysis of the basic relationships of the various physical quantities involved in the motion and in the dynamic action of the fluid is called *dimensional analysis*. It is the purpose of this chapter to give briefly these basic principles and methods of analysis as they relate particularly to hydraulic model testing.

Few if any important hydraulic structures are now designed and built without more or less extensive preliminary model studies. Such studies may have for their purpose: (1) the reduction of head losses in canal or pipe intakes or at transition sections; (2) the development of effective methods of dissipating the energy in the stream at the foot of overflow dams or at the outlet end of culverts, thus reducing stream bed erosion; (3) the determination of discharge coefficients for overflow dams and weirs; (4) the development of the best design of shaft and siphon spillways and of other outlet structures at reservoirs; (5) harbor design, involving a determination of the best cross section, height, and location of breakwater, as well as the size and location of opening; (6) the design of locks, including the effect, upon ships, of the currents set up in the operation of those locks; and (7) innumerable other problems that arise in engineering work but perhaps occur with less frequency than those mentioned above.

River model studies may have for their purpose the determination of: (1) the time of travel of flood waves through river channels; (2) methods of improving channels for transmission of floods with less danger of overflow of banks; (3) the effect of river bend cut-offs on the regimen of the stream both above and below the point of cut-off; (4) the effect of bends, levees, spur dikes, and training walls on the formation of bars or on the erosion of the bed; (5) the height of backwater caused by permanent or temporary structures built in a stream; and (6) the direction and force of currents in rivers and harbors and their effect on navigation.

Hydraulic model studies may also be classified with respect to the character of equipment or the type of laboratory that is needed. For instance, in the river hydraulics laboratory, designed to study river problems, one of the principal features is a circulating system capable of providing a constant rate of discharge that may be accurately measured and controlled between the desired limits. On the other hand, in a lake hydraulics laboratory the outstanding characteristics are equipment capable of creating and measuring waves of any desired height and frequency.

An important phase of many model studies is the determination of the resistance offered to motion of objects through fluids. In hydraulic models this study usually takes the form of determining the force required to move a ship or submarine model through still water at various speeds. Practically all modern airplane design is based on wind tunnel experiments, in which models of planes or parts of planes are fixed in a conduit through which air is propelled at high velocities. A recent development in hydraulic research is the water tunnel, in which water or other liquid is moved past a fixed model.

160. Principles of Similitude. Similarity between model and prototype may take three different forms: geometric, kinematic, dynamic.

1. Geometric similarity implies similarity of form. A model is geometrically similar to the prototype if the ratios of all homologous lengths in model and prototype are equal.

2. Kinematic similarity implies similarity of motion. Kinematic similarity of model to prototype is attained if the paths of

homologous moving particles are geometrically similar and if the ratios of velocities of the various homologous particles are equal.

3. Dynamic similarity implies similarity of forces. A model is dynamically similar to the prototype if it is kinematically similar, and if the ratios of homologous moving masses and of the forces producing motion are respectively equal.

In some model studies, particularly of open channel flow, strict geometric similarity would result in too small a depth of flow in the model. It then becomes necessary to make the vertical scale of the model larger than the horizontal scale. The model is then said to be distorted.

The interrelationship of the various quantities involved in the three forms of similarity can be derived from a consideration of the units in which the quantities are expressed. Let the subscripts m and p indicate respectively model and prototype, and let the subscript r indicate, for each quantity, the ratio between model and prototype. This ratio is ordinarily expressed as a fraction. For instance, a model in which all linear dimensions are one-thirtieth of the homologous dimensions of the prototype is said to be built on a scale of $1:30$, or the scale ratio is $1/30$. It is usually desirable in hydraulic model studies to express the ratio of all quantities involved in terms of the geometric scale ratio.

161. Geometric Similarity. The quantities involved in geometric similarity are length, area, and volume. The ratio of homologous lengths in model and prototype is expressed as

$$\frac{L_m}{L_p} = L_r \tag{1}$$

An area A is equal to the square of a characteristic length; hence, the ratio of homologous areas can be expressed as

$$\frac{A_m}{A_p} = \frac{L_m{}^2}{L_p{}^2} = L_r{}^2 \tag{2}$$

A volume is equal to the cube of a characteristic length; hence, the ratio of homologous volumes can be expressed as

$$\frac{\mathrm{Vol}_m}{\mathrm{Vol}_p} = \frac{L_m{}^3}{L_p{}^3} = L_r{}^3 \tag{3}$$

162. Kinematic Similarity. Kinematic similarity introduces the concept of time as well as of length. The ratio of the times required for homologous particles to travel homologous distances in model and prototype is

$$\frac{T_m}{T_p} = T_r \tag{4}$$

Kinematic quantities usually involved in model studies are linear velocity and acceleration, discharge, and angular velocity and acceleration. Linear velocity V is expressed in terms of length per unit of time; thus

$$\frac{V_m}{V_p} = \frac{L_m/T_m}{L_p/T_p} = \frac{L_m/L_p}{T_m/T_p} = \frac{L_r}{T_r} \tag{5}$$

Linear acceleration a is expressed as a length per unit of time squared. Thus

$$\frac{a_m}{a_p} = \frac{L_m/T_m{}^2}{L_p/T_p{}^2} = \frac{L_m/L_p}{T_m{}^2/T_p{}^2} = \frac{L_r}{T_r{}^2} \tag{6}$$

The units of discharge Q are volume per unit of time; thus

$$\frac{Q_m}{Q_p} = \frac{\mathrm{Vol}_m/T_m}{\mathrm{Vol}_p/T_p} = \frac{L_m{}^3/L_p{}^3}{T_m/T_p} = \frac{L_r{}^3}{T_r} \tag{7}$$

Angular velocity ω is expressed in radians per unit of time and is equal to the tangential linear velocity divided by the length of radius R of the curve at the point of tangency. The units of ω are thus

$$\frac{\omega_m}{\omega_p} = \frac{V_m/R_m}{V_p/R_p} = \frac{V_m/V_p}{R_m/R_p} = \frac{L_r/T_r}{L_r} = \frac{1}{T_r} \tag{8}$$

Since revolutions per minute, N, is a measure of angular velocity,

$$\frac{N_m}{N_p} = \frac{1}{T_r} \tag{9}$$

Angular acceleration α is expressed in radians of angle per unit of time squared; hence, from equation 8

$$\frac{\alpha_m}{\alpha_p} = \frac{1}{T_r{}^2} \tag{10}$$

163. Dynamic Similarity. For dynamic similarity it is necessary that the ratios of homologous forces in model and prototype be equal, or

$$\frac{F_m}{F_p} = F_r \tag{11}$$

Since force equals mass, M, times acceleration, a,

$$F_r = \frac{F_m}{F_p} = \frac{M_m a_m}{M_p a_p} = M_r \frac{L_r}{T_r^2} \tag{12}$$

The force defined by the equation $F = Ma$ has been called *inertia* force, and equation 12 therefore defines the ratio of homologous inertia forces in model and prototype. Since mass $= \rho \times$ Vol, equation 12 can be written

$$F_r = (\rho_r L_r^3) \frac{L_r}{T_r^2} = \rho_r L_r^2 \left(\frac{L_r}{T_r}\right)^2 = \rho_r A_r V_r^2 \tag{12a}$$

This equation, which should be compared with equation 34 (page 305), has been called the Newtonian law of similitude since it is based on the relation, $F = Ma$, and expresses the general law of dynamic similarity between model and prototype.

From equation 12 the ratio of homologous masses in model and prototype becomes

$$M_r = \frac{F_r T_r^2}{L_r} \tag{13}$$

Work being equal to force times distance, the ratio of the work done by homologous forces is

$$\frac{\text{Work}_m}{\text{Work}_p} = \frac{F_m L_m}{F_p L_p} = F_r L_r \tag{14}$$

Power is the rate of doing work; hence

$$\frac{\text{Power}_m}{\text{Power}_p} = \frac{F_m L_m / T_m}{F_p L_p / T_p} = \frac{F_r L_r}{T_r} \tag{15}$$

Unit weight w is the force of attraction of the earth on each unit volume of the substance; hence

$$\frac{w_m}{w_p} = \frac{F_m / \text{Vol}_m}{F_p / \text{Vol}_p} = \frac{F_m / F_p}{\text{Vol}_m / \text{Vol}_p} = \frac{F_r}{L_r^3} \tag{16}$$

Mass density ρ is the mass per unit volume; hence

$$\frac{\rho_m}{\rho_p} = \frac{(F_m T_m{}^2/L_m) \div \text{Vol}_m}{(F_p T_p{}^2/L_p) \div \text{Vol}_p} = \frac{F_r T_r{}^2}{L_r{}^4} \quad (17)$$

164. Gravitational Forces Predominant — Froude's Law. When the force of inertia and the force of gravity can be considered to be the only forces which control the motion, the ratio of forces acting on homologous particles in model and prototype is defined not only by equation 12,

$$F_r = M_r \frac{L_r}{T_r{}^2} \quad (12)$$

but also by the fact that the force of gravity acting on a particle is equal to the weight W of the particle. Hence, for homologous particles,

$$F_r = \frac{W_m}{W_p} = \frac{w_m L_m{}^3}{w_p L_p{}^3} = w_r L_r{}^3 \quad (18)$$

Since mass $= (w/g)$ times volume, M_r in equation 12 can be written

$$M_r = \frac{w_r}{g_r} L_r{}^3 \quad (19)$$

and thence

$$F_r = \frac{w_r}{g_r} \cdot \frac{L_r{}^4}{T_r{}^2} \quad (20)$$

Equating values of F_r from equations 18 and 20,

$$w_r L_r{}^3 = \frac{w_r}{g_r} \cdot \frac{L_r{}^4}{T_r{}^2} \quad (21)$$

from which

$$T_r = \sqrt{\frac{L_r}{g_r}} \quad (22)$$

This equation, derived for the condition under which it can be assumed that the forces of inertia and gravity control the flow, is known as the Froude model law.

Ordinarily the value of g in equation 22 can be considered to be the same in model and prototype. Then $g_r = 1$ and equation 22 becomes

$$T_r = \sqrt{L_r} \quad (23)$$

If $\sqrt{L_r}$ is substituted for T_r, the ratios of the quantities given by equations 1 to 17 can be expressed in terms of the length ratio L_r and the unit weight ratio w_r, as shown in the accompanying table. If also $w_r = 1$, as, for example, if water is used in both model and prototype, the ratios of the quantities involved in dynamic similarity can also be expressed in terms of L_r.

SCALE RATIOS FOR THE FROUDE MODEL LAW, WHEN $g_r = 1$

Geometric Similarity		Kinematic Similarity		Dynamic Similarity	
Length	L_r	Time	$L_r^{1/2}$	Unit weight	w_r
Area	L_r^2	Velocity	$L_r^{1/2}$	Mass density	w_r
Volume	L_r^3	Acceleration	1	Force	$w_r L_r^3$
		Discharge	$L_r^{5/2}$	Mass	$w_r L_r^3$
		Angular velocity	$L_r^{-1/2}$	Work	$w_r L_r^4$
		Angular acceleration	L_r^{-1}	Power	$w_r L_r^{7/2}$

165. Froude's Number. If the value of T_r from equation 22 is substituted in equation 5, the following expression for the velocity ratio in model and prototype is obtained:

$$\frac{V_m}{V_p} = \frac{L_r}{T_r} = \sqrt{L_r g_r} = \frac{\sqrt{L_m g_m}}{\sqrt{L_p g_p}} \qquad (24)$$

This equation can be written

$$\frac{V_m}{\sqrt{L_m g_m}} = \frac{V_p}{\sqrt{L_p g_p}} \qquad (25)$$

The general expression $V/\sqrt{Lg}$ is a dimensionless ratio called Froude's number. The dimensionless nature of the ratio can be shown by substituting proper units for V, L, and g, as follows:

$$\text{ft/sec} \div \sqrt{\text{ft} \times \text{ft/sec}^2} = \text{ft/sec} \div \text{ft/sec} = 1$$

Alternate forms in which the Froude number has been expressed are V^2/Lg, $\sqrt{Lg}/V$, and Lg/V^2.

Froude's number, denoted by N_F, is a significant ratio in model studies in which gravitational forces, together with inertia forces, control the motion. When this condition exists, complete similarity between model and prototype is attained if the Froude number is the same for both.

The Froude model law applies to most model tests of hydraulic structures, particularly those in which there is a relatively large loss of energy, such as occurs at dams, weirs, spillways, and other outlet structures, or in which relatively large surface waves play an important part, as with surface ships and beach and harbor structures.

PROBLEMS

1. A caisson, rectangular in plan, 32 ft wide, 80 ft long, and 30 ft high, is to be sunk to the bed of a river at a point where the depth is 24 ft and the velocity of flow is 6.5 ft per sec. It is desired to determine the probable force of the current against the upstream end of the caisson by means of a test on a 1:12 scale model in a stream of water. Assume that the Froude law applies. Determine the proper width and length of the model caisson and the depth and velocity of flow in the model. Under test, the force against the end of the model caisson was found to be 21.5 lb. Determine the corresponding force to be expected on the prototype. Determine the coefficient of drag resistance.

2. The spillway section of a dam is to have piers and gates by means of which the upstream pond level can be controlled A 1:40 scale model is to be built and tested to determine the approximate discharge coefficients to be expected under various conditions of head and gate openings and to assist in the design of downstream apron and baffles to dissipate energy and protect the stream bed. The length of each prototype crest gate between piers is to be 60 ft and the expected discharge through each gate will be 7500 cfs. Compute the corresponding quantities in the model. The weir head on the crest of the dam in the model is found to be 0.256 ft. Compute the weir coefficient, neglecting velocity of approach. The velocity at a given point on the apron in the model is 8.50 ft per sec and the time required for a particle to travel from the crest of the dam to a point on the apron is 1.08 sec. Compute the corresponding quantities to be expected in the prototype.

3. A ship having a hull length of 450 ft is to be propelled at a speed of 16 knots. (1 knot = 1 nautical mile per hour; 1 nautical mile = 6087 ft.) Compute the Froude number. At what velocity should a 1:30 scale model be towed through water, if the Froude number is to be the same for model as for prototype?

4. An object is to be towed submerged through sea water. Studies of the flow conditions in the prototype are to be made by towing a 1:9 scale model through fresh water. The conditions are such that the Froude model law applies. If the velocity of the prototype is to be 15 ft per sec, what should be the velocity of the model? If the towing force on the model is 25 lb, what would be the corresponding force on the prototype? The measured pressure difference between two points on the surface of the model is found to be 10.6 lb per sq ft. What will be the pressure difference between corresponding points on the prototype?

166. Viscous Forces Predominant — Reynolds' Law. If viscous forces influence the motion or action of a fluid to so marked a degree that they can be considered predominant to the exclusion of gravitational forces, the force of viscosity as well as the force of inertia simultaneously govern the motion of any particle.

The unit shearing stress resulting from the viscous resistance of a fluid in motion is expressed as (equation 2, page 5)

$$\tau = \mu \frac{dV}{dx} \tag{26}$$

If this unit force is applied over area A, the total force

$$F = \mu \frac{dV}{dx} A \tag{27}$$

The ratio of homologous viscous forces in model and prototype can be expressed as

$$F_r = \frac{\mu_m (dV_m/dx_m) A_m}{\mu_p (dV_p/dx_p) A_p} = \frac{\mu_m (L_m/T_m L_m) L_m{}^2}{\mu_p (L_p/T_p L_p) L_p{}^2} = \mu_r \frac{L_r{}^2}{T_r} \tag{28}$$

SCALE RATIOS FOR THE REYNOLDS MODEL LAW, WHEN $g_r = 1$

Geometric Similarity		Kinematic Similarity		Dynamic Similarity	
Length	L_r	Time	$L_r{}^2/\nu_r$	Unit weight	$\rho_r \nu_r{}^2/L_r{}^3$
Area	$L_r{}^2$	Velocity	ν_r/L_r	Mass density	ρ_r
Volume	$L_r{}^3$	Acceleration	$\nu_r{}^2/L_r{}^3$	Force	$\rho_r \nu_r{}^2$
		Discharge	$L_r \nu_r$	Mass	$\rho_r L_r{}^3$
		Angular velocity	$\nu_r/L_r{}^2$	Work	$\rho_r \nu_r{}^2 L_r$
		Angular acceleration	$\nu_r{}^2/L_r{}^4$	Power	$\rho_r \nu_r{}^3/L_r$

Equating 12 and 28, and substituting $\rho_r L_{r3}$ for M_r,

$$\mu_r \frac{L_r{}^2}{T_r} = \rho_r L_r{}^3 \frac{L_r}{T_r{}^2} \tag{29}$$

Solving for T_r,

$$T_r = \frac{L_r{}^2}{\mu_r/\rho_r} = \frac{L_r{}^2}{\nu_r} \tag{30}$$

This equation, derived for the condition under which it can be assumed that the forces of inertia and viscosity control the flow, is known as the Reynolds model law.

If L_r^2/ν_r is substituted for T_r in equations 1 to 17, the ratios of the various quantities involved have the values shown in the table on page 326. If also the same fluid is used in model and prototype so that $w_r = \rho_r = 1$ and $\nu_r = 1$, all the ratios can be expressed in terms of the length ratio.

167. Reynolds' Number. If the value of T_r from equation 30 is substituted in equation 5, the following expression for velocity ratio in model and prototype is obtained:

$$\frac{V_m}{V_p} = \frac{L_r}{T_r} = \frac{L_r \nu_r}{L_r^2} = \frac{\nu_r}{L_r} = \frac{\nu_m/\nu_p}{L_m/L_p} \tag{31}$$

This equation can be written

$$\frac{V_m L_m}{\nu_m} = \frac{V_p L_p}{\nu_p} \tag{32}$$

The general expression VL/ν is a dimensionless ratio and is the Reynolds number, N_R, which has been referred to in previous chapters. In pipe flow the characteristic length used is the diameter. The equation of units showing that the Reynolds number is dimensionless is

$$(\text{ft/sec}) \ (\text{ft}) \div (\text{ft}^2/\text{sec}) = 1$$

Reynolds' number is a significant ratio in model studies in which viscous forces, together with inertia forces, control the motion. When this condition exists, complete similarity between model and prototype is attained if the Reynolds number is the same for both.

The Reynolds model law is usually followed in model studies of the flow of fluids in pipes and in river channels, of the motion of submerged objects through liquids, and of the relative motion of air past airplanes, automobiles, and trains at velocities which are not excessive.

168. Surface Tension Forces Predominant—Weber's Law. The surface tension of liquids may affect the flow over weirs under low heads and the propagation of small ripples on the surface of a liquid. Surface tension, denoted by σ, is measured in terms of force per unit of length (Art. 4). Hence, the force of surface tension is $F = \sigma L$. The ratio of homologous surface tension forces in model and prototype is thus

$$\frac{F_m}{F_p} = \frac{\sigma_m L_m}{\sigma_p L_p} = \sigma_r L_r \qquad (33)$$

Equating this surface tension force ratio to the inertia force ratio

$$\sigma_r L_r = \rho_r \frac{L_r{}^4}{T_r{}^2} \qquad (34)$$

from which

$$T_r = \sqrt{\frac{L_r{}^3 \rho_r}{\sigma_r}} \qquad (35)$$

If the same liquid is used in model and prototype, $\rho_r = 1$, $\sigma_r = 1$, and

$$T_r = L_r{}^{3/2} \qquad (36)$$

Substituting the value of T_r from equation 35 in equation 5

$$\frac{V_m}{V_p} = \frac{L_r}{\sqrt{L_r{}^3 \rho_r / \sigma_r}} = \sqrt{\frac{\sigma_r}{L_r \rho_r}} = \frac{\sqrt{\sigma_m / L_m \rho_m}}{\sqrt{\sigma_p / L_p \rho_p}} \qquad (37)$$

This equation can be written

$$\frac{V_m{}^2 L_m \rho_m}{\sigma_m} = \frac{V_p{}^2 L_p \rho_p}{\sigma_p} \qquad (38)$$

The expression $V^2 L \rho / \sigma$ is a dimensionless ratio known as Weber's number, N_W, and is a significant ratio in model studies in which surface tension and inertia forces control the motion.

169. Summary. The relationship of the various physical quantities involved in most tests with scale models has been developed for the conditions under which the fundamental equation, force = mass × acceleration, is combined with (1) gravitational forces, (2) viscous forces, or (3) surface tension forces, to control the motion. Frequently, however, more than one of these latter three forces has an important effect on the motion. A brief summary of the problems involved in model tests in this event is given in the following quotation from Manual 25 of the American Society of Civil Engineers[1]:

[1] *Hydraulic Models*, prepared by the Committee on Hydraulic Research, 1942.

If the occurrence be such that several equally important forces produce the motion, the problem of attaining similitude becomes more involved. Similitude can be secured, when two forces exist, by the use of different fluids in model and prototype. Once the two fluids are chosen, the scale ratio immediately becomes fixed by the properties of the chosen fluids.

In general practice, however, the concern of the investigator is with the effect of only one force, which is considered dominant. The neglect of the other forces is responsible for inaccuracies in only the final result. The endeavor, therefore, should be to choose scales and to build and operate models in such a manner that the effect of the non-dominant forces is compensating or negligible. If this is done, such model studies will produce sufficiently accurate information to predict major occurrences in the prototype. Thus, in the case of the flow past a bridge pier, gravitation ordinarily will be found to be the predominating force in the motion occurrence, both in the prototype and in the model, provided the latter is not too small. Fluid friction and surface tension will also influence the occurrence, and in a case where the model is made too small, the influence of the latter forces upon the flow pattern may result in important dissimilarity of the two occurrences between model and prototype.

It has been stated that the best results which have been obtained from model tests are qualitative rather than quantitative. That means, for instance, in a study of the discharge characteristics of a spillway, that properly conducted model tests of different spillway designs should indicate quite reliably which design would provide the most favorable flow characteristics in the prototype, but that scale effects resulting from the impossibility of obtaining complete similarity of model to prototype may cause numerical values of coefficients of discharge or velocity in the prototype to be somewhat different from those obtained with the model. Considerable success has, however, been achieved with quantitative tests of scale models, and values of coefficients obtained in such tests have been closely substantiated by measurements made on the prototype.

EXAMPLE. It is desired to determine by model tests how far the outlet pipe in the side of a large oil tank should be below the oil surface in order that air will not be drawn into the pipe when oil is discharged from the tank. The oil has a kinematic viscosity of 0.00080 sq ft per sec.

(a) Determine the proper kinematic viscosity of the model fluid if the scale of the model is 1:4.

(b) If a glycerol solution with a kinematic viscosity of 0.000096 sq ft per sec is used as the model fluid, determine·the proper scale ratios of length, velocity, and discharge.

Solution. Both gravitational and viscous forces are important in this problem. Therefore both the Froude and the Reynolds numbers must be the same for model and prototype. Surface tension effects will be minimized by using a fairly large-scale model.

Equating the velocity scale ratios for the Froude law (page 324) and the Reynolds law (page 326), assuming $g_r = 1$,

$$\sqrt{L_r} = v_r/L_r$$

whence

$$v_r = L_r^{3/2}$$

If $L_r = \frac{1}{4}$, $v_r = \frac{1}{8}$, and the kinematic viscosity of the model fluid should be 0.00010.

For $v_m = 0.000096$, $v_r = 0.12$, and

$$L_r = v_r^{2/3} = 0.243$$

The velocity and discharge ratios can be determined by either the Froude or the Reynolds model law.

By Froude: $\qquad\qquad V_r = \sqrt{L_r} = 0.493$

$$Q_r = L_r^{5/2} = 0.0291$$

By Reynolds: $\qquad\quad V_r = v_r/L_r = 0.493$

$$Q_r = L_r v_r = 0.0291$$

170. Dimensional Analysis. The attention òf the student has been repeatedly called to the units in which the various physical quantities used in this book are expressed, and equations of units are given for the more important formulas. A number of these formulas with their unit equations in the foot-pound-second system are repeated in the table on page 332. The column at the right gives the algebraic equations of units with F (force) substituted for pounds, L (length) for feet, and T (time) for seconds.

It should be noted that, with three exceptions, the equations of units in the table reduce algebraically to the identity $1 = 1$, indicating that the equations are dimensionally correct. The exceptions are the rectangular weir discharge formula and the Chezy and the Manning formulas for flow in pipes and open channels. If, however, C in the weir formula and C in the Chezy formula are considered to include $\sqrt{g}$, these two equations become dimensionally correct, as is shown by multiplying the right sides of the unit equations by $\sqrt{\text{ft/sec}^2}$ and $\sqrt{L/T^2}$. This is not true of the Manning formula, which therefore remains dimensionally incorrect, though it is applied with much success in hydraulic engineering because of the rather complete information available concerning the proper values of roughness coefficient n over a wide range of flow conditions.

In order for the equations of units in the right-hand column of the table to reduce to the identity $1 = 1$, it is evidently necessary that the algebraic sum of the exponents of each of the independent quantities F, L, and T, respectively, be the same on each side of the equation. For example, the equation of units for power $= QwH$, can be written

$$LFT^{-1} = L^3 T^{-1} \times FL^{-3} \times L$$

The equations of the exponents of F, L, and T are

$$\text{For } F: \qquad 1 = 1$$

$$\text{For } L: \qquad 1 = 3 - 3 + 1$$

$$\text{For } T: \qquad -1 = -1$$

Each of these equations being an identity, the original equation for power is dimensionally correct.

Dimensional analysis provides a method of developing the general form of an equation to express the relationship of the various physical quantities involved in force or motion. It is assumed that if any such relationship is completely defined by a mathematical equation, that equation must be dimensionally correct.

Each physical quantity involved in force or motion can be expressed in units of force, length, and time (or mass, length, and time, as some prefer). A number of these quantities and their corresponding F-L-T units are shown in the table. Other quantities commonly occurring in hydraulics (Art. 4) are μ, dynamic viscosity (FT/L^2); ρ, mass density (FT^2/L^4) and σ, surface tension (F/L).

The method of dimensional analysis will be illustrated by its application to a few important problems of hydraulics.

1. Let it be assumed that the discharge Q over a rectangular weir varies directly with the length L of the weir and is also a function of the head H on the weir and g, the acceleration due to

UNIT EQUATIONS FOR IMPORTANT HYDRAULIC FORMULAS

Subject	Formula	Unit Equations	
Intensity of pressure	$p = wh$	$\dfrac{lb}{ft^2} = \dfrac{lb}{ft^3} \times ft$	$\dfrac{F}{L^2} = \dfrac{F}{L^3} \times L$
Hydrostatic force	$P = whA$	$lb = \dfrac{lb}{ft^3} \times ft \times ft^2$	$F = \dfrac{F}{L^3} \times L \times L^2$
Equation of continuity	$Q = AV$	$\dfrac{ft^3}{sec} = ft^2 \times \dfrac{ft}{sec}$	$\dfrac{L^3}{T} = L^2 \times \dfrac{L}{T}$
Power	$G = QwH$	$\dfrac{ft \cdot lb}{sec} = \dfrac{ft^3}{sec} \times \dfrac{lb}{ft^3} \times ft$	$\dfrac{LF}{T} = \dfrac{L^3}{T} \times \dfrac{F}{L^3} \times L$
Velocity of discharge	$V = \sqrt{2gH}$	$\dfrac{ft}{sec} = \sqrt{\dfrac{ft}{sec^2} \times ft}$	$\dfrac{L}{T} = \sqrt{\dfrac{L}{T^2} \times L}$
Orifice discharge	$Q = CA\sqrt{2gH}$	$\dfrac{ft^3}{sec} = ft^2\sqrt{\dfrac{ft}{sec^2} \times ft}$	$\dfrac{L^3}{T} = L^2\sqrt{\dfrac{L}{T^2} \times L}$
Rectangular weir discharge	$Q = CLH^{3/2}$	$\dfrac{ft^3}{sec} \neq ft \times ft^{3/2}$	$\dfrac{L^3}{T} \neq L \times L^{3/2}$
Head lost in pipe flow (general)	$h_f = f\dfrac{L}{D}\dfrac{V^2}{2g}$	$ft = \dfrac{ft}{ft} \times \dfrac{(ft/sec)^2}{ft/sec^2}$	$L = \dfrac{L}{L} \times \dfrac{(L/T)^2}{L/T^2}$
Head lost in laminar flow in pipes	$h_f = \dfrac{32L\nu V}{gD^2}$	$ft = \dfrac{ft \times (ft^2/sec) \times (ft/sec)}{(ft/sec^2) \times ft^2}$	$L = \dfrac{L \times (L^2/T) \times (L/T)}{(L/T^2) \times L^2}$
Chezy formula	$V = C\sqrt{RS}$	$\dfrac{ft}{sec} \neq \sqrt{ft \times (ft/ft)}$	$\dfrac{L}{T} \neq \sqrt{L \times (L/L)}$
Manning formula	$V = \dfrac{1.486}{n}R^{2/3}S^{1/2}$	$\dfrac{ft}{sec} \neq ft^{2/3} \times (ft/ft)^{1/2}$	$\dfrac{L}{T} \neq L^{2/3} \times (L/L)^{1/2}$
Critical velocity	$V = \sqrt{gd}$	$\dfrac{ft}{sec} = \sqrt{\dfrac{ft}{sec^2} \times ft}$	$\dfrac{L}{T} = \sqrt{\dfrac{L}{T^2} \times L}$
Drag resistance	$D_f = C_D w \dfrac{V^2}{2g} A$	$lb = \dfrac{lb}{ft^3} \times \dfrac{(ft/sec)^2}{ft/sec^2} \times ft^2$	$F = \dfrac{F}{L^3} \times \dfrac{(L/T)^2}{L/T^2} \times L^3$

gravity. It is further assumed that Q can be expressed by an equation of the form

$$Q = kLH^a g^b \qquad (39)$$

where a and b are unknown exponents and k is a non-dimensional factor of proportionality. Substituting for each variable its corresponding units of force, length, and time, the following equation is obtained:

$$L^3 T^{-1} = LL^a(LT^{-2})^b \qquad (40)$$

For equation 39 to be dimensionally correct the equation of exponents of each independent quantity in equation 40 must be an identity. These equations are

$$\text{For } L: \qquad 3 = 1 + a + b$$

$$\text{For } T: \qquad -1 = -2b$$

Solving for a and b, $b = 1/2; a = 3/2$. Substituting these values in equation 39,

$$Q = kLH^{3/2} g^{1/2} \qquad (41)$$

If $k\sqrt{g}$ is denoted by a general coefficient C, the equation of rectangular weir discharge without correction for velocity of approach is obtained:

$$Q = CLH^{3/2} \qquad (42)$$

If the only quantities affecting weir discharge were those assumed above, C would be constant for all heads and lengths of weir. That this is not true was brought out in Chapter VI; therefore, other physical quantities must be involved in the laws governing weir flow. The application of dimensional analysis to the problem of determining the relationship of other physical quantities to flow through a V-notch weir is shown in the following problem.

2. Assume that the velocity of flow V through a V-notch weir is a function of H, the head, of ρ, μ, and σ, respectively the mass density, dynamic viscosity, and surface tension of the liquid, and of g, the acceleration due to gravity; and that the relationship of these quantities may be expressed in the following form:

$$V = kH^a \rho^b \mu^c \sigma^d g^e \qquad (43)$$

Expressing each quantity in its respective F-L-T units,

$$\frac{L}{T} = L^a \left(\frac{FT^2}{L^4}\right)^b \left(\frac{FT}{L^2}\right)^c \left(\frac{F}{L}\right)^d \left(\frac{L}{T^2}\right)^e \tag{44}$$

Writing the equations of exponents,

For F: $0 = b + c + d$

For L: $1 = a - 4b - 2c - d + e$

For T: $-1 = 2b + c - 2e$

Since there are five unknown exponents and only three equations, three of the exponents must be expressed in terms of the other two. Solving for a, b, and e in terms of c and d,

$$b = -c - d$$

$$e = \frac{1}{2} - \frac{c}{2} - d$$

$$a = \tfrac{1}{2} - \tfrac{3}{2}c - 2d$$

Substituting these values in equation 43 and reducing,

$$V = kH^{\frac{1}{2}-3c/2-2d}\rho^{-c-d}\mu^c\sigma^d g^{\frac{1}{2}-c/2-d}$$

$$= k\frac{H^{\frac{1}{2}}\mu^c\sigma^d g^{\frac{1}{2}}}{H^{3c/2}H^{2d}\rho^c\rho^d g^{c/2}g^d} = kg^{\frac{1}{2}}H^{\frac{1}{2}}\left(\frac{\mu}{H^{\frac{3}{2}}\rho g^{\frac{1}{2}}}\right)^c\left(\frac{\sigma}{H^2\rho g}\right)^d \tag{45}$$

Since V is proportional to $\sqrt{gH}$, it can be substituted therefor in the ratios in parentheses by making a corresponding change in the proportionality factor k. The equation can therefore be written

$$V = k'\left(\frac{\mu}{HV\rho}\right)^c\left(\frac{\sigma}{HV^2\rho}\right)^d \sqrt{gH} \tag{46}$$

The ratios in parentheses are identified as the reciprocal of Reynolds' number (Art. 167) and Weber's number (Art. 168), the characteristic length being H, the head on the weir.

Multiplying the velocity by the area of the weir, $H^2 \tan \theta/2$, the equation for discharge through a V-notch weir becomes

$$Q = k'\left(\frac{1}{N_R}\right)^c\left(\frac{1}{N_W}\right)^d \sqrt{g} \tan\frac{\theta}{2} H^{\frac{5}{2}} \tag{47}$$

This equation is in a form comparable with equation 48 on page 161. Values of k', c, and d must be determined by experiment.[1]

If any three exponents, other than a, b, and e, had been expressed in terms of the other two in the foregoing development, equation 46 would have contained a different function of the Reynolds and Weber numbers.

3. It is shown in Chapter VI that in equation $Q = CA \sqrt{2gH}$, C is not constant but varies with the head and with the size of orifice. It will now be shown by dimensional analysis that C is a function of Froude's and Reynolds' numbers.

Let it be assumed that the discharge Q through an orifice varies with the dynamic viscosity μ and the mass density ρ of the liquid, with the area A of the orifice, with the head H on the orifice, and with g, the acceleration due to gravity; and that the discharge can be expressed as

$$Q = k\mu^a \rho^b A^c H^d g^e \tag{48}$$

Introducing the F-L-T units for each quantity

$$\frac{L^3}{T} = \left(\frac{FT}{L^2}\right)^a \left(\frac{FT^2}{L^4}\right)^b L^{2c} L^d \left(\frac{L}{T^2}\right)^e \tag{49}$$

Writing the equations of exponents

For F: $0 = a + b$

For L: $3 = -2a - 4b + 2c + d + e$

For T: $-1 = a + 2b - 2e$

Solving for b, c, and e, in terms of a and d:

$$b = -a$$

$$c = \frac{5}{4} - \frac{3}{4}a - \frac{d}{2}$$

$$e = \frac{1}{2} - \frac{a}{2}$$

Substituting these values in the original equation and reducing,

$$Q = k\mu^a \rho^{-a} A^{5/4 - 3a/4 - d/2} H^d g^{1/2 - a/2} = k\left(\frac{\mu}{\rho A^{3/4} g^{1/2}}\right)^a \left(\frac{H}{A^{1/2}}\right)^d A^{5/4} g^{1/2} \tag{50}$$

[1] See footnote 3 on page 162.

The quantity $A^{1/2}$, being the square root of an area, can be replaced by a characteristic length L of the orifice provided the coefficient k is changed accordingly. Making this substitution and multiplying numerator and denominator by $\sqrt{2H}$, equation 50 becomes

$$Q = \frac{k'}{\sqrt{2}}\left(\frac{\mu}{\rho L\sqrt{Lg}}\right)^{a}\left(\frac{H}{L}\right)^{d-\frac{1}{2}} A\sqrt{2gH} \qquad (51)$$

The first quantity in parentheses can be written

$$\frac{1}{\sqrt{Lg}}\cdot\frac{\mu}{\rho L}$$

Multiplying numerator and denominator by some characteristic velocity V, the expression becomes

$$\frac{V}{\sqrt{Lg}}\cdot\frac{\mu}{\rho L V}$$

which is seen to be the ratio of Froude's number to Reynolds' number. The discharge coefficient C in equation 10 (page 122) therefore has the value

$$C = k''\left(\frac{N_F}{N_R}\right)^{a}\left(\frac{H}{L}\right)^{d-\frac{1}{2}} \qquad (52)$$

provided the discharge is governed by the quantities included in equation 48.

4. The drag force of resistance acting on an object moving through a fluid was shown in Art. 154 to be given by the equation

$$D_f = C_D\,\frac{\rho V^2}{2}\,A$$

It will now be shown by dimensional analysis that C_D is a function of Reynolds' number as indicated by the experimental data illustrated in Fig. 148 (page 306).

Let it be assumed that the drag force D varies with the mass density ρ and dynamic viscosity μ of the fluid, with the projected area A of the object, and with the velocity V of the object relative to the fluid; and that the drag force can be expressed in the form

$$D = k\rho^{a}A^{b}V^{c}\mu^{d} \qquad (53)$$

Substituting F-L-T units

$$F = \left(\frac{FT^2}{L^4}\right)^a (L^2)^b \left(\frac{L}{T}\right)^c \left(\frac{FT}{L^2}\right)^d \tag{54}$$

Writing the equations of exponents

Of F: $1 = a + d$

Of L: $0 = -4a + 2b + c - 2d$

Of T: $0 = 2a - c + d$

Solving for a, b, and c in terms of d,

$$a = 1 - d$$

$$b = 1 - \frac{d}{2}$$

$$c = 2 - d$$

Substituting in equation 53 and reducing,

$$D = k\rho^{1-d}A^{1-d/2}V^{2-d}\mu^d = k\left(\frac{\mu}{\rho A^{1/2}V}\right)^d \rho A V^2 \tag{55}$$

Replacing $A^{1/2}$ by some characteristic length L of the object, with corresponding change in the proportionality factor k, and multiplying and dividing by 2, equation 55 becomes

$$D = k'\left(\frac{\mu}{\rho L V}\right)^d \frac{\rho V^2}{2} A \tag{56}$$

The drag coefficient C_D in equation 56 therefore has the value

$$C_D = \frac{k'}{N_R{}^d} \tag{57}$$

5. Another problem, the solution of which has been provided by dimensional analysis, is that of the resistance to motion of a ship through water. This resistance is considered to be composed of two parts: (1) a force which sets up surface waves of more or less complicated patterns, called the wave-making resistance; (2) a force produced by so-called " skin friction " on the submerged part of the hull.

Let it be assumed that the wave-making resistance D_w varies with the density ρ of the water, with the volume of displacement Δ

and the velocity V of the ship, and with g, the acceleration due to gravity; and that it can be expressed by an equation of the form

$$D_w = k\rho^a \Delta^b V^c g^d \tag{58}$$

Introducing the units of the various quantities,

$$F = \left(\frac{FT^2}{L^4}\right)^a (L^3)^b \left(\frac{L}{T}\right)^c \left(\frac{L}{T^2}\right)^d \tag{59}$$

Writing the equation of exponents,

Of F: $1 = a$

Of L: $0 = -4a + 3b + c + d$

Of T: $0 = 2a - c - 2d$

Solving for a, b, and c in terms of d

$$a = 1$$

$$b = \frac{2}{3} + \frac{d}{3}$$

$$c = 2 - 2d$$

Substituting in equation 58 and reducing,

$$D_w = k\rho \Delta^{2/3 + d/3} V^{2-2d} g^d = k\left(\frac{\Delta^{1/3} g}{V^2}\right)^d \rho \Delta^{2/3} V^2 \tag{60}$$

The quantity $\Delta^{1/3}$ can be replaced by some characteristic dimension L, usually the ship's length. Substituting this value and multiplying numerator and denominator by 2,

$$D_w = k'\left(\frac{Lg}{V^2}\right)^d \frac{\rho V^2}{2} \Delta^{2/3} = \frac{k'}{(N_F)^{2d}} \frac{\rho V^2}{2} \Delta^{2/3} \tag{61}$$

This equation can be written

$$D_w = C_w \frac{\rho V^2}{2} \Delta^{2/3} \tag{62}$$

where C_w is a function of Froude's number.

The skin friction drag[1] on a ship's hull is assumed to be equal to the skin friction drag on one side of a thin flat plate of equal area

[1] For a more extended discussion of skin friction resistance see Karl E. Schoenherr, "Resistance of Flat Surfaces Moving through a Fluid," *Trans. Soc. Naval Architects and Marine Engrs.*, Vol. 40, 1932, pp. 279–313.

which is moving through a fluid in a direction parallel to the surface of the plate.   The skin friction drag is given by the equation

$$D_f = C_f \frac{\rho V^2}{2} A \tag{63}$$

where A is the area of the plate and C_f is a coefficient which has been shown to be a function of Reynolds' number, VL/ν, L being usually the length of the plate (or ship).   Schlichting gives the following value for C_f in terms of Reynolds' number:

$$C_f = \frac{0.455}{(\log N_R)^{2.58}} \tag{64}$$

Answers

Chapter I

Chapter II

Chapter III

Page 56. **1.** 13,640 lb per sq in. **2.** 0.15 in. **3.** 3.8 in. **4.** 0.71, 1.37 **in.** **5.** Top, 3120 lb; bottom, 6240 lb. **6.** Top, 2370 lb; bottom, 7110 lb.

Page 61. **1.** (a) 4.78 ft from toe; 1720, 2240 lb per sq ft; (b) 4.20 ft from toe; 780, 2240 lb per sq ft. **2.** 2.41 ft from toe.

Page 63. **1.** 2.99, 2.92 ft. **2.** 100. **3.** 353, 530 lb. **5.** 1750, 4710 lb. **6.** 6550 cu yd. **7.** 880 lb. **8.** 13,250 lb. **9.** 174 lb. **10.** 2560 lb. **11.** 1440 lb. **12.** 3.56 ft.

Page 68. **1.** 1.98 ft. **2.** 4.38 ft, 932,000 ft-lb. **3.** 3.25 ft, 221,000 ft-lb (approximately). **4.** 16.8 ft. **5.** 1,880,000 ft-lb. **6.** 0.064 ft.

Chapter IV

Page 81. **3.** (a) 108; (b) 133. **5.** 5.75 ft. **6.** 108.5. **8.** 1.02 ft . **10.** (a) 1.21 ft. **11.** 62.6 rpm. **13.** 2.18 cu ft. **14.** 1.43. **15.** 34.2.

Chapter V

Page 87. **1.** 0.417 cfs. **2.** 45.7. **3.** At 3 ft, 6.37 ft per sec. **4.** 7.34, 11.4, 16.5 ft per sec. **5.** At 3 ft, 8.23 in. **6.** $59\frac{1}{2}$ in., $34\frac{1}{2}$ in. **7.** 0.764 in. **8.** At top, 11.16 ft per sec.

Page 93. **1.** 28.1, 31.0, 24.2, 18,017. **2.** 15.1 ft per sec., 1.83. **3.** 224 ft, 600. **4.** 7.8 lb per sq in. **5.** −10.9 ft. **6.** 1.046.

Page 99. **1.** 8 ft, 1 ft. **2.** 26.1, 26.3, 25.7 lb per sq in. **3.** 11.0 ft. **4.** 10.9 lb per sq in. **5.** 24 lb per sq in. **6.** 96.3 ft. **7.** 6.8. **8.** 0.16. **9.** 1.9 ft. **10.** 2.1 ft. **11.** 0.39 cfs. **12.** 1.33 cfs; at 3, $p/w = 17.3$ ft. **13.** 0.62, 0.59, cfs. **14.** 28.7. **15.** 23.3 ft. **16.** 8.2, 3.15, 1410. **17.** 1.67 cfs; at 2, $p/w =$ −24.1 ft. **18.** (a) 12.6, 19.7 ft, 267.7 ft; (b) 47.7, 17.1 ft, 329.0 ft. **19.** 553, 179.1 ft, −15.8 ft.

Page 106. **1.** 6.30, 6.08, 0.967. **2.** 9.0 in. **3.** 24.7 gpm. **4.** 0.972. **5.** 16,600. **6.** 1.82 cfs.

Page 108. **1.** 229, 270 ft. **2.** 43.2 lb per sq in. **3.** 17 ft.

Chapter VI

Page 125. **1.** 32.1, 31.1 ft per sec; 0.678 cfs. **2.** 1.575 in., 31.4 ft per sec, 0.425 cfs. **3.** 1.33 cfs, 1.24 ft. **4.** 59.6 lb per sq in. **5.** 3.28 ft. **6.** 29.3 ft per sec, 2.55 cfs. **7.** 53. **8.** 7.20 ft, 2.73 in. **9.** 0.330 cfs, 1.587 in., 24.0 ft per sec, 0.56 ft. **10.** 1.77 cfs. **11.** 36.8. **12.** 0.715, 1.82, 2.60, 3.57, cfs.

Page 132. **1.** 0.0211, 0.0779, 0.1243 cfs. **2.** 0.0943, 0.1330, 0.1862 cfs. **3.** 0.630, 0.024 cfs. **4.** 3.77 in. **5.** 1.12, 28.6, 115 ft. **6.** 0.885 cfs, 0.985. **7.** 27.6 ft. **8.** 0.619, 0.979, 0.605. **9.** 0.595, 0.960, 2.0 ft. **10.** 43.8 cfs. **11.** 5.68 ft, 0.0632 cfs. **12.** 0.278, 0.461, 0.874 cfs. **13.** 8.94 cfs. **14.** 19.0 gpm. **15.** 3.96 gpm. **16.** 1560 cfm.

Page 140. **3.** 157 sec. **4.** $6\frac{1}{2}$ min. **5.** $14\frac{1}{4}$ min. **6.** 1 hr. **7.** 9 min 23 sec. **8.** 21 sec.

Page 149. **1.** 2.57, 1,660,000. **2.** 15.7 ft. **3.** 2.5, 4.3 in. **4.** 0.97 ft, 9.0. **5.** 0.333 cfs. **6.** 167 cfs. **7.** 68 in.

Page 158. **1.** Combination b-k-p-w, 98.7 cfs; c-i-p-v, 18.94 cfs; e-j-n-t, 10.12 cfs. **2.** With $H = 1.0$ ft, 3.365, 3.400, 3.349. **3.** 3.17, 25.2 cfs. **4.** 0.437,

0.417, 4.86, 4.79 cfs. **5.** Rehbock, -0.3 per cent; Harris, -0.4 per cent.
6. 45.5 cfs. **7.** 79.6, 80.7 cfs. **8.** 6.7 cfs. **9.** $3\frac{1}{3}$ ft. **10.** 4.3 ft.

Page 164. **1.** 36.6 cfs. **2.** 22.2, 21.7 cfs. **3.** 1.78 ft. **4.** 2.37 ft. **5.**
0.444, 2.035 cfs. **6.** 0.868 ft. **7.** 0.558 cfs. **8.** 0.393 ft. **9.** 1.12, 1.117,
1.112 cfs. **10.** 13.2 cfs. **11.** 1.37 ft. **12.** 19.4 ft.

Chapter VII

Page 186. **1.** 50 ft. **2.** 1.05 ft. **3.** 11.4 ft. **4.** 4.3 cfs. **5.** 42 cfs. **6.**
109. **7.** 12 in. **8.** 3.13 cfs. **9.** 43 in. **10.** 64 in. **11.** 5.2 lb per sq in.
12. 164 lb per sq in. **13.** 37 lb per sq in. **14.** 8.3 cfs.

Page 194. **1.** 134. **2.** 1350. **3.** 6.2. **4.** 650,000. **5.** 4650, 3100. **6.**
11.5, 11.1, 10.7. **7.** 0.036, 3650. **8.** 1.86 in. **9.** 8 in., 20.3. **10.** 1510.
11. 123. **12.** 2. **13.** 0.023, 3.57, 0.342. **14.** 1.75 in. **15.** 0.70. **16.** 24.
17. 610 cfs. **18.** 25.

Page 205. (a) 0.23 ft, 0.58 lb per sq in.

Page 208. (a) 0.31 ft, 0.34 lb per sq in.

Page 216. **1.** 14.8 ft. **2.** 2.6 cfs. **3.** 21.2 ft. **4.** 34 in. **5.** 45 in. **6.**
3.7 ft.

Page 219. **1.** With $n = 0.011$, 50 ft. **2.** 27.3 ft, 33.2 ft lower. **3.** With
$n = 0.011$, 0.74 cfs. **4.** 2.1 cfs. **5.** 10.3 in. **6.** 28 cfs. **7.** 27.2 ft.

Page 223. **1.** 5 to 3, with $n = 0.011$, 30 ft. **2.** 8.9 cfs. **3.** With $n =$
0.011, 40 ft. **4.** 11. **5.** 18 in. **6.** With $n = 0.011$, 93 ft. **7.** 27 cfs. **8.**
With $C_1 = 120$, 6 in.

Page 227. **1.** ad, 2.3 cfs; cd, 0.6 cfs; fd, 1.1 cfs. **2.** de, 1070 gpm; fe, 550
gpm; ie, 380 gpm; 76 ft. **3.** cf, 1180 gpm; ef, 530 gpm; hf, 1290 gpm; 52.1 lb
per sq in.

Page 231. **1.** 1.6, 1.0, 0.6 cfs. **2.** 11.0, 1.6, 2.6, 6.8 cfs. **3.** 9.7, 2.3, 7.4 cfs.
4. With $n = 0.011$, 55 cfs. **5.** With $n = 0.011$, 13.5 ft below A. **6.** With C_1
$= 120$, 0.25 cfs. **7.** 56 ft. **8.** With $n = 0.011$, 16 in.

Page 233. **1.** 48.2. **2.** 45.2 lb per sq in. **3.** 72 per cent. **4.** -10.8,
96.5 ft; 108.1 ft; 34.2. **5.** 12 lb per sq in. **6.** 1.73, 33.3, 221. **7.** 108, 14.6.
8. 8.7, 233 ft; 227 ft, 257.

Chapter VIII

Page 243. **1.** (a) 39.3 sq ft, 15.7 ft, 2.50 ft. **2.** (a) 5.09 ft per sec. **3.** (b)
$B = 4.08$ ft, $d = 2.04$ ft. **5.** (b) 7.93, 1.26 ft.

Page 251.* **1.** 50 cfs. **2.** 330 cfs. **3.** 36 cfs. **4.** 3380 cfs. **5.** 379 cfs,
0.0253, 2.24. **6.** 0.00080. **7.** 0.00041. **8.** 0.00018. **9.** 0.0095. **10.** $B =$
4.92 ft, $d = 2.46$ ft. **11.** 12.3 ft. **12.** $D = 14.1$ ft. **13.** 4.0 ft. **14.** 7.9 ft.

Page 255.* **1.** 3.68 ft. **2.** 120 cfs. **3.** 0.0020, 1.20 ft, 0.0037. **4.** 4.25 ft.
5. 0.00245.

Page 263.* **1.** $D = 10.45$ ft. **2.** $D = 19.4$ ft. **3.** $B = 5.75$ ft, $d = 4.98$
ft. **4.** 6 ft, 3 ft, 5. **5.** 7.76, 6.28 ft. **6.** $B = 3.77$ ft, $d = 7.96$ ft, 0.00026.
7. $B = d = 8.58$, 9.3 ft per sec. **8.** 0.00299. **9.** 6.47 ft, 350 cfs. **10.** $B =$
12.18 ft, $d = 6.09$ ft, 590 cfs. **11.** 3900 cfs. **12.** 30,500, 20,600 cfs.

Page 274.* **1.** 5.9, 340 ft. **2.** 360 cfs. **3.** 4.81 ft. **4.** 1.26 ft. **5.** 2.80 ft.

*By Manning formula.

6. 3.2 ft. **7.** 3.43 ft. **8.** 48 ft per sec. **9.** 6.2 ft. **10.** 3.75 ft. **11.** 0.26 ft below. **12.** 440 cfs.

Page 279. **1.** 2.35 ft. **2.** 17.5 ft per sec. **3.** 69.5 cfs. **4.** 96 ft. **5.** 77 ft.

Page 286. **1.** 4.86 ft. **2.** 450 cfs. **3.** 19.9, 3.06 ft per sec. **4.** 235 cfs.

Chapter IX

Page 298. **1.** 6.6, 2.4 lb. **2.** 34, 59 lb. **3.** 71, 265 lb. **4.** 19, 33 lb. **5.** 226, 52 lb. **6.** 190 lb. **7.** 45 lb, 1.9. **8.** 15.5 ft per sec, 11.8.

Page 303. **1.** $13\frac{1}{3}$ ft per sec, 100. **2.** 2170, 930, 3100 ft-lb. **3.** 262 lb, 7860 ft-lb per sec. **4.** 2550 ft-lb per sec, 5.23 ft per sec. **5.** 3220 lb. **6.** 34,700 lb. **7.** 390, 1350 lb; 9870, 3170 lb.

Page 308. **1.** At 40 miles per hr, 55.1 lb, 5.9. **2.** 1.94 lb, 52.5 ft per sec². **3.** 29.4 ft per sec. **4.** 9.7. **5.** 4680 lb, 187.

Chapter X

Page 325. **1.** $F_p = 37,200$ lb, $C_D = 1.17$. **2.** $C = 3.82$, $T_p = 6.84$ sec. **3.** $V_m = 4.95$ ft per sec. **4.** $F_p = 18,700$ lb, $\Delta p_p = 98$ lb per sq ft.

Index

Numbers refer to pages.